Six Sigma Statistics with Excel and Minitab

Issa Bass

New York Chicago San Francisco Lisbon London Madrid
Mexico City Milan New Delhi San Juan Seoul
Singapore Sydney Toronto

The McGraw-Hill Companies

Library of Congress Cataloging-in-Publication Data

Bass, Issa.
Six sigma statistics with Excel and Minitab / Issa Bass. – 1st ed.
p. cm.
Includes index.
ISBN-13: 978-0-07-148969-0–ISBN-10: 0-07-148969-X (alk. paper).
1. Six sigma (Quality control standard) 2. Quality control–data processing.
3. Quality control–Statistical methods–Computer programs. 4. Microsoft Excel
(Computer file) 5. Minitab. I. Title
TS156.B432 2007
658.4′013028554—dc22

2007001157

9 10 11 12 13 14 15 DOC/DOC 1 9 8 7 6 5 4

ISBN 13: P/N 978-0-07-149646-9 of set
978-0-07-148969-0
ISBN 10: P/N 0-07-149646-7 of set
0-07-148969-X

This book is printed on acid-free paper.

Sponsoring Editor
Kenneth P. McCombs

Editing Supervisor
David E. Fogarty

Project Manager
Vageesh Sharma

Copy Editor
Paul Hightower

Indexer
Paul Hightower

Production Supervisor
Pamela A. Pelton

Composition
Aptara, Inc.

Art Director Cover
Jeff Weeks

I dedicate this book to

My brother, Demba who left me too soon
Rest in peace

President Leopol Sedar Senghor and Professor Cheikh Anta Diop
For showing me the way

Monsieur Cisse, my very first elementary school teacher.
I will never forget how you used to hold my little five year old fingers to teach me how to write.

Thank you!
You are my hero.

ABOUT THE AUTHOR

ISSA BASS is a Six Sigma Master Black Belt and Six Sigma project leader for the Kenco Group, Inc. He is the founding editor of SixSigmaFirst.com, a portal where he promotes Six Sigma methodology, providing the public with accessible, easy-to-understand tutorials on quality management, statistics, and Six Sigma.

Contents

Preface

The role of statistics in quality management in general and Six Sigma in particular has never been so great. Quality control cannot be dissociated from statistics and Six Sigma finds its definition in that science.

In June 2005, we decided to create sixsigmafirst.com, a website aimed at contributing to the dissemination of the Six Sigma methodology. The site was primarily focusing on tutorials about Six Sigma. Since statistical analysis is the fulcrum of that methodology, a great deal of the site was slated to enhance the understanding of the science of Statistics. The site has put us in contact with a variety of audiences that range from students who need help with their homework to quality control managers who seek to better understand how to apply some statistics tools to their daily operations.

Some of the questions that we receive are theoretical while others are just about how to use some statistics software to conduct an analysis or how to interpret the results of a statistical testing.

The many questions that we have been getting have brought about the idea of writing a comprehensive book that covers both statistical theory and helps to better understand how to utilize the most widely used software in statistics.

Minitab and Excel are currently the most preponderant software tools for statistical analysis; they are easy to use and provide reliable results. Excel is very accessible because it is found on almost any Windows-based operating system and Minitab is widely used in corporations and universities.

But we believe that without a thorough understanding of the theory behind the analyses that these tools provide, any interpretation made of results obtained from their use would be misleading.

That is why we have elected to not only use hundreds of examples in this book, with each example, each study case being analyzed from a theoretical standpoint, using algebraic demonstrations, but we also graphically show step by step how to use Minitab and Excel to come to the same conclusions we obtained from our mathematical reasoning.

This comprehensive approach does help better understand how the results are obtained and best of all, it does help make a better interpretation of the results.

We hope that this book will be a good tool for a better understanding of statistics theory through the use of Minitab and Excel.

Acknowledgments

I would like to thank all those without whom, I would not have written this book. I would especially thank my two sisters Oumou and Amy. Thank you for your support.

I am also grateful to John Rogers, my good friend and former Operations' Director at the Cingular Wireless Memphis Distribution Centre in Memphis, Tennessee.

My thanks also go to my good friend Jarrett Atkinson from the Kenco group, at KM Logistics in Memphis.

Thank you all for your constant support.

Chapter

1

Introduction

Learning Objectives:

- Clearly understand the definition of Six Sigma
- Understand the Six Sigma methodology
- Understand the Six Sigma project selection methods
- Understand balanced scorecards
- Understand how metrics are selected and integrated in scorecards
- Understand how metrics are managed and aligned with the organization's strategy
- Understand the role of statistics in quality control and Six Sigma
- Understand the statistical definition of Six Sigma

A good business performance over a long period of time is never the product of sheer happenstance. It is always the result of a well-crafted and well-implemented strategy. A *strategy* is a time-bound plan of structured actions aimed at attaining predetermined objectives. Not only should the strategy be clearly geared toward the objectives to be attained, but it should also include the identification of the resources needed and the definition of the processes used to reach the objectives.

Over the last decades, several methodologies have been used to improve on quality and productivity and enhance customer satisfaction. Among the methodologies used for these purposes, Six Sigma has so far proved to be one of the most effective.

1.1 Six Sigma Methodology

Six Sigma is a meticulous, data-driven methodology that aims at generating quasi-perfect production processes that would result in no more than 3.4 defects per 1 million opportunities. By definition, Six Sigma is rooted in statistical analysis because it is data-driven and is a strict approach that drives process improvements through statistical measurements and analyses.

The Six Sigma approach to process improvements is project driven. In other words, areas that show opportunities for improvements are identified and projects are selected to proceed with the necessary improvements. The project executions follow a rigorous pattern called the *DMAIC (Define, Measure, Analyze, Improve and Control)*. At every step in the DMAIC roadmap, specific tools are used, and most of these tools are statistical.

Even though Six Sigma is a project-driven strategy, the initiation of a Six Sigma deployment does not start with project selections. It starts with the overall understanding of the organization in terms of how it defines itself, in terms of what its objectives are, how it measures itself, what performance metrics are crucial for it to reach its objectives and how those metrics are analyzed.

1.1.1 Define the organization

Defining an organization means putting it precisely in its context; it means defining it in terms of its objectives, in terms of its internal operations and in terms of its relations with its customers and suppliers.

Mission statement. Most companies' operational strategies are based on their mission statements. A *mission statement* (sometimes called *strategic intent*) is a short inspirational statement that defines the purpose of the organization and its core values and beliefs. It tells why the organization was created and what it intends to achieve in the future.

Mission statements are in general very broad in perspective and not very precise in scope. They are mirrors as well as rudders: they are mirrors because they reflect what the organization is about, and they are rudders because they point the direction that the organization should be heading. Even though they do not help navigate through obstacles and certainly do not fix precise quarterly or annual objectives, such as the projected increase of Return On Investment (ROI) by a certain percentage for a coming quarter, mission statements should clearly define the company's objective so that management can align its strategy with that objective.

What questions should an organization ask? Every organization's existence depends on the profits derived from the sales of the goods or services to its customers. So to fulfill its objectives, an organization must elect to produce goods or services for which it has a competitive advantage and it must produce them at the lowest cost possible while still satisfying its customers. The decision on what to produce raises more questions and addresses the nature of the organization's internal processes and its relations with its suppliers, customers, and competitors.

So to define itself, an organization must answer the following questions:

- How to be structured?
- What to produce?
- How to produce its products or services?
- Who are its customers?
- Who are its suppliers?
- Who are its competitors?
- Who are its competitors' customers and suppliers?

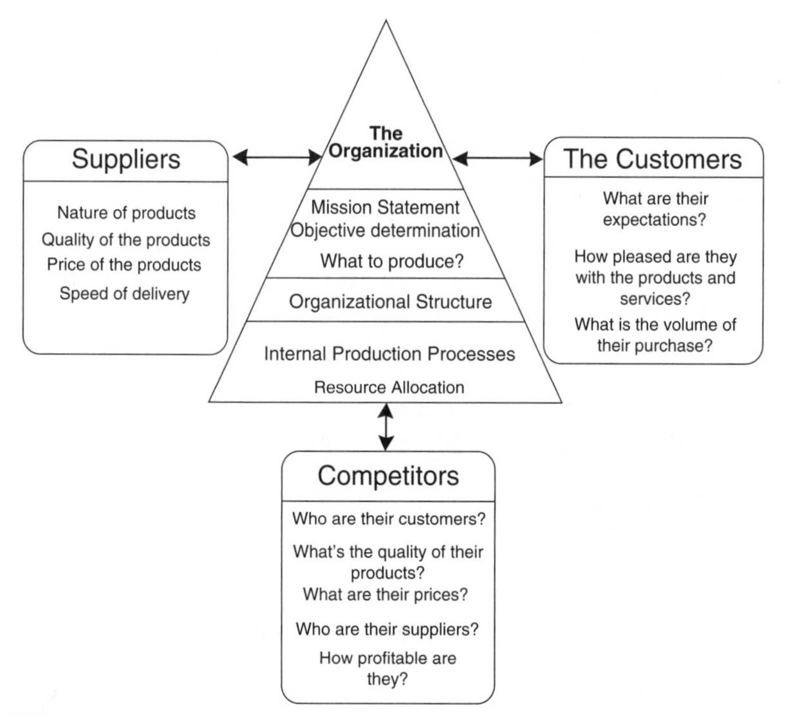

Figure 1.1

What does an organization produce? Years ago, a U.S. semiconductor supply company excelled in its operations and was the number one in its field in America. It won the Malcolm Baldrige National Quality Award twice and was a very well-respected company on Wall Street; its stocks were selling at about $85 a share in 1999. At that time, it had narrowed the scope of its operations to mainly manufacturing and supplying electronic components to major companies.

After it won the Baldrige Award for the second time, the euphoria of the suddenly confirmed success led its executives to decide to broaden the scope of its operations and become an end-to-end service provider—to not only supply its customers (who were generally computer manufacturers) with all the electronic components for their products but also to provide the aftermarket services, the repair services, and customer services for the products. The company bought repair centers where the end users would send their damaged products for repair and also bought call centers to handle the customer complaints. About a year after it broadened the scope of its operations, it nearly collapsed. Its stocks plunged to $3 a share (where they still are), it was obliged to sell most of the newly acquired businesses and had to lay off thousands of employees and is still struggling to redefine itself and gain back its lost market share.

At one point, Daimler-Benz, the car manufacturer, decided to expand its operations to become a conglomerate that would include computers and information technology services and aeronautics and related activities. That decision shifted the company's focus from what it does best and it started to lose its efficiency and effectiveness at making and selling competitive cars. Under Jac Nasser, Ford Motor Company went through the same situation when it decided to expand its services and create an end-to-end chain of operations that would range from the designing and manufacturing of the cars to distribution networks to the servicing of the cars at the Ford automobile repair shops. And there again, Ford lost the focus to its purpose, which was just to design, manufacture, and sell competitive cars.

What happened to these companies shows how crucial it is for an organization to not only elect to produce goods or services for which it is well suited, because it has the competence and the capabilities to produce, but it must also have the passion for it and must be in a position to constantly seek and maintain a competitive advantage for those products.

How does the organization produce its goods or services? One of the essential traits that make an organization unique is its production processes. Even though competing companies usually produce the same products, they seldom use the exact same processes. The production processes determine the quality of the products and the cost of production; therefore, they also determine who the customers are and the degree to which they can be retained.

Who are the organization's customers? Because an organization grows through an increase in sales, which is determined by the number of customers it has and the volume of their purchases, a clear identification and definition of the customers becomes a crucial part of how an organization defines itself. Not only should an organization know who its customers are but, to retain them and gain their long term loyalty and increase them in numbers and the volume of their purchases, it should strive to know why those customers choose it over its competitors.

Who are the organization's suppliers? In global competitive markets, the speeds at which the suppliers provide their products or services and the quality of those products and services have become vital to the survival of any organization. Therefore, the selection of the suppliers and the type of relationship established with them is as important as the selection of the employees who run the daily operations because, in a way, the suppliers are nothing but extensions of an organization's operations. A supplier that provides a car manufacturer with its needed transmission boxes or its alternators may be as important to the car manufacturer's operations as its own plant that manufactures its doors. The speed of innovation for a major manufacturer can be affected by the speed at which its suppliers can adapt to new changes.

Most companies have understood that fact and have engaged in long-term relationships founded on a constant exchange of information and technologies for a mutual benefit. For instance, when Toyota Motor Company decided to put out the Prius, its first hybrid car, if its suppliers of batteries had not been able to meet its new requirements and make the necessary changes to their operations to meet Toyota's demands on time, this would have had negative impacts on the projected date of release of the new cars and their cost of production. Toyota understood that fact and engaged in a special relationship with Matsushita Electric's Panasonic EV Energy to get the right batteries for the Prius on time and within specifications. Therefore, the definition of an organization must also include who its suppliers are and the nature of their relationship.

Who are the organization's competitors? How your competitors perform, their market share, the volume of their sales, and the number of their customers are gauges of your performance. An organization's rank in its field is not necessarily a sign of excellence or poor performance;

some companies deliberately choose not be the leaders in the products or services they provide but still align their production strategies with their financial goals and have excellent results.

Yet in a competitive global market, ignoring your competitors and how they strive to capture your customers can be a fatal mistake. Competitors are part of the context in which an organization evolves and they must be taken into account. They can be used for benchmarking purposes.

Who are the competitors' customers and suppliers? When Carlos Ghosn became the CEO of Nissan, he found the company in total disarray. One of the first projects he initiated was to compare Nissan's cost of acquisition of parts from its suppliers to Renault's cost of acquisition of parts. He found that Nissan was paying 20 percent more than Renault to acquire the same parts. At that time, Nissan was producing about two million cars a year. Imagine the number of parts that are in a car and think about the competitive disadvantage that such a margin could cause for Nissan.

An organization's competitors' suppliers are its potential suppliers. Knowing what they produce, how they produce it, the speed at which they fulfill their orders, and the quality and the prices of their products must be relevant to the organization.

1.1.2 Measure the organization

The overall performance of an organization is generally measured in terms of its financial results. This is because ultimately profit is the life blood of an enterprise. When an organization is being measured at the highest level—as an entity—financial metrics such as the ROI, the net profit, the Return On Assets (ROA), and cash flow are used to monitor and assess performance. Yet, these metrics cannot explain why the organization is performing well or not; they are just an expression of the results, indicators of what is happening. They do not explain the reason why it is happening.

Good or bad financial performance can be the result of non-financial factors such as customer retention, how the resources are managed, how the internal business processes are managed or with how much training the employees are provided. How each one of these factors contributes to the financial results can be measured using specific metrics. Those metrics that are called *mid-level metrics* in this book (to differentiate from the *high-level metrics* used to measure financial results) are also just indicators of how each one of the factors they measure is performing without explaining why they are doing so. For instance, suppose that the Days' Supply of Inventory (DSI) is a mid-level metric used to monitor how many days worth of inventory are kept in a warehouse. DSI can tell us "there is three or four days' worth of inventory in the warehouse" but it will not tell us why.

How high or low the mid-level metrics are is also explained by still lower-level factors that contribute to the performance of the factors measured by the mid-level metrics. The lower-level metrics can range from how often employees are late to work to the sizes of the samples taken to measure the quality of the products. They are factors that explain the fluctuations of mid-level metrics such as the Customer Satisfaction Index (CSI). A high or low CSI only indicates that the customers are satisfied or unsatisfied, but it does not tell us why. The CSI level is dependent on still other metrics such as the speed of delivery and the quality of the products. So there is a vertical relationship between the factors that contribute to the financial results of an organization.

A good example of correlation analysis between metrics in a manufacturing or distribution environment would be the study of how all the different areas of operations in those types of industries relate to the volume of held inventory. The higher the volume of held inventory, the more money will be needed for its maintenance. The money needed for its maintenance comes under the form of expenses for the extra direct labor needed to stock, pick, and transfer the products, which requires extra employees; extra equipment such as forklifts, extra batteries, and therefore more electricity and more trainers to train the employees on how to use the equipment; more RF devices, therefore more IT personnel to maintain the computer systems. A high volume of physical inbound or outbound inventory will also require more transactions in the accounting department because not only are the movements of products for production in progress financially tracked but the insurance paid on the stock of inventory is also a proportion of its value and the space the inventory occupies is also rented real estate.

The performance of every one of the areas mentioned above is measured by specific metrics, and as their fluctuations can be explained by the variations in the volume of inventory, it becomes necessary to find ways and means to quantify their correlations to optimize the production processes.

Measuring the organization through balanced scorecards. *Metrics* are measurements used to assess performance. They are very important for an organization because not only do they show how a given area of an organization performs but also because the area being measured performs according to the kind of metric used to assess its performance. Business units perform according to how they are measured; therefore,

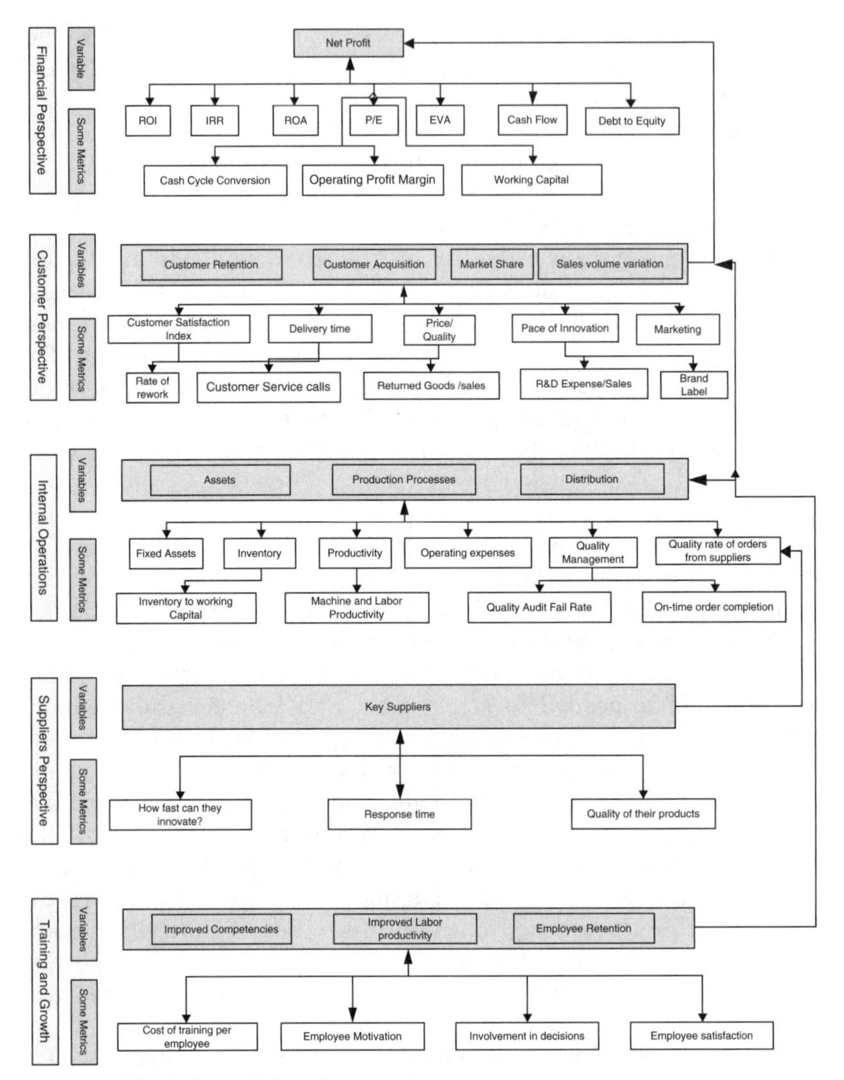

Figure 1.2 Metrics correlation diagram

selecting the correct metrics is extremely important because it ultimately determines performance.

Many organizations tabulate the most important metrics used to monitor their performance in scorecards. *Scorecards* are organized and structured sets of metrics used to translate strategic business objectives into reality. They are report cards that consist of tables containing sets of metrics that measure the performance of every area of an organization. Every measurement is expected to be at a certain level at a given time. Scorecards are used to see if the measured factors are meeting expectations.

In their book, *The Balanced Scorecard*, Robert S. Kaplan and David P. Norton show how a balance must be instilled in the scorecards to go beyond just monitoring the performance of the financial and non-financial measures to effectively determine how the metrics relate to one another and how they drive each other to enhance the overall performance of an enterprise. Balanced scorecards can help determine how to better align business metrics to the organization's long and short-term strategies and how to translate business visions and strategies into actions.

There is not a set standard number of metrics used to monitor performance for an organization. Some scorecards include hundreds of metrics while others concentrate on the few critical ones. Kaplan and Norton's approach to balanced scorecards is focused on four crucial elements of an organization's operations:

- Financial
- Customer
- Internal business processes
- Learning and growth

Finance	2/15/2005	2/27/2005	3/4/2005	3/12/2005	YTD	Target by..
Net profit						
Cost of goods sold						
Operating Expenses						
EVA						
P/E						
Return On Assest						
ROIC						
Net Sales						
DSO						

Learning and Growth

	2/15/2005	2/27/2005	3/4/2005	3/12/2005	YTD	Target by..
Employee Satisfaction Index						
Employee Involvement						
Internal promotions						
Number of employees with a degree						

Internal Operations

	2/15/2005	2/27/2005	3/4/2005	3/12/2005	YTD	Target by..
Productivity						
Rework						
On Time delivery						
Order Cycle time						
Cost per unit produced						
Inventory turn over						

Customer

	2/15/2005	2/27/2005	3/4/2005	3/12/2005	YTD	Target by..
Customer Calls						
Returned goods/Sales						
Customer retention						
New customers						
Customer Satisfaction index						
Market share						

Even though Kaplan and Norton did extensively elaborate on the importance of the suppliers, they did not include them in their balanced scorecards. Suppliers are a crucial element for an organization's performance—their speed of delivery and the quality of their products can have very serious repercussions on an organization's results.

Solectron supplies Dell, IBM, and Hewlett-Packard with computer parts and also repairs their end users' defective products, so in a way, Solectron is an extension to those companies operations. Should it supply them with circuit boards with hidden defects that are only noticeable after extensive use, this can cost the computer manufacturers customers and profit. Motorola supplies Cingular Wireless with mobile phones but Cingular's customers are more likely to blame Cingular than they would blame Motorola for poor reception even when the defects are due to a poor manufacturing of the phones. So suppliers must be integrated into the balanced score cards.

1.1.3 Analyze the organization

If careful attention is not given to how metrics relate to one another in both vertical and horizontal ways, scorecards can end up being nothing but a stack of metrics that may be a good tool to see how the different areas of a business perform but not an effective tool to align those metrics to a business strategy. If the vertical and horizontal contingence between the metrics is not established and made obvious and clear, it would not be right to qualify the scorecards as balanced and some of the metrics contained in them may not be adequate and relevant to the organization.

A distribution center for a cellular phone service provider used Quality Assurance (QA) audit fail rate as a quality metric in its scorecard. They took a sample of the cell phones and accessories and audited them at the end of the production line; if the fail rate was two percent, they would multiply the volume of the products shipped by 0.02 to determine the projected volume of defective phones or accessories sent to their customers. The projected fail rate is used by customer services to plan for the volume of calls that will be received from unhappy customers and allocate the necessary human and financial resources to respond to the customers' complaints.

The projected volume of defective products sent to the customers has never come anywhere close to the volume of customer complaints, but they were still using the two metrics in the same scorecard. It is obvious that there should be a correlation between these two metrics. If there is none, one of the metrics is wrong and should not be used to explain the other.

The purpose of analyzing the organization is primarily to determine if the correct metrics are being used to measure performance and, if they are, to determine how the metrics relate to one another, to quantify that relationship to determine what metrics are performance drivers, and how they can be managed to elevate the organization's performance and improve its results.

What metrics should be used and what is the standard formula to calculate metrics? Every company has a unique way of selecting the metrics it uses to measure itself, and there is no legal requirement for companies to measure themselves, let alone to use a specific metric. Yet at the highest level of an enterprise, financial metrics are generally used to measure performance; however, the definition of these metrics may not be identical from one company to another. For instance, a company might use the ROA to measure the return it gets from the investments made in the acquisition of its assets. The questions that come to mind would be: "What assets? Do the assets include the assets for which the accounting value has been totally depleted but are still being used, or is it just the assets that have an accounting value? What about the revenue—does it include the revenue generated from all the sales or is it just the sales that come from the products generated by a given set of assets?"

The components of ROA for one company may be very different from the components of the same metrics in a competing company. So in the "Analyze the Organization" phase, not only should the interactions between the metrics be assessed but the compositions of the metrics themselves must be studied.

How to analyze the organization. Most companies rely on financial analysts to evaluate their results, determine trends, and make projections. In a Six Sigma environment, the Master Black Belt plays that role. Statistical tools are used to determine what metrics are relevant to the organization and in what area of the organization they are appropriate; those metrics and only those are tracked in the scorecards. Once the metrics are determined, the next step will consist in establishing correlations between the metrics. If the relationships between the relevant measurements are not established, management will end up concentrating on making local improvements that will not necessarily impact the overall performance of the organization.

These correlations between the measurements can be horizontal when they pertain to metrics that are at the same level of operations. For instance, the quality of the product sent to the customers and on-time delivery are both factors that affect the CSI. And a correlation can be found between quality and on-time delivery because poor quality can cause more rework, which can affect the time it takes to complete customer orders.

An example of a vertical correlation would be the effect of training on productivity and the effect of productivity and customer satisfaction and the effect of those factors on profit. In a nutshell, the composition of every metric, how the metrics measure the factors they pertain to and how they affect the rest of the organization must be understood very well.

This understanding is better obtained through statistical analysis. Historic data are analyzed using statistics to measure the organic composition of the metrics, the interactions between them and how they are aligned with respect to the overall organizational strategy. The statistical measurement of the metrics also enables the organization to forecast its future performance and better situate itself with regard to its strategic objectives.

1.1.4 Improve the organization

The theory of constraints is founded on the notion that all business operations are sequences of events, interrelated processes that are linked to one another like a chain, and at any given time one process will act as the weakest link—the bottleneck—and prevent the whole "chain" from achieving its purpose. To improve on the organization's performance, it is necessary to identify the weakest link and improve it. Any improvement made on any process other than the one that happens to be the bottleneck may not improve the overall performance of the organization; it may even result in lowering its performance.

One company claimed to extensively use Six Sigma projects to drive performance. The results of the projects were posted on a board and they seemed to always be excellent and the plant seemed to be saving millions of dollars a year, but were the project savings really reflected in the company's quarterly financial results? They seemed too good to be true and because the projects that the Black Belts work on tend to concentrate on local optima without considering the contingence between departments, improving one process or department in the organization may not necessarily positively impact the overall organizational performance.

For instance, if the company improves on its shipping department while it still fails to better the time it takes to fill customer orders, the improvement made in the shipping process will not have any impact in the overall operations. The correlations between departments' measurements and processes must be taken into account when selecting a project. Once the Master Black belt analyzes the organization, he or she determines what the areas that show opportunity for improvement are and selects a Black Belt to work on a project to address the bottleneck.

Once the Black belt is selected, he or she works on the project following the same DMAIC process:

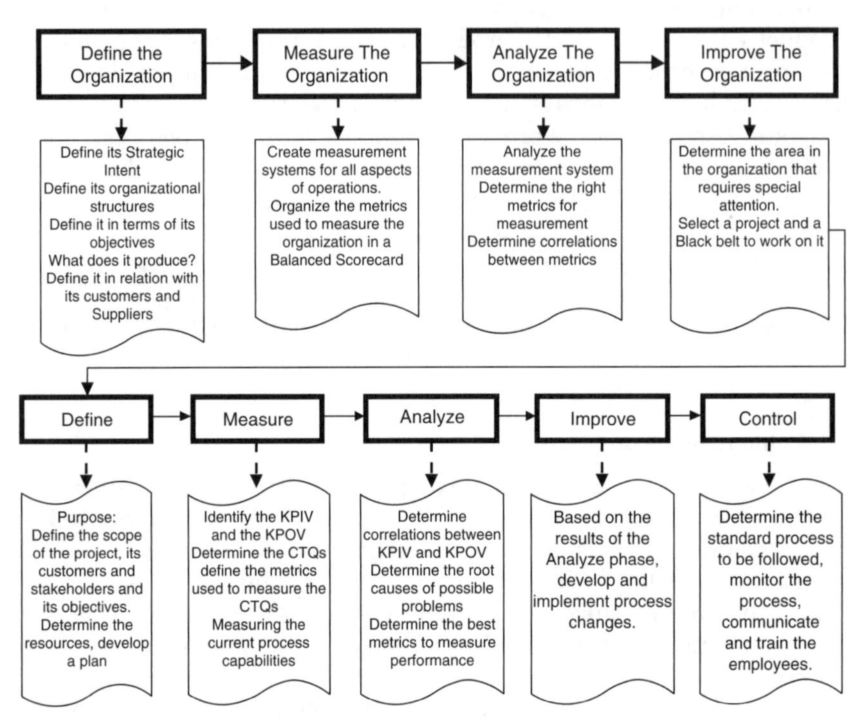

Figure 1.3 Six Sigma project selection process

Notice that at the organization level, we did not include a Control phase. This is because improvement is a lifelong process and after improvements are made, the organization still needs to be continuously measured, analyzed, and improved again.

1.2 Statistics, Quality Control, and Six Sigma

The use of statistics in management in general, and quality control in particular, did not begin with Six Sigma. Statistics has started to play an increasingly important role in quality management since Walter Shewhart from the Bell Laboratories introduced the Statistical Process Control (SPC) in 1924.

Shewhart had determined in the early years of the twentieth century that the variations in a production process are the causes of products' poor quality. He invented the control charts to monitor the production processes and make the necessary adjustments to keep them under control. At that time, statistics as a science was still in its infancy, and most of its developments took place in the twentieth century. As it was becoming more and more refined, its application in quality control became more intense. The whole idea was not to just control quality but to measure it and accurately report it to ultimately reduce defects and rework and improve on productivity.

From W. Edward Deming to Genichi Taguchi, statisticians found ways to use more and more sophisticated statistical tools to improve on quality. Statistics has since been so intrinsically integrated into quality control that it is difficult to initiate quality improvement without the use of statistics. In a production process, there is a positive correlation between the quality of a product and productivity. Improving the production process to where the production of defective parts is reduced will lead to a decrease in rework and returned products.

1.2.1 Poor quality defined as a deviation from engineered standards

The quality of a product is one of the most important factors that determine a company's sales and profit. *Quality* is measured in relation to the characteristics of the products that customers' expect to find, so the quality level of the products is ultimately determined by the customers.

The customers' expectations about a product's performance, reliability, and attributes are translated into Critical-To-Quality (CTQ) characteristics and integrated into the products' design by the design engineers. While designing the products, engineers must also take into account the resources' (machines, people, materials) capabilities—that is, their ability to produce products that meet the customers' expectations. They specify exactly the quality targets for every aspect of the products.

But quality comes with a cost. The definition of the *Cost Of Quality* (COQ) is contentious. Some authors define it as the cost of nonconformance, that is, how much producing nonconforming products would cost a company. This is a one-sided approach because it does not consider the cost incurred to prevent nonconformance and, above all in a competitive market, the cost of improving the quality targets.

For instance, in the case of an LCD (liquid crystal display) manufacturer, if the market standard for a 15-inch LCD with a resolution of 1024 × 768 is 786,432 pixels and a higher resolution requires more pixels, improving the quality of the 15-inch LCDs and pushing the company's specifications beyond the market standards would require the engineering of LCDs with more pixels, which would require extra cost.

In the now-traditional quality management acceptance, the engineers integrate all the CTQ characteristics in the design of their new products and clearly specify the target for their production processes as they define the characteristics of the products to be sent to the customers. But because of unavoidable common causes of variation (variations that are inherent to the production process and that are hard to eliminate) and the high costs of conformance, they are obliged to allow some variation or tolerance around the target. Any product that falls within the specified tolerance is considered to meet the customers' expectations, and any product outside the specified limits would be considered as nonconforming.

Even after the limits around the targets are specified, it is impossible to eliminate variations from the production processes. And since the variations are nothing but deviations from the engineered targets, they eventually lead to the production of substandard products. Those poor-quality products end up being a financial burden for organizations. The deviations from the engineered standards are statistically quantified in terms of standard deviation, or sigma (σ).

1.2.2 Sampling and quality control

The high volumes of mass production, whether in manufacturing or services, make it necessary to use samples to test the conformance of production outputs to engineered standards. This practice raises several questions:

- What sample size would reflect the overall production?
- What inferences do we make when we use the samples to test the whole production?
- How do we interpret the results and what makes us believe that the samples reflect the whole population?

These are practical questions that can only be answered through statistical analysis.

1.3 Statistical Definition of Six Sigma

Six Sigma is defined as a methodology that aims at a quasi-perfect production process. Some authors define it as a methodology that aims at a rate of 3.4 defects per million opportunities (DPMO), but the 3.4 DPMO remains very controversial among the Six Sigma practitioners.

Why 6? Why σ? And why 3.4 DPMO? To answer these questions, we must get acquainted with at least three statistical tools: the mean, the standard deviation, and the normal distribution theory.

In the design phase of their manufacturing processes, businesses correctly identify their customers' needs and expectations. They design products that can be consistently and economically manufactured to meet those expectations. Every product exhibits particular characteristics, some of which are CTQ because their absence or their lack of conformance to the customers' requirement can have a negative impact on the reliability of the product and on its value. Because of the importance of the CTQ characteristics, after deciding what to produce the design engineers set the nominal values and the design parameters of the products. They decide on what would be the best design under current circumstances.

For the sake of discussion, consider a rivet manufacturer. Rivets are pins that are used to connect mating parts. Each rivet is manufactured for a given size of hole, so the rivet must exhibit certain characteristics such as length and diameter to properly fit the holes it is intended for and correctly connect the mating parts. If the diameter of the shaft is too big, it will not fit the hole and if it is too small, the connection will be too loose. To simplify the argument, only consider one CTQ characteristic—the length of the rivet, which the manufacturer sets to exactly 15 inches.

1.3.1 Variability: the source of defects

But a lot of variables come into action when the production process is started, and some of them can cause variations to the process over a period of time. Some of those variables are inherent to the production process itself (referred to as *noise factors* by Taguchi) and they are unpredictable sources of variation in the characteristics of the output. The sources of variation are multiple and can be the result of untrained operators, unstable materials received from suppliers, poorly designed production processes, and so on. Because the sources of variation can be unpredictable and uncontrollable, when it is acceptable to the customers, businesses specify tolerated limits around the target.

For instance, our rivet manufacturer would allow ± 0.002 inches added to the 15-inch rivets it produces; therefore, 15 inches becomes the length of the mean of the acceptable rivets.

The *mean* is just the sum of all the scores divided by their number,

$$\bar{x} = \frac{\sum x}{N}$$

Because all the output will not necessarily match the target, it becomes imperative for the manufacturer to be able to measure and control the variations.

The most widely used measurements of variation are the range and the standard deviation. The *range* is the difference between the highest and the lowest observed data. Because the range does not take into account the data in between, the standard deviation will be used in the attempt to measure the level of deviation from the set target. The standard deviation shows how the data are scattered around the mean, which is the target.

The *standard deviation* (s) is defined as

$$s = \sqrt{\frac{\sum_{i=1}^{n} (x_i - \overline{x})^2}{n - 1}}$$

for small samples, where $\overline{x}$ is the mean, x_i is ith rivet observed, n is the number of rivets observed, and $n - 1$ is the degrees of freedom. It is used to derive an unbiased estimator of the population's standard deviation.

If the sample is greater than or equal to 30 or the whole population is being studied, there would be no need for a population adjustment and the Greek letter σ will be used instead of s. Therefore, the standard deviation becomes

$$\sigma = \sqrt{\frac{\sum (x - \mu)^2}{N}}$$

where μ is the arithmetic mean and N represents the population observed.

Suppose that the standard deviation in the case of our rivet manufacturer is 0.002, and $\pm 3\sigma$ from the mean are allowed. In that case, the specified limits around a rivet would be 15 ± 0.006 inches ($0.002 \times 3 = 0.006$). So any rivet that measures between 14.994 inches ($15 - 0.006$) and 15.006 inches ($15 + 0.006$) would be accepted, and anything outside that interval would be considered as a defect.

1.3.2 Evaluation of the process performance

Once the specified limits are determined, the manufacturer will want to measure the process performance to know how the output compares to the specified limits. They will therefore be interested in two aspects of the process, the process capabilities and the process stability. The *process capability* refers to the ability of the process to generate products that are within the specified limits, and the *process stability* refers to the manufacturer's ability to predict the process performance based on past experience. In most cases, the SPC is used for that purpose and control charts are used to interpret the production patterns.

Because it would be costly to physically inspect every rivet that comes off the production lines, a sample will be taken and audited at specified intervals of time and an estimation will be derived for the whole production to determine the number of defects.

1.3.3 Normal distribution and process capability

A distribution is said to be *normal* when most of the observations are clustered around the mean. In general, manufactured products are normally distributed and when they are not, the Central Limit Theorem usually applies. So the normal distribution is used when samples are taken from the production line and the probability for a rivet being defective is estimated.

The density function of the normal distribution is

$$f(x) = \frac{1}{\sigma\sqrt{2\pi}} e^{-(x-\mu)^2/2\sigma^2}$$

The curve associated with that function is a bell-shaped curve that spreads from $-\infty$ to $+\infty$ and never touches the horizontal line. The area under the curve represents the probabilities for an event to occur, and the whole area under the curve is estimated to be equal to 1.

In the graph in Figure 1.4, the area between the USL and the LSL represents the products in conformance and the darkened areas at the tails of the curve represent the defective ones.

If the manufacturer uses the sigma scale and sets the specifications to $\pm 3\sigma$, how many rivets should we expect to be within specification?

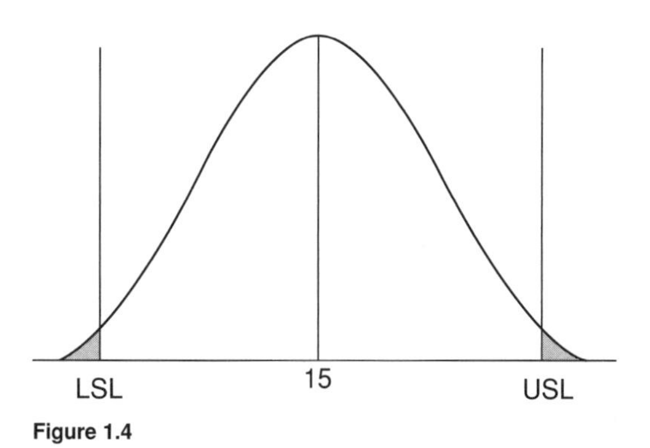

Figure 1.4

TABLE 1.1

Range around μ	Percentage of products in conformance	Percentage of nonconforming products	Nonconformance out of a million
-1σ to $+1\sigma$	68.26	31.74	317,400
-2σ to $+2\sigma$	95.46	4.54	45,400
-3σ to $+3\sigma$	99.73	0.27	2700
-4σ to $+4\sigma$	99.9937	0.0063	63
-5σ to $+5\sigma$	99.999943	0.000057	0.57
-6σ to $+6\sigma$	99.9999998	0.00000002	0.002

Because the area under the normal curve that uses σ scale has already been statistically estimated (see Table 1.1), we can derive an estimation of the quantity of the products that are in conformance.

The probability for a rivet to be between $\mu - 3\sigma$ and $\mu + 3\sigma$ is 0.9973, and the probability for it to be outside the limits will be 0.0027 $(1 - 0.9973)$. In other words, 99.73 percent of the rivets will be within the specified limit, or 2700 out of 1 million will be defective.

Suppose that the manufacturer improves the production process and reduces the variation to where the standard deviation is cut in half and it becomes 0.001. Bear in mind that a higher standard deviation implies a higher level of variation and that the further the specified limits are from the target μ, the more variation is tolerated and therefore the more poor-quality products are tolerated (a 15.0001-inch long rivet is closer to the target than a 15.005-inch long rivet).

Table 1.2 shows the level of quality associated with σ and the specified limits $(\mu + z\sigma)$. Clearly, the quality level at $\pm 6\sigma$ after improvement is the same as the one at $\pm 3\sigma$ when σ was 0.002 (14.994, 15.006) but the quantity of conforming products has risen to 99.9999998 percent and the defects per million have dropped to 0.002. An improvement of the process has lead to a reduction of the defects.

TABLE 1.2

$(\mu = 15)$	0.002	0.001
$(\mu + 1\sigma)$	15.002	15.001
$(\mu - 1\sigma)$	14.998	14.999
$(\mu + 2\sigma)$	15.004	15.002
$(\mu - 2\sigma)$	14.996	14.998
$(\mu + 3\sigma)$	15.006	15.003
$(\mu - 3\sigma)$	14.994	14.997
$(\mu + 4\sigma)$	15.008	15.004
$(\mu - 4\sigma)$	14.998	14.996
$(\mu + 5\sigma)$	15.010	15.005
$(\mu - 5\sigma)$	14.990	14.995
$(\mu + 6\sigma)$	15.012	15.006
$(\mu - 6\sigma)$	14.988	14.994

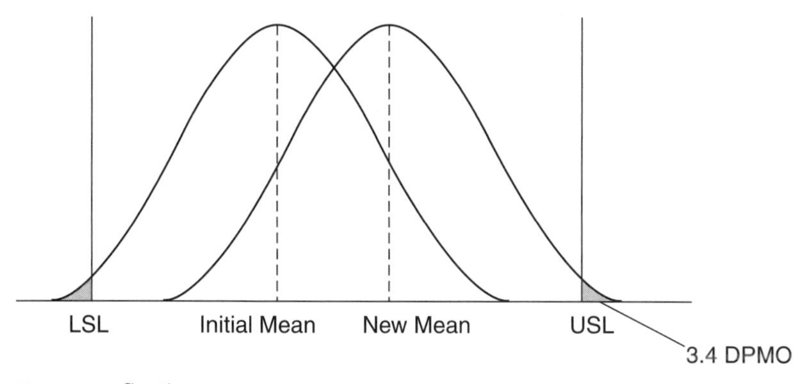

Figure 1.5 Caption

What does 3.4 DPMO have to do with all this? We see in Table 1.2 that a 6σ-level corresponds to 0.002 defects per 1 million opportunities. In fact, 3.4 DPMO is obtained at about $\pm 4.5\sigma$. But this only applies to a static process — in other words, to a short-term process.

According to the Motorola Six Sigma advocates, small shifts that are greater than 1.5σ will be detected and corrective actions taken, but shifts smaller than 1.5σ can go unnoticed over a period of time. In the long run, an accumulation of small shifts in the process average will lead to a drift in the standard deviation of the process. So in the worst case, the noise factors will cause a process average shift that will result in it being 1.5σ away from the target, therefore only 4.5σ will be the distance between the new average process and the closest specified limit. And 4.5σ corresponds to 3.4 DPMO (Figure 1.5).

Note that manufacturers seldom aim at 3.4 DPMO. Their main objective is to use Six Sigma for the sake of minimizing defects to the lowest possible rates and increase customer satisfaction. 3.4 DPMO and the 1.5 sigma shift remain very controversial among the six sigma practitioners.

Chapter

2

An Overview of Minitab and Microsoft Excel

Learning Objectives:

- Understanding the primary tools used in Minitab and Excel

The complexity involved in manipulating some statistics formulae and the desire to get the results quickly have resulted in the creation of a plethora of statistical software. Most of them are very effective tools at solving problems quickly and are very easy to use.

Minitab has been around for many years and has proven to be very sophisticated and easy to use. It has also been adopted by most Six Sigma practitioners as a preferred tool. The statistics portion of Microsoft Excel's functionalities does not match Minitab's capabilities but because it is easy to access and easy to use, Excel is widely used by professionals.

2.1 Starting with Minitab

From a Microsoft desktop, Minitab is opened like any other program, either by double-clicking on its icon or from the Windows Start Menu. Once the program is open, we obtain a window that resembles the one shown in Figure 2.1. The top part of the window resembles most Microsoft programs. It has a title bar, a menu bar, and a tool bar.

The session window displays the output of the statistical analysis while the worksheets store the data to be analyzed. The worksheets resemble Excel's spreadsheets but their functionalities are totally different. Some operations can be directly performed on the Excel

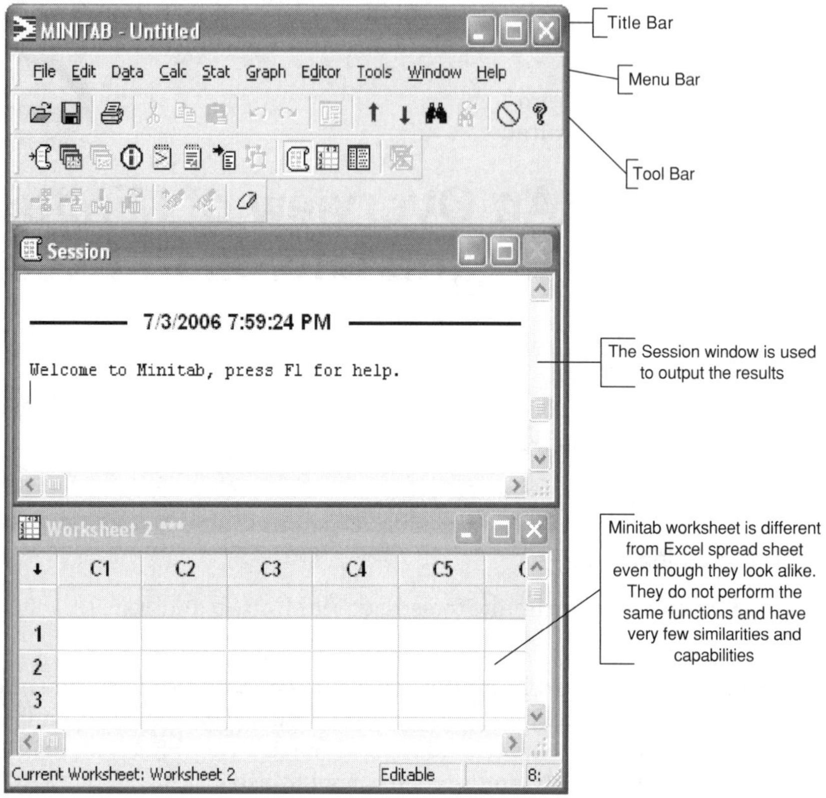

Figure 2.1

spreadsheet but would require extra steps when using Minitab. Minitab does not offer the same flexibility as Excel outside of statistical analysis.

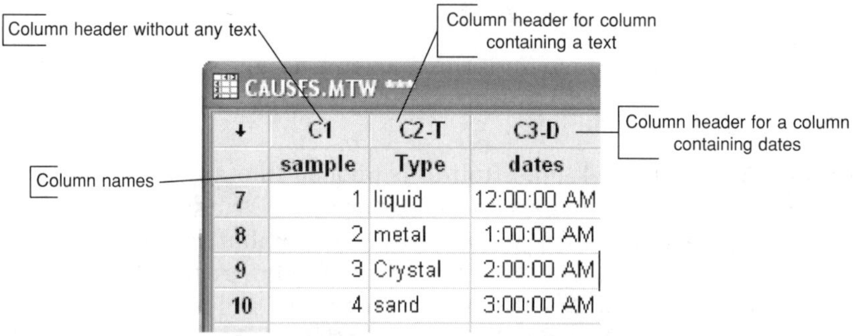

↓	C1	C2-T	C3-D
	sample	Type	dates
7	1	liquid	12:00:00 AM
8	2	metal	1:00:00 AM
9	3	Crystal	2:00:00 AM
10	4	sand	3:00:00 AM

The worksheets come with default column headers. The default headers start with the letter "C" and a number. If the columns are filled with numeric data only, their headers remain unchanged; if the columns contain text data, "-T" will be added to the default header; if the data contained in the columns are dates or times, "-D" will be added to the default header.

Underneath the column headers we have the column names, which are blank when the Minitab window is first opened. The column names are entered or pasted with the data. These names are important because they will follow the data throughout the analysis and will be part of the output.

2.1.1 Minitab's menus

Minitab uses some namings on the menu bar that are common to most Microsoft products. The content and the layout of the File menu are close to Excel's File menu with a few exceptions. The shortcuts are the same in both Excel and Minitab: *Ctrl+n* is for a new file and *Ctrl+p* is for the print function.

File menu. The "New. . . " option of the File menu will prompt the user to choose between a new project and a new worksheet. Choosing a new project will generate a new session window and a new worksheet, whereas choosing a new worksheet will maintain the current session window but create a new worksheet.

The "Open Project . . . " option will list all the previously saved Minitab projects so that the user can make a selection from prior work. The "Project Description . . . " option enables the user to save a project under a new name along with comments and the date and time.

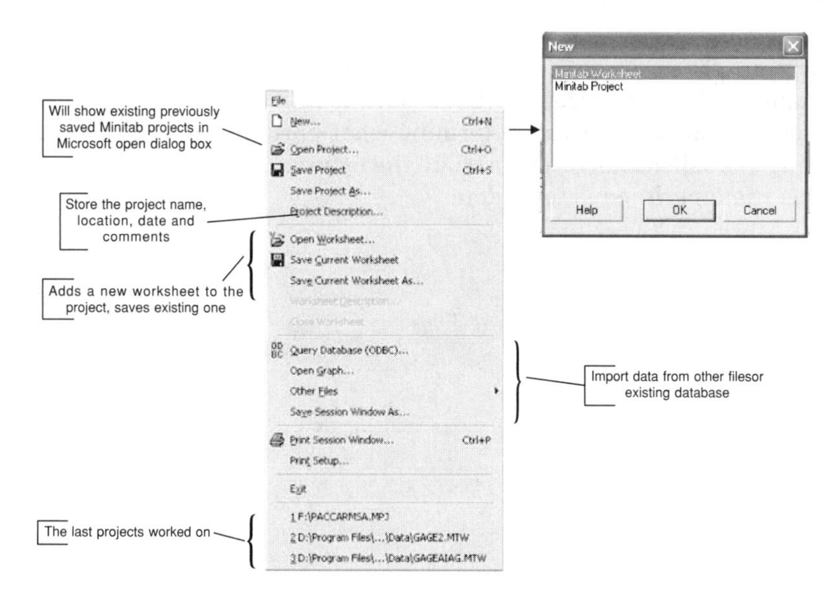

Edit menu. Except for "Command Line Editor," "Edit Last Dialog" and "Worksheet Links," all the other options are the usual options found on Microsoft Edit menus. The "Command Line Editor" opens a dialog box in which the user can type in the commands they would want to execute.

When a value is missing from a set of data in a column, Minitab uses an asterisk (*) to fill out the empty cell. The "Edit Last Dialog" option enables the user to modify the default symbol. The symbols may not change on the worksheet, but once the data are copied and pasted to a spreadsheet, the selected symbols will show.

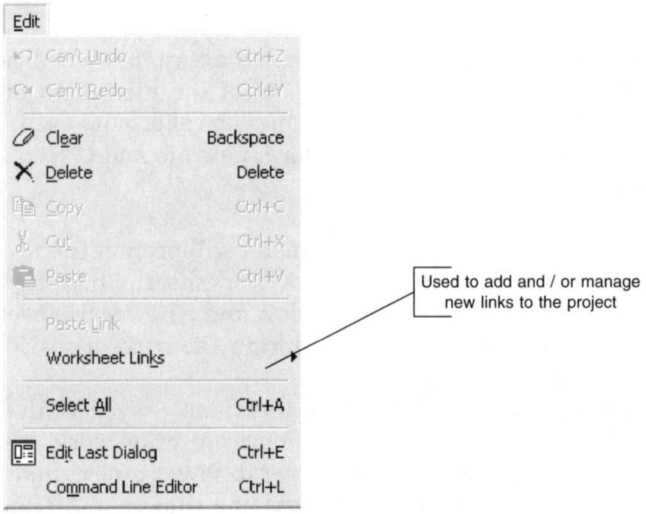

Edit
Can't Undo Ctrl+Z
Can't Redo Ctrl+Y
Clear Backspace
Delete Delete
Copy Ctrl+C
Cut Ctrl+X
Paste Ctrl+V
Paste Link
Worksheet Links
Select All Ctrl+A
Edit Last Dialog Ctrl+E
Command Line Editor Ctrl+L

Used to add and / or manage new links to the project

Data menu. The Data menu helps organize the data before the analyses are conducted. The first option, "Subset Worksheet . . . ," generates a new worksheet and the user can determine what part of the data from the existing worksheet can be exported to the new one and under what conditions that operation is to be done.

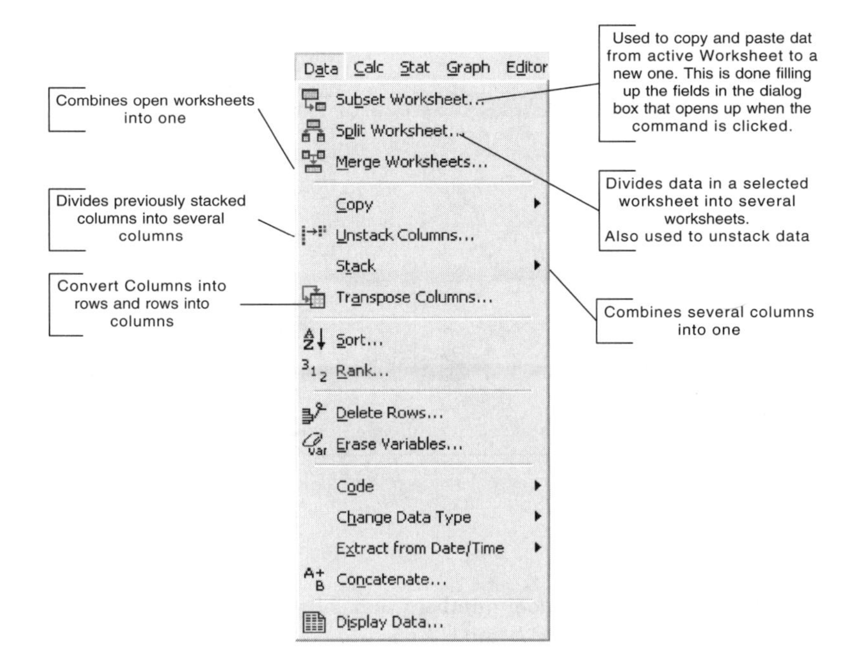

Calc menu. The Calc menu enables options such as the Minitab Calculator and basic functions such as adding or subtracting columns.

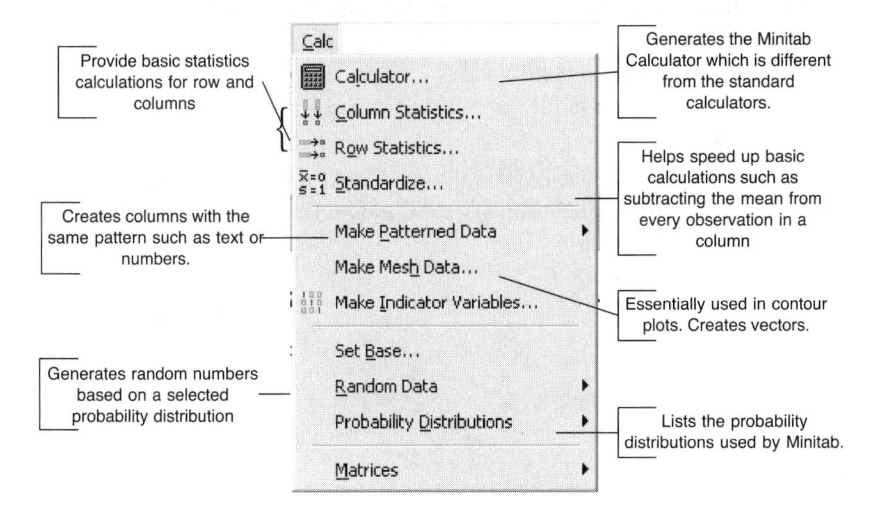

Basic functionalities of the Minitab Calculator. The Minitab Calculator is very different from a regular calculator. It is closer to Excel's Insert Function dialog box.

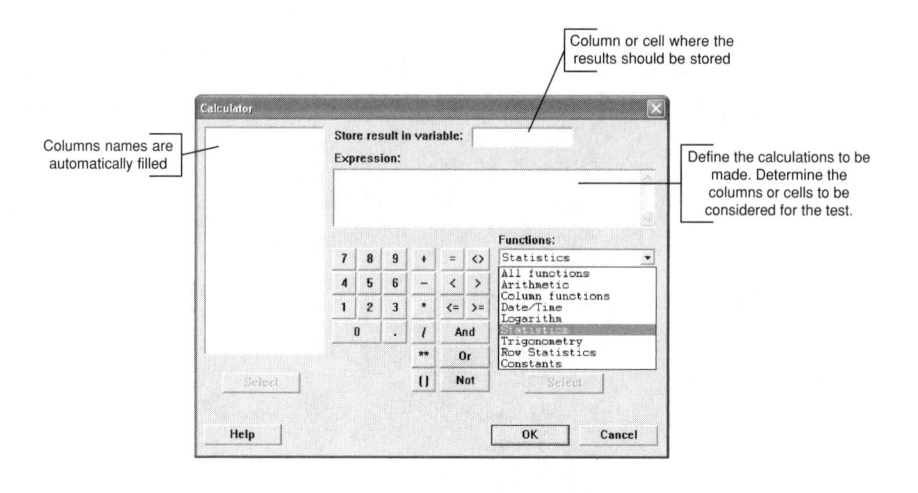

Example

1. Generate 25 rows of random numbers that follow a Poisson distribution with a mean of 25 in C1, C2, and C3.
2. Store the square root of C1 in C4.
3. After generating the numbers, sum the three columns and store the result in C5.
4. Find the medians for each row and store the results in C6.
5. Find all cells in C1 that have greater values than the ones in C2 on the same rows. Store the results in C7.
6. Find the total value of all the cells in C1 and store the results in C8.

Solution

1. *Generating random numbers*. Open a new Minitab worksheet. From the menu bar, select "Calc," then select "Random Data" and then select "Poisson . . . "

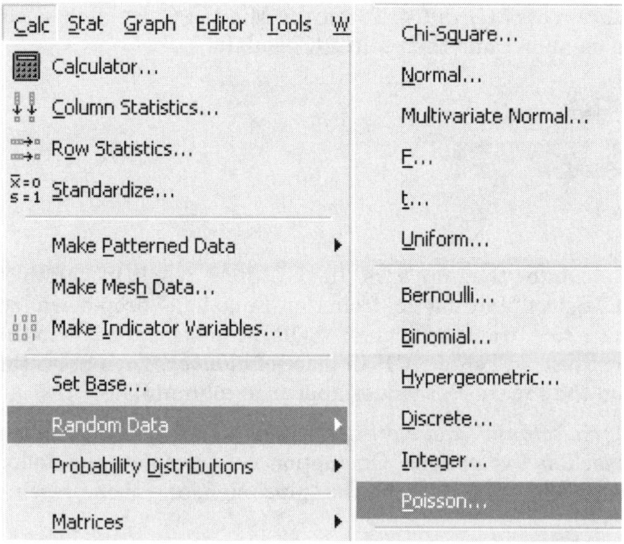

Fill out the fields in the Poisson distribution dialog box, then press the "OK" button.

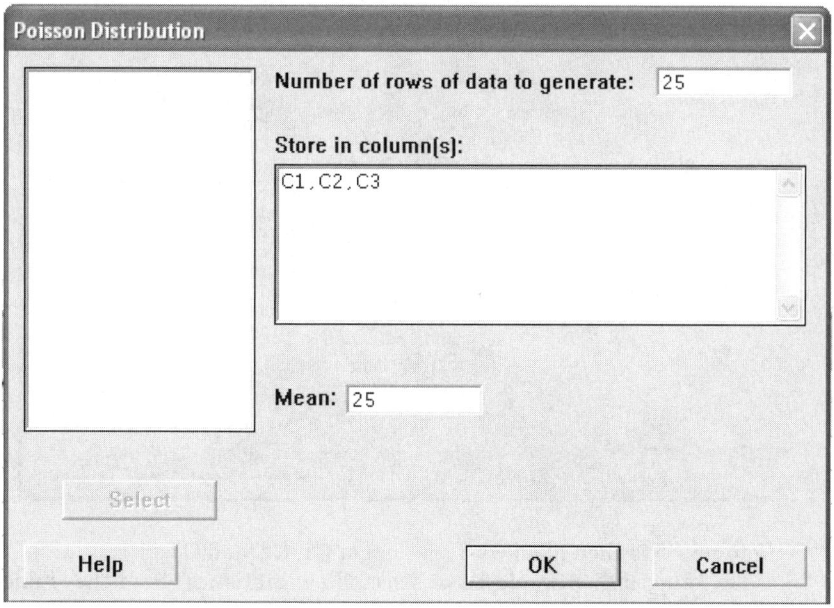

The three first columns will be filled with 25 random numbers following the Poisson distribution.

2. *Storing the square root of C1 in C4.* To get the Minitab Calculator, select "Calc" from the menu bar and then select "Calculator . . . "

The Minitab Calculator then appears. In the "Store result in variable" field, enter "C4." Select "Arithmetic" from the "Functions" drop-down list and select "Square root" from the list box. "SQRT(number)" appears in the "Expression" text box, and enter "C1" in place of *number*. Then press the "OK" button and the square root values appear in column C4.

3. *Summing the three columns and storing the results in C5.* To add up the columns, the user has two options. One option is to complete the dialog box as indicated in Figure 2.11, using the "plus" sign, and then pressing the "OK" button.

Calculator
C1
C2
C3
Store result in variable: C5
Expression:
C1 + C2 + C3
Functions:
All functions
7 8 9 + = <>
4 5 6 – < >
1 2 3 * <= >=
0 . / And
** Or
() Not
Absolute value
Antilog
Arcsine
Arccosine
Arctangent
Ceiling
Cosine
Current time
Select
Select
Help
OK
Cancel

Column C5 is then filled with the sum of C1, C2, and C3.

The other option would be to select "Row Statistics" from the "Functions" drop-down list and choose "Sum" from the list box below it. "RSUM(number, number, . . .)" appears in the "Expression" text box, replace *number* with each of the column names, and then press "OK" and C5 will be filled with sum of the columns.

4. *Finding the medians for each row and store the results in C6.* Remember that we are not looking for the median within an individual column but rather across columns. From the "Functions" drop-down list, select "Row Statistics." "RMEDIAN(number,number, . . .)" then appears in the "Expression" text box, and replace each *number* with C1, C2, and C3, and then press "OK." The median values across the three columns appear in C6.
5. *Finding all the cells in column C1 that have greater values than the ones in column C2 on the same rows.* In the field "Store result in variable," enter "C7." In the "Expression" text box, enter "C1 > C2," then press "OK" and the column C7 will be filled with values of 0 and 1. The zeros represent the cells in C1 whose values are lower than the corresponding cells in C2.
6. *Find the total value of all the cells in C1 and store the results in C8.* In the field "Store result in variable," enter "C8." From the "Functions" drop-down list, select "Statistics" and then choose "Sum" in the list box beneath it. "SUM(number)" appears in the "Expression" text box, replace *number* with "C1," and then press "OK" and the total should appear in the first cell of column C8.

Help menu. Minitab has spared no effort to build solid resources under the Help menu. It contains very rich tutorials is and easy to access. The Help menu can be accessed from the menu bar but it can also be accessed from any dialog box. The Help menu contains examples and solutions with the interpretations of the results. The tutorials can be accessed by selecting "Help" and then selecting "Tutorials" or by pressing the "Help" button on any dialog box.

Example The user is running a regression analysis and would like to know how to interpret the results. To understand how the test is conducted and how to interpret the results, all that must be done is to open a Regression dialog box and select the Help button. Select "Stat," then select "Regression" and from the drop-down list, select "Regression" again. Once the dialog box opens, press the "Help" button.

Regression

Response:

Predictors:

Select

Help

Graphs...

Options...

Results...

Storage...

OK

Cancel

After pressing "Help," the tutorial window will appear. The tutorials contain not only an overview of the topic but also practical examples on how to solve problems and how to interpret the results obtained.

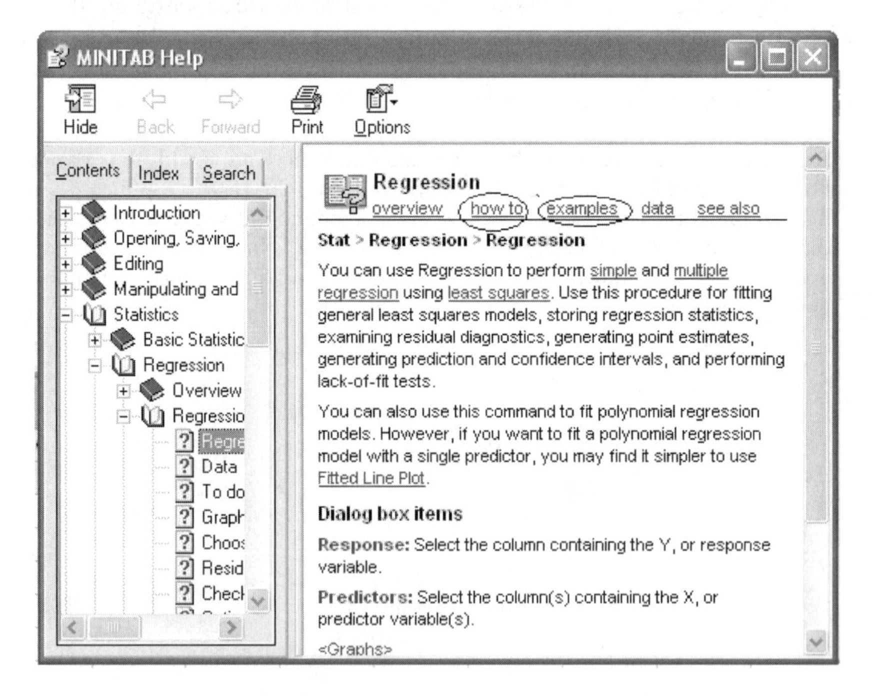

Another way to access the Help menu and get interpretations of the results that are actually generated after conducting an analysis is to right-click on the output in the session window and select "StatGuide," or by pressing *Shift+F1*.

```
C3     9  15.556                      *------------)
                                 ---------+---------+
                                       17.5      20.0

Pooled StDev =

Tukey 95% Simul
All Pairwise Co

Individual conf

C1 subtracted f

     Lower  Cent                   -+---------+---------+
C2  -5.854  -0.                    *------------)
C3  -3.743   1.                    -----*-------------)
                                   -+---------+---------+
                                   .0       4.0       8.0

C2 subtracted f

     Lower  Cent                   -+---------+---------+
C3  -3.299   2.                    ------*-------------)
                                   -+---------+---------+
                         -4.0     0.0       4.0       8.0
```

```
Can't Undo                  Ctrl+Z
Can't Redo                  Ctrl+Y
Append Section to Report
Delete                      Delete
Copy                        Ctrl+C
Cut                         Ctrl+X
Paste                       Ctrl+V
Next Command                F2
Previous Command            Alt+F2
Bring Graph to Front
Find...                     Ctrl+F
Replace...                  Ctrl+H
Apply I/O Font              Alt+1
Apply Title Font            Alt+2
Apply Comment Font          Alt+3
StatGuide                   Shift+F1
```

The StatGuide appears under two windows: MiniGuide and StatGuide. The MiniGuide contains the links to the different topics that are found on the StatGuide.

2.2 An Overview of Data Analysis with Excel

Microsoft Excel is the most powerful spreadsheet software on the consumer market today. Like any other spreadsheet software, it is used to enter, organize, store, and analyze data and display the results in a comprehensive way. Most of the basic probabilities and descriptive statistical analyses can be done using built-in tools that are preloaded in Excel. Some more complex tools required to perform further analyses are contained in the add-ins found in the "Tools" menu. Excel's statistical functionalities are not a match for Minitab. However, Excel is very flexible and allows the creation of macros to enable the user to personalize the program and add more capabilities to it.

The flexibility of Excel's macros has made it possible to create very rich and powerful statistical programs that have become widely used. In this book, we will only use the capabilities built into the basic Excel package. This will reduce the ability to perform some analyses with Excel but the use of a specific additional macro-generated program will require the user to purchase that program.

The basic probability and descriptive statistics analyses are performed through the "Insert Function" tool. The "Insert Function" is accessed either by selecting its shortcut on the tool bar,

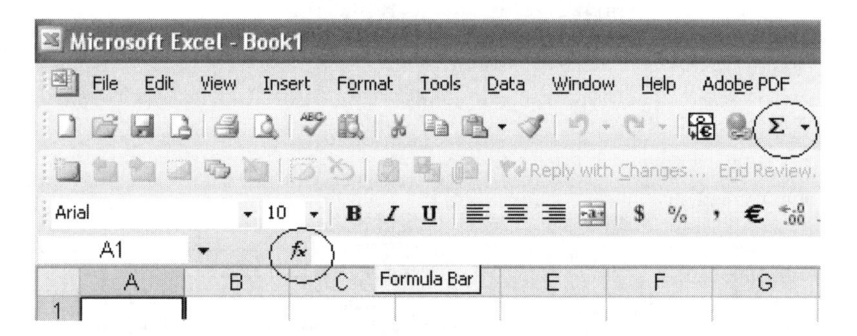

or by selecting "Insert" on the menu bar and then selecting "Function..."

Insert Format Tools Da
Cells...
Rows
Columns
Worksheet
Chart...
Symbol...
Page Break
Function...
Name
Comment
Picture
Diagram...
Object...
Hyperlink... Ctrl+K

In either case, the "Insert Function" dialog box will appear.

This box contains more than just statistical tools. To view all the available functions, select "All" from the "Select a Category" drop-down list and all the options will appear in alphabetical order in the "Select a function" list box. To only view the statistics category, select "Statistical."

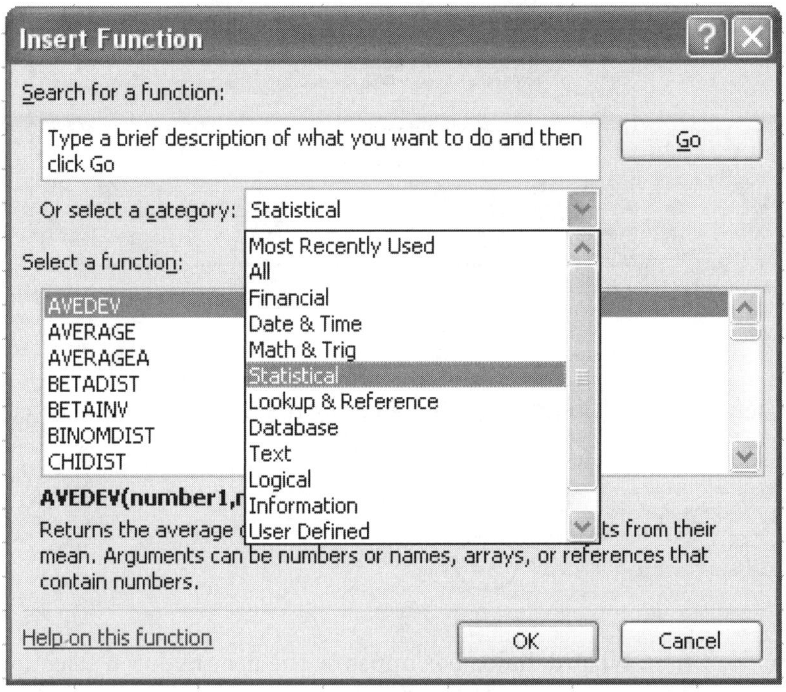

Once the category has been selected, press "OK" and the "Function Arguments" dialog box appears. The forms that this dialog box takes depend on the category selected, but areas of the different parts can be reduced to 5. The dialog box shown in Figure 2.2 pertains to the binomial distribution.

Once the analysis is completed, it is displayed either on a separate worksheet or in a preselected field.

2.2.1 Graphical display of data

Both Minitab and Excel display graphics in windows separate from the worksheets. Excel's graphs are obtained from the Chart Wizard. The Chart Wizard is obtained either by selecting "Insert" on the menu bar

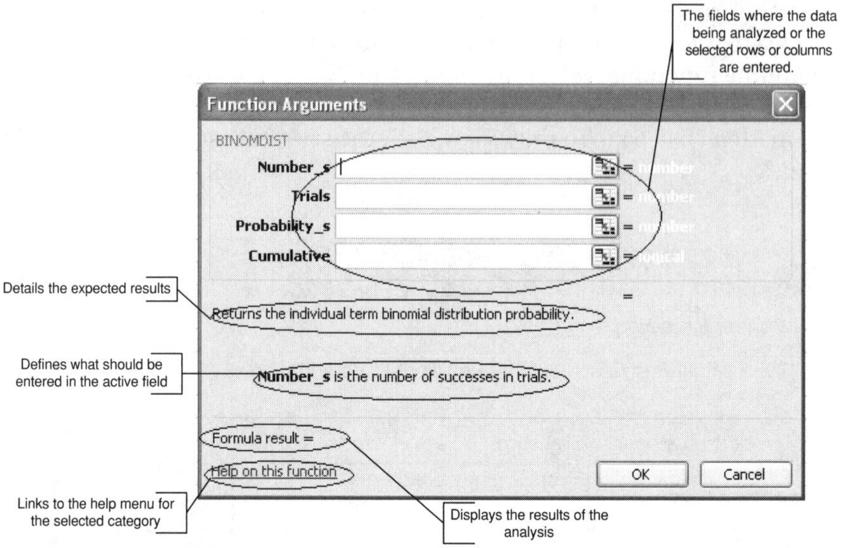

The fields where the data being analyzed or the selected rows or columns are entered.

Function Arguments

BINOMDIST

Number_s

Trials

Probability_s

Cumulative

Details the expected results

Returns the individual term binomial distribution probability.

Defines what should be entered in the active field

Number_s is the number of successes in trials.

Formula result =

Help on this function

OK Cancel

Links to the help menu for the selected category

Displays the results of the analysis

Figure 2.2

and then selecting "Chart . . . " or by selecting the Chart Wizard shortcut on the tool bar.

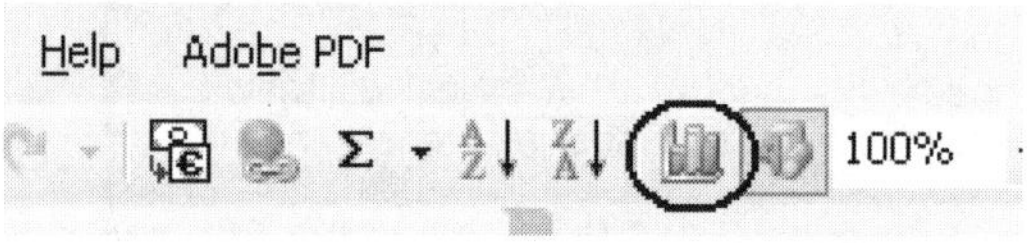

Help Adobe PDF

100%

When the Chart Wizard dialog box appears, the user makes a selection from the "Chart type" list menu and follows the four subsequent steps to obtain the graph.

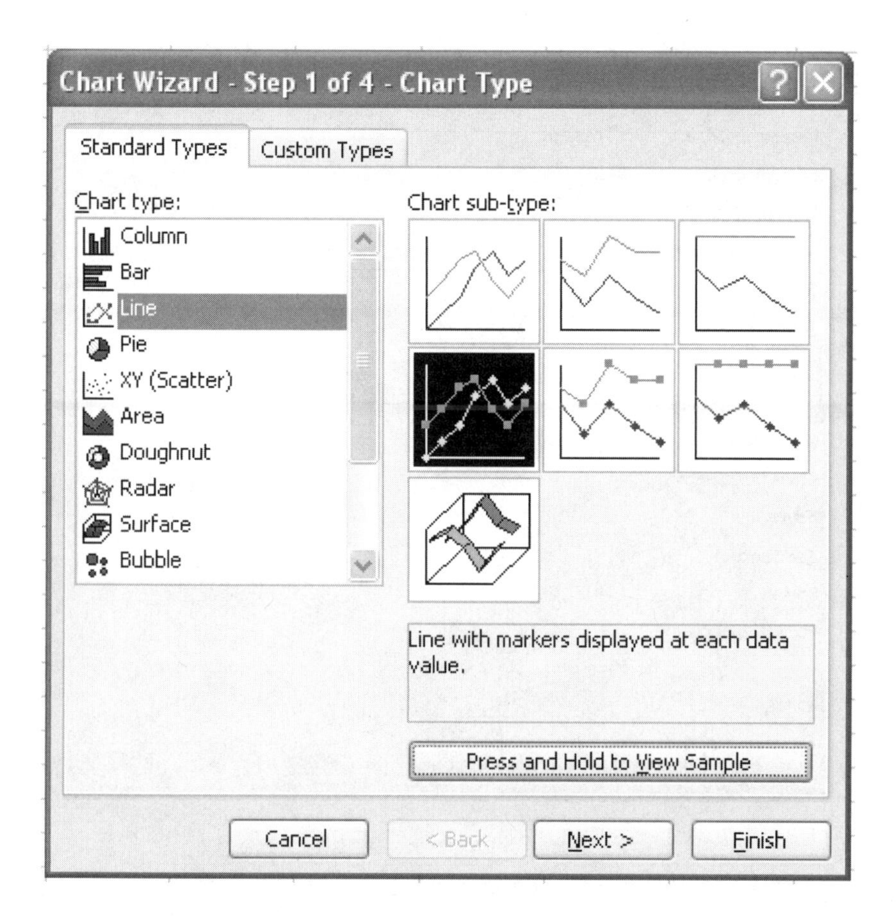

2.2.2 Data Analysis add-in

The tools under the Statistical category in the "Insert Function" dialog are only for basic probability and descriptive statistics; they are not fit for more complex data analysis. Analyses such as regression or ANOVA cannot be performed using the Insert Function tools. These are done through Data Analysis, which is an add-in that can be easily installed from the Tools menu.

To install Data Analysis, select "Tools" from the menu bar, then select "Add-Ins . . . "

The Add-Ins dialog box then appears.

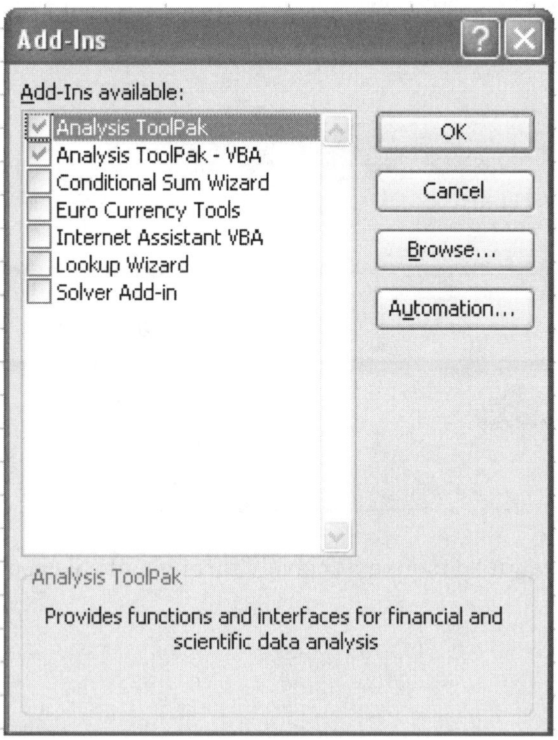

Check the options "Analysis ToolPack" and "Analysis ToolPack — VBA" and the press the "OK" button. This action will add the "Data Analysis . . . " option to the Tools menu.

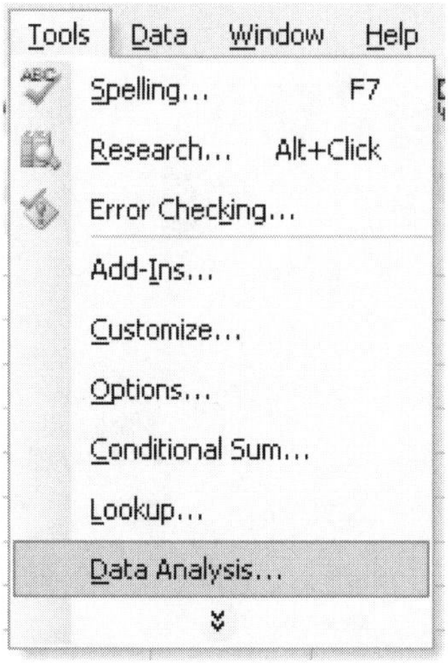

To browse all the capabilities offered by Data Analysis, select that option from the Tools menu and then scroll down the Analysis Tools list box.

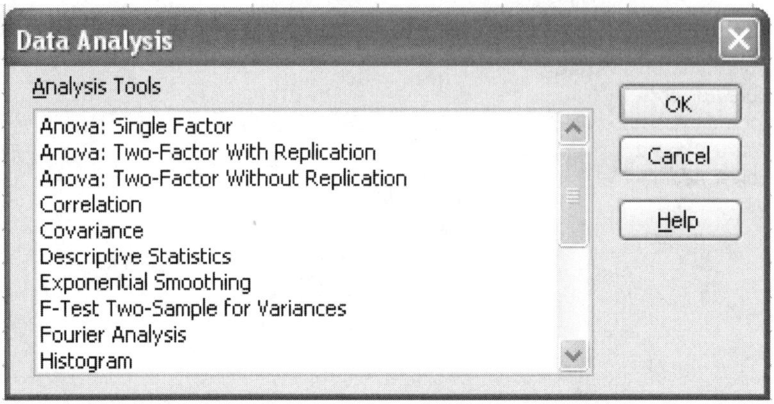

Data Analysis

Analysis Tools

Anova: Single Factor
Anova: Two-Factor With Replication
Anova: Two-Factor Without Replication
Correlation
Covariance
Descriptive Statistics
Exponential Smoothing
F-Test Two-Sample for Variances
Fourier Analysis
Histogram

OK
Cancel
Help

These options will be examined more extensively throughout this book.

Chapter

3

Basic Tools for Data Collection, Organization and Description

Learning Objectives:

- Understand the fundamental statistical tools needed for analysis
- Understand how to collect and interpret data
- Be able to differentiate between the measures of location
- Understand how the measures of variability are calculated
- Understand how to quantify the level of relatedness between variables
- Know how to create and interpret histograms, stem-and-leaf and box plot graphs

Statistics is a science of collecting, analyzing, representing, and interpreting numerical data. It is about how to convert raw numerical data into informative and actionable data. As such, it applies to all spheres of management. The science of statistics is in general divided into two areas:

- *Descriptive statistics*, which deals with the analysis and description of a particular population in its entity. *Population* in statistics is just a group of interest. That group can be composed of people or objects of any kind.
- *Inference statistics*, which seeks to make a deduction based on an analysis made using a sample of a population.

This chapter is about descriptive statistics. It shows the basic tools needed to collect, analyze, and interpret data.

3.1 The Measures of Central Tendency Give a First Perception of Your Data

In most cases, a single value can help describe a set of data. That value can give a glimpse of the magnitude or the location of a measurement of interest. For instance, when we say that the average diameter of bolts that are produced by a given machine is 10 inches, even though we know that all the bolts may not have the exact same diameter, we expect their dimension to be close to 10 inches if the machine that generated them is well calibrated and the sizes of the bolts are normally distributed.

The single value used to describe data is referred to as a *measure of central tendency* or *measure of location*. The most common measures of central tendency used to describe data are the arithmetic mean, the mode, and the median. The geometric mean is not often used but is useful in finding the mean of percentages, ratios, and growth rates.

3.1.1 Arithmetic mean

The *arithmetic mean* is the ratio of the sum of the scores to the number of the scores.

Arithmetic mean for raw data. For ungrouped data—that is, data that has not been grouped in intervals—the arithmetic mean of a population is the sum of all the values in that population divided by the number of values in the population:

$$\mu = \sum_{i=1}^{N} \frac{X_i}{N}$$

where

μ is the arithmetic mean of the population
X_i is the ith value observed
N is the number of items in the observed population
Σ is the sum of the values

Example Table 3.1 shows how many computers are produced during five days of work. What is the average daily production?

TABLE 3.1

Day	Production
1	500
2	750
3	600
4	450
5	775

Solution

$$\mu = \frac{500 + 750 + 600 + 450 + 775}{5} = 615$$

Using Minitab. After having pasted the data into a worksheet, select "Calc" and then "Calculator..."

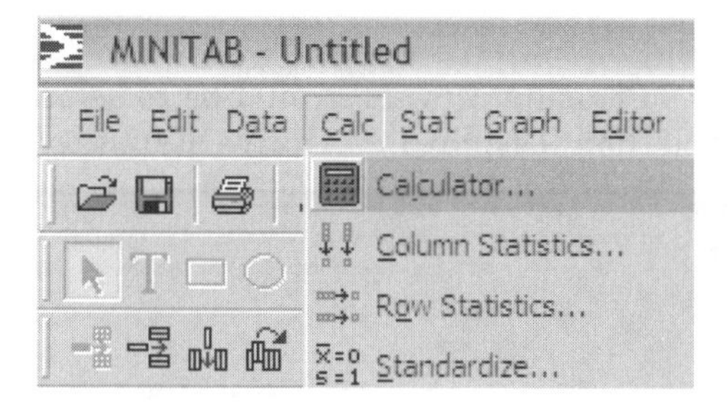

Select the field where you want to store the results. In this case, we elected to store it in C3. Select the "Functions" drop-down list, select "Statistical," and then select "Mean." Notice that "MEAN" appears in the Expression text box with *number* in parenthesis and highlighted. Double-click on "Production" in the left-most text box and then press "OK."

Then the result appears in C3.

Worksheet 1 ***

↓	C1-T	C2	C3
	Days	Production	
1	Day 1	500	615
2	Day 2	750	

Another way to get the same result using Minitab would be to select "Stat," then "Basic Statistics," and then "Display Descriptive Statistics."

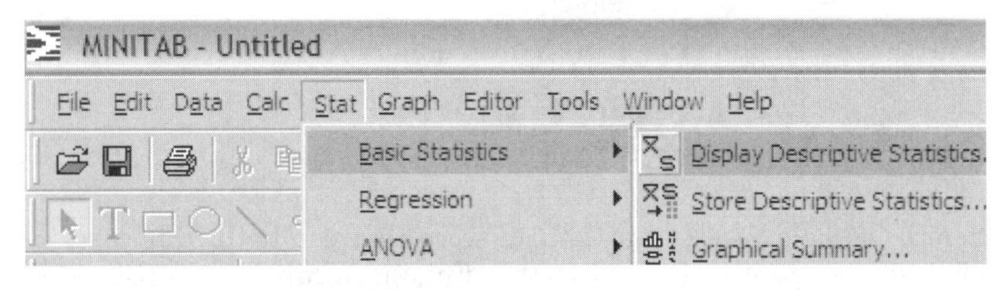

After selecting "Production" for the Variables field, select "Statistics" and check the option "Mean." Then press "OK," and press "OK" again.

The result will display as shown in Figure 3.1.

Descriptive Statistics: Production

```
Variable    Mean
production 615.0
```

Figure 3.1

Using Excel. After having pasted the data into a worksheet, either select "Insert" and then "Function" or just select the "Insert Function" shortcut.

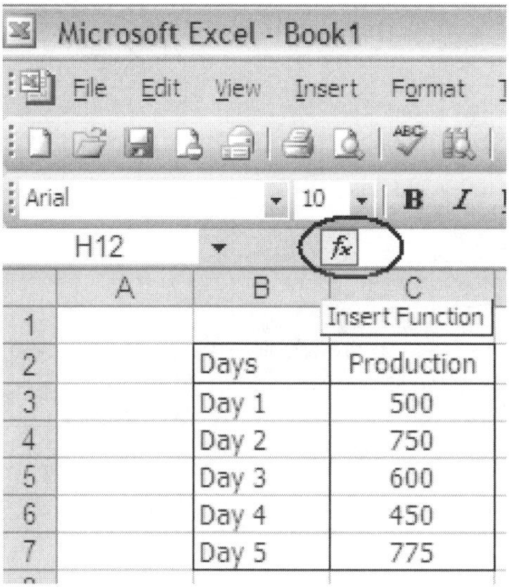

Days	Production
Day 1	500
Day 2	750
Day 3	600
Day 4	450
Day 5	775

In the "Insert Function" dialog box, select "Statistical" and then "AVERAGE."

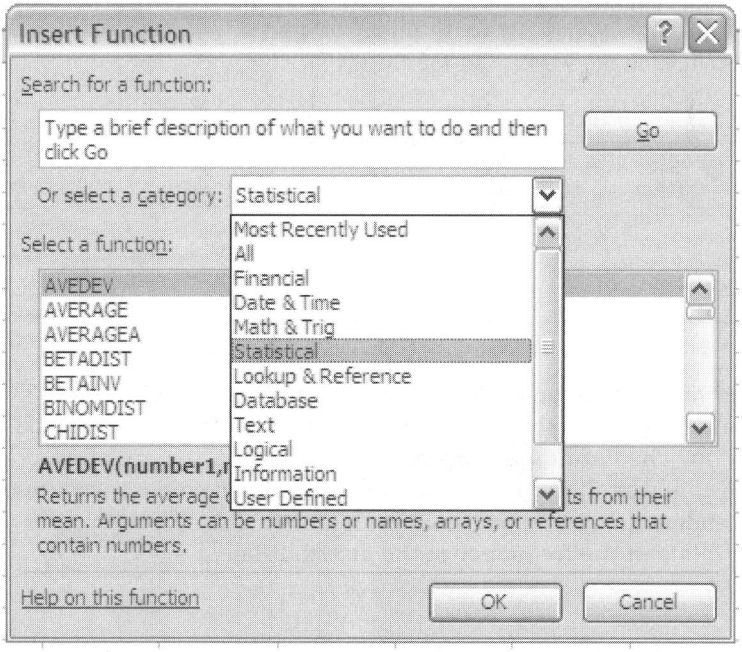

In the "Function Arguments" dialog box, enter the range of the numbers in the *Number1* field. Notice that the result of the formula shows on the box even before you press "OK."

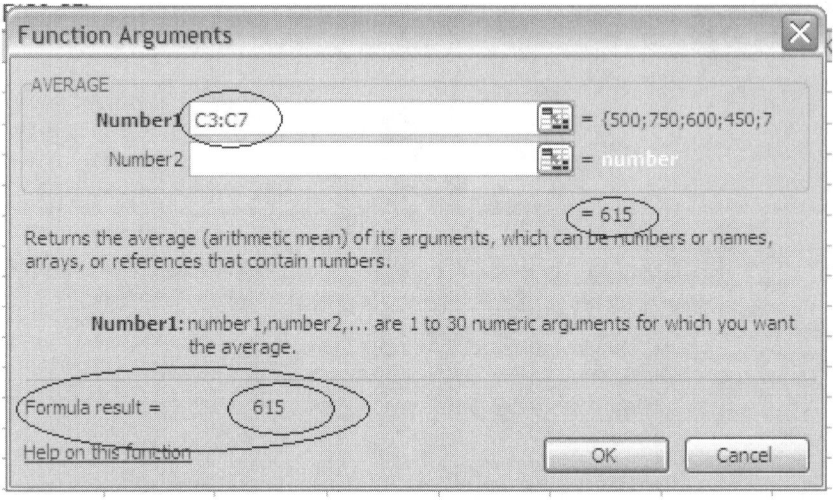

Function Arguments

AVERAGE

Number1 C3:C7 = {500;750;600;450;7

Number2 = number

= 615

Returns the average (arithmetic mean) of its arguments, which can be numbers or names, arrays, or references that contain numbers.

Number1: number1,number2,... are 1 to 30 numeric arguments for which you want the average.

Formula result = 615

Help on this function

OK Cancel

Example Table 3.2 shows the daily production of five teams of workers over a period of four days. Each team has a different number of workers. What is the average production per worker during that period?

Total production over the four days = 750 + 400 + 700 + 600 = 2450
Total number of workers = 15 + 13 + 12 + 10 = 50
Mean production per worker = 2450/50 = 49

TABLE 3.2

Day	Team	Number of workers per team	Production
1	1	15	750
2	2	13	400
3	3	12	700
4	4	10	600

Arithmetic mean of grouped data. Sometimes the available data are grouped in intervals or classes and presented in the form of a frequency distribution. The data on income or age of a population are often presented in this way. It is impossible to exactly determine a measure of central tendency, so an approximation is done using the midpoints of the intervals and the frequency of the distribution:

$$\mu = \frac{\sum fX}{N}$$

where μ is the arithmetic mean, X is the midpoint, f is the frequency in each interval, and N is the total number of the frequencies.

Example The net revenues for a group of companies are organized as shown in Table 3.3. Determine the estimated arithmetic mean revenue of the companies.

TABLE 3.3

Revenues ($ millions)	Number of companies
18–22	3
23–27	17
28–32	10
33–37	15
38–42	9
43–47	3
48–52	14
53–57	5

Solution

Revenues ($ millions)	Number of companies	Midpoint of revenues	fX
18–22	3	20	60
23–27	17	25	425
28–32	10	30	300
33–37	15	35	525
38–42	9	40	360
43–47	3	45	135
48–52	14	50	700
53–57	5	55	275
Total:	**76**	—	**2780**

$$\mu = \frac{\sum fX}{N} = \frac{2780}{76} = 36.579$$

So the mean revenue per company is $36.579 million.

3.1.2 Geometric mean

The *geometric mean* is used to find the average of ratios, indexes, or growth rates. It is the nth root of the product of n values:

$$GM = \sqrt[n]{(x_1)(x_2).....(x_n)}$$

Suppose that a company's revenues have grown by 15 percent last year and 25 percent this year. The average increase will not be 20 percent, as with an arithmetic mean, but instead will be

$$GM = \sqrt{15 \times 25} = \sqrt{375} = 19.365$$

Using Excel. After having pasted the data into a worksheet and selecting the field to store the result, select the "Insert Function" shortcut and in the "Function" dialog box, select "Statistical" and then "GEOMEAN."

In the "Function Argument" dialog box, enter the range into the *Number1* text box. The results appear by "Formula result."

3.1.3 Mode

The *mode* is not a very frequently used measure of central tendency but it is still an important one. It represents the value of the observation that appears most frequently. Consider the following sample measurement:

$$75, 60, 65, 75, 80, 90, 75, 80, 67$$

The value 75 appears most frequently, thus it is the mode.

3.1.4 Median

The *median* of a set of data is the value of x such that half the measurements are less than x and half are greater. Consider the following set of data:

$$12, 25, 15, 19, 40, 17, 36$$

The total $n = 7$ is odd. If we rearrange the data in order of increasing magnitude, we obtain:

$$12, 15, 17, 19, 25, 36, 40$$

The median would be the fourth value, 19.

3.2 Measures of Dispersion

The measures of central tendency only locate the center of the data; they do not provide information on how the data are spread. The *measures of dispersion* or *variability* provide that information. If the values of the measures of dispersion show that the data are closely clustered around the mean, the mean would be a good representation of the data and a good and reliable average.

Variation is very important in quality control because it determines the level of conformance of the production process to the set standards. For instance, if we are manufacturing tires, an excessive variation in the depth of the treads of the tires would imply a high rate of defective products.

The study of variability also helps compare the spread in more than one distribution. Suppose that the arithmetic mean of a daily production of cars in two manufacturing plants is 1000. We can conclude that the two plants produce the same number of cars every day. But an observation over a certain period of time might show that one produces between 950 and 1050 cars a day and the other between 450 and 1550.

So the second plant's production is more erratic and has a less stable production process.

The most widely used measures of dispersion are the range, the variance, and the standard deviation.

3.2.1 Range

The *range* is the simplest of all measures of variability. It is the difference between the highest and the lowest values of a data set.

$$Range = highest\ value - lowest\ value$$

Example The weekly output on a production line is given in Table 3.4.

TABLE 3.4

Day	Production
1	700
2	850
3	600
4	575
5	450
6	900
7	300

The range is 900 − 300 = 600.

The concept of range will be investigated more closely when we study the Statistical Process Control (SPC).

3.2.2 Mean deviation

The range is very simple; in fact, it is too simple because it only considers two values in a set of data. It is not informative about the other values. If the highest and the lowest values in a distribution are both outliers (i.e., extremely far from the rest of the observations), then the range would be a very bad measure of spread. The mean deviation, the variance, and the standard deviation provide more information about all the data observed.

Single deviations from the mean for a given distribution measure the difference between every observation and the mean of the distribution. The deviation indicates how far an observation is away from a mean and it is denoted $X - \mu$. The sum of all the deviations from the mean is given as $\Sigma(X - \mu)$, and that sum is always equal to zero.

In Table 3.5, the production for Day 1 deviates from the mean by 75 units. Consider the example in Table 3.5 and find the sum of all the deviations from the production mean.

TABLE 3.5

Day	Production	$X-\mu$
1	700	75
2	850	225
3	600	−25
4	575	−50
5	450	−175
6	900	275
7	300	−325
Total	**4375**	**0**
Mean	**625**	—

The *mean deviation* measures the average amount by which the values in a population deviate from the mean. Because the sum of the deviations is always equal to zero, it cannot be used to measure the mean deviation; another method should be used instead.

The mean deviation is the sum of the absolute values of the deviations from the mean divided by the number of observations in the population. The absolute value of the sum of the deviations from μ is used because $\sum(X-\mu)$ is always equal to zero. The mean deviation is written as

$$MD = \frac{\sum_{i=1}^{N} |x_i - \mu|}{N}$$

where x_i is the value of each observation, μ is the arithmetic mean of the observation, $|x_i - \mu|$ is the absolute value of the deviations from the mean, and N is the number of observations.

Example Use Table 3.4 to find the mean deviation of the weekly production.

Solution We need to find the arithmetic mean first.

$$\mu = \frac{700 + 850 + 600 + 575 + 450 + 900 + 300}{7} = 625$$

We will add another column for the absolute values of the deviations from the mean.

TABLE 3.6

Day	Production	$\|X - \mu\|$
1	700	75
2	850	225
3	600	25
4	575	50
5	450	175
6	900	275
7	300	325
Total	—	**1150**

$$MD = \frac{1150}{7} = 164.29$$

The mean deviation is 164.29 items produced a day. In other words, on average 164.29 items produced deviated from the mean every day during that week.

3.2.3 Variance

Because $\Sigma(X - \mu)$ equals zero and the use of absolute values does not always lend itself to easy manipulation, the square of the deviation from the mean is used instead. The *variance* is the average of the squared deviation from the arithmetic mean. (For the remainder of this chapter, whenever we say "mean," we will understand arithmetic mean.) The variance for the population mean is denoted by σ^2 for whole populations or for samples greater than 30:

$$\sigma^2 = \frac{\sum_{i=1}^{N} (X_i - \mu)^2}{N}$$

For samples, the letter s will be used instead and the sum of square of the deviations will be divided by $n - 1$.

$$s^2 = \frac{\sum_{i=1}^{n} (x_i - \overline{x})^2}{n - 1}$$

If we want to find the variance for the example in Table 3.4, we will add a new column for the squared deviation.

TABLE 3.7

Day	Production	$(X-\mu)^2$
1	700	5625
2	850	50,625
3	600	625
4	575	2500
5	450	30,625
6	900	75,625
7	300	105,625
Total	—	**271,250**

$$\sigma^2 = \frac{271,250}{7} = 38,750$$

The variance is not only a high number but it is also difficult to interpret because it is the square of a value. For that reason, we will consider the variance as a transitory step in the process of obtaining the standard deviation.

Using Excel, we must to distinguish between the variance based on a sample (ignoring logical values and text in the sample), variance based on a sample (including logical values and text), and variance based on a population. We will use the latter.

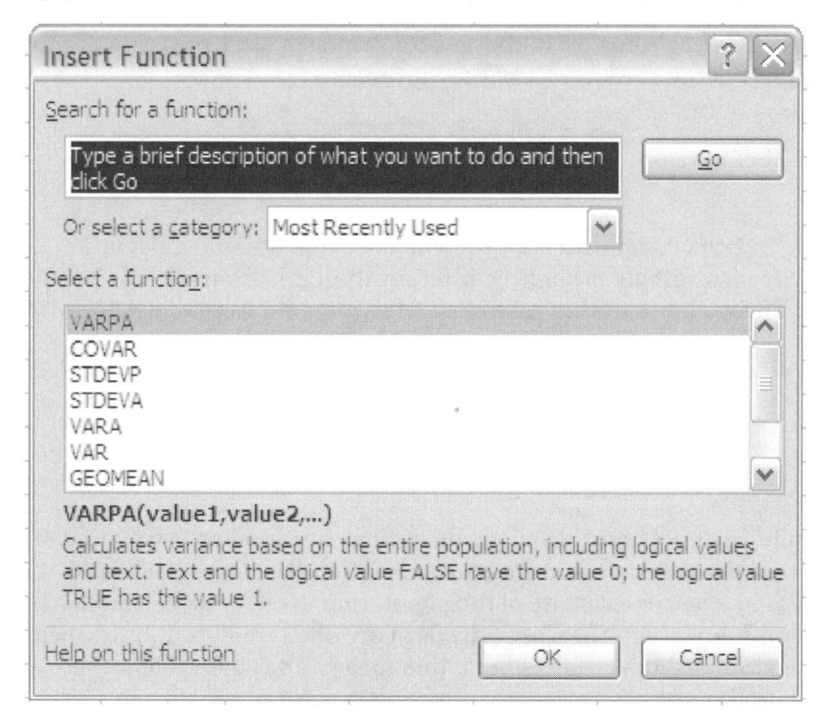

Select "OK," then select the "Function Arguments" dialog box, and then select the fields under "Production."

Function Arguments

VARPA

Value1 E5:E17 = {700;0;850;0;600;0

Value2 = number

= 38750

Calculates variance based on the entire population, including logical values and text. Text and the logical value FALSE have the value 0; the logical value TRUE has the value 1.

Value1: value1,value2,... are 1 to 30 value arguments corresponding to a population.

Formula result = 38750

Help on this function

OK Cancel

3.2.4 Standard deviation

The *standard deviation* is the most commonly used measure of variability. It is the square root of the variance:

$$\sigma = \sqrt{\sigma^2} = \sqrt{\frac{\sum (X - \mu)^2}{N}}$$

Note that the computation of the variance and standard deviation derived from a sample is slightly different than it is from a whole population. The variance in that case is noted as s^2 and the standard deviation as s.

$$s = \sqrt{s^2} = \sqrt{\frac{\sum_{i=1}^{n} (x_i - \overline{x})^2}{n - 1}}$$

Sample variances and standard deviations are used as estimators of a population's variance and standard deviation. Using $n - 1$ instead of N results in a better estimate of the population. Note that the smaller the standard deviation, the closer the data are scattered around the mean. If the standard deviation is zero, this means all the data observed are equal to the mean.

3.2.5 Chebycheff's theorem

Chebycheff's theorem allows us to determine the minimum proportion of the values that lie within a specified number of the standard deviation of the mean. Given the number k greater than or equal to 1 and a set of n measurements $a_1, a_2, \ldots a_n$, at least $\left(1 - \frac{1}{k^2}\right)$ of the measurements lie within k standard deviations of their mean.

Example A sample of bolts taken out of a production line has a mean of 2 inches in diameter and a standard deviation of 1.5. At least what percentage of the bolts lie within ± 1.75 standard deviations from the mean?

Solution

$$1 - \frac{1}{k^2} = 1 - \frac{1}{1.75^2} = 1 - \frac{1}{3.0625} = 0.6735$$

At least 67.35 percent of the bolts are within ± 1.75 standard deviations from the mean.

3.2.6 Coefficient of variation

A comparison of one or more measures of variability is not possible using the variance or the standard deviation. We cannot compare the standard deviation of the production of bolts to one of the availability of parts. If the standard deviation of the production of bolts is 5 and that of the availability of parts is 7 for a given time frame, we cannot conclude that the standard deviation of the availability of parts is greater than that of the production of bolts, and therefore the variability is greater with the parts. For a meaningful comparison to be made, a relative measure called the coefficient of variation is used.

The *coefficient of variation* is the ratio of the standard deviation to the mean:

$$cv = \frac{\sigma}{\mu}$$

for a population and

$$cv = \frac{s}{\overline{X}}$$

for a sample.

Example A sample of 100 students was taken to compare their income and expenditure on books. The standard deviations and means are summarized in Table 3.8. How do the relative dispersions for income and expenditure on books compare?

TABLE 3.8

Statistics	Income ($)	Expenditure on books
$\overline{X}$	750	70
s	15	9

Solution For the students' income:

$$cv = \left(\frac{15}{750}\right) \times 100 = 2\%$$

For their expenditure on books:

$$cv = \left(\frac{9}{70}\right) \times 100 = 12.86\%$$

The students' expenditure on books is more than six times as variable as their income.

3.3 The Measures of Association Quantify the Level of Relatedness between Factors

Measures of association are statistics that provide information about the relatedness between variables. These statistics can help estimate the existence of a relationship between variables and the strength of that relationship. The three most widely used measures of association are the covariance, the correlation coefficient, and the coefficient of determination.

3.3.1 Covariance

The *covariance* shows how the variable y reacts to a variation of the variable x. Its formula is given as

$$\text{cov}\,(X, Y) = \frac{\sum (x_i - \mu_x)(y_i - \mu_y)}{N}$$

for a population and

$$\text{cov}\,(X, Y) = \frac{\sum (x_i - \bar{x})(y_i - \bar{y})}{n - 1}$$

for a sample.

Example Based on the data in Table 3.9, how does the variable y react to a change in x?

TABLE 3.9

x	y
9	10
7	9
6	3
4	7

Solution

x	y	$x - \mu_x$	$y - \mu_y$	$(x - \mu_x)(y - \mu_y)$
9	10	2.5	2.75	6.875
7	9	0.5	1.75	0.875
6	3	−0.5	−4.25	2.125
4	7	−2.5	−0.25	0.625
$\mu_x = 6.5$	$\mu_y = 7.25$			10.5

$$\text{cov}\,(X, Y) = \frac{10.5}{4} = 2.625$$

Using Excel In the "Insert Function" dialog box, select "COVAR" and then select "OK."

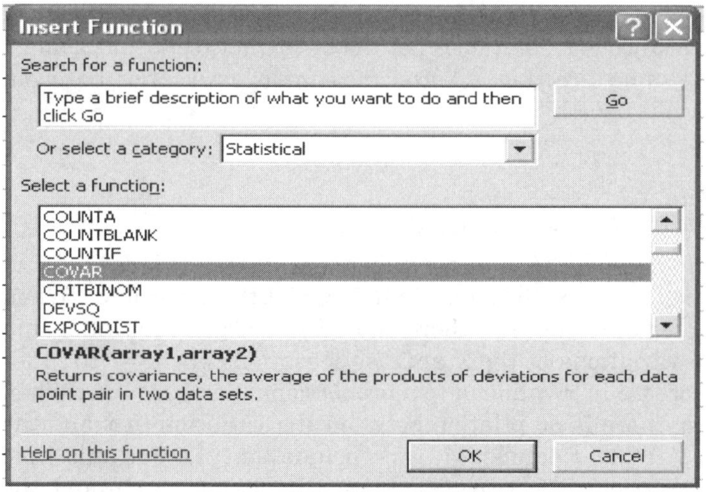

Insert Function

Search for a function:

Type a brief description of what you want to do and then click Go

Go

Or select a category: Statistical

Select a function:

COUNTA
COUNTBLANK
COUNTIF
COVAR
CRITBINOM
DEVSQ
EXPONDIST

COVAR(array1,array2)

Returns covariance, the average of the products of deviations for each data point pair in two data sets.

Help on this function

OK

Cancel

Fill in *Array1* and *Array2* accordingly and the results are obtained.

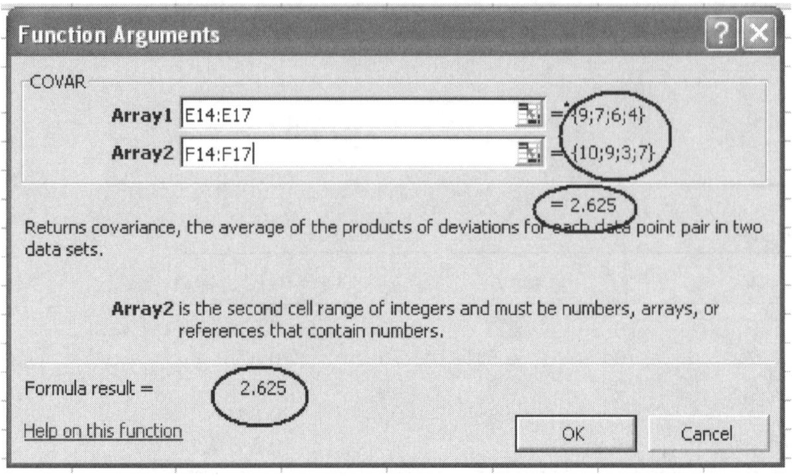

Function Arguments

COVAR

Array1 E14:E17 = {9;7;6;4}

Array2 F14:F17 = {10;9;3;7}

= 2.625

Returns covariance, the average of the products of deviations for each data point pair in two data sets.

Array2 is the second cell range of integers and must be numbers, arrays, or references that contain numbers.

Formula result = 2.625

Help on this function

OK Cancel

A result of 2.625 suggests that x and y vary in the same direction. As x increases, so does y, and when x is greater than its mean, so is y.

The covariance is limited in describing the relatedness of x and y. It can show the direction in which y moves when x changes but it does not show the magnitude of the relationship between x and y. If we say that the covariance is 2.65, it does not tell us much except that x and y change in the same direction. A better measure of association based on the covariance is used by statisticians.

3.3.2 Correlation coefficient

The *correlation coefficient* (r) is a number that ranges between -1 and $+1$. The sign of r will be the same as the sign of the covariance. When r equals -1, we conclude that there is a perfect negative relationship between the variations of the x and the variations of the y. In other words, an increase in x will lead to a proportional decrease in y. When r equals zero, there is no relation between the variation in x and the variation in y. When r equals $+1$, we conclude that there is a positive relationship between the two variables—the changes in x and the changes in y are in the same direction and in the same proportion. Any other value of r is interpreted according to how close it is to -1, 0, or $+1$.

The formula for the correlation coefficient is

$$\rho = \frac{Cov\ (X, Y)}{\sigma_x \sigma_y}$$

for a population and

$$r = \frac{\text{cov}\ (X, Y)}{s_x s_y}$$

for a sample.

Example Given the data in Table 3.10, find the correlation coefficient between the availability of parts and the level of output.

TABLE 3.10

Week	Parts	Output
1	256	450
2	250	445
3	270	465
4	265	460
5	267	462
6	269	465
7	270	466

Solution

TABLE 3.11

Week	Parts (x)	Output (y)	$x - \mu_x$	$(x - \mu_x)^2$	$y - \mu_y$	$(y - \mu_y)^2$	$(x - \mu_x)(y - \mu_y)$
1	256	450	−7.85714	61.73469	−9	81	70.71429
2	250	445	−13.8571	192.0204	−14	196	194
3	270	465	6.142857	37.73469	6	36	36.85714
4	265	460	1.142857	1.306122	1	1	1.142857
5	267	462	3.142857	9.877551	3	9	9.428571
6	269	465	5.142857	26.44898	6	36	30.85714
7	270	466	6.142857	37.73469	7	49	43
Total	**1847**	**3213**	**0**	**366.8571**	**0**	**408**	**386**
Mean	**263.8571**	**459**					
Stdev	**7.239348**	**7.634508**					
Cov	**55.14286**						

We will have to find the covariance and the standard deviations for the Parts and the Output to find the correlation coefficient.

$$\rho = \frac{\text{cov}\,(X, Y)}{\sigma_x \sigma_y}$$

The covariance will be

$$Cov(x, y) = \frac{(x - \mu_x)(y - \mu_y)}{N} = \frac{386}{7} = 55.143$$

The standard deviation for the Parts will be

$$\sigma_x = \sqrt{\frac{\sum (X - \mu_x)^2}{N}} = \sqrt{\frac{366.8571}{7}} = 7.2393$$

The standard deviation for the Output will be

$$\sigma_Y = \sqrt{\frac{\sum (y - \mu_Y)^2}{N}} = \sqrt{\frac{408}{7}} = 7.635$$

Therefore, the correlation coefficient will be

$$r = \frac{55.143}{7.2393 \times 7.635} = 0.9977$$

Using Minitab. After entering the data in a worksheet, from the Stat menu select "Basic Statistics" and then "Correlation..."

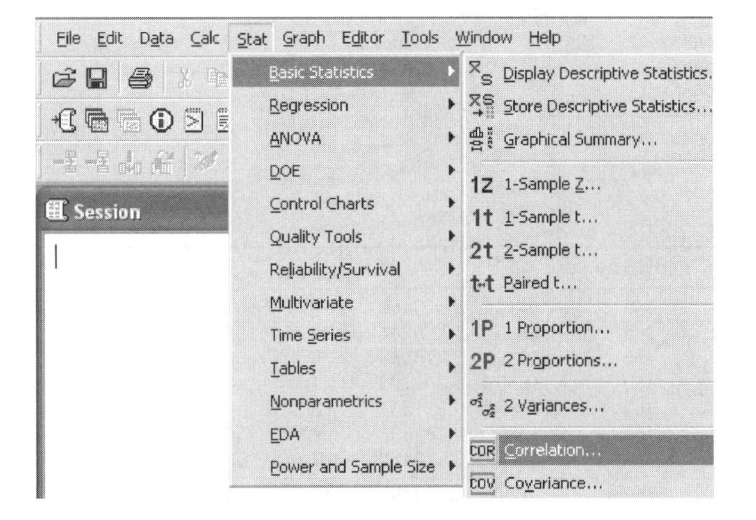

In the "Correlation" dialog box, insert the variables in the "Variables" textbox and then select "OK."

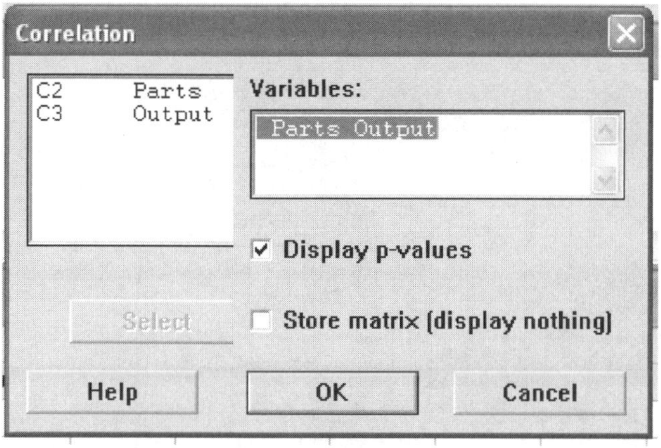

The output shows that $r = 0.998$

Correlations: Parts, Output

```
Pearson correlation of Parts and Output = 0.998
P-Valve = 0.000
```

The correlation coefficient $r = 0.9977201$, which is very close to 1, so we conclude that there is a strong positive correlation between the availability of parts and the level of the output.

Using Excel. We can also determine the correlation coefficient using Excel. After selecting the cell to store the results, select the "Insert Function" button. In the subsequent dialog box, select "CORREL."

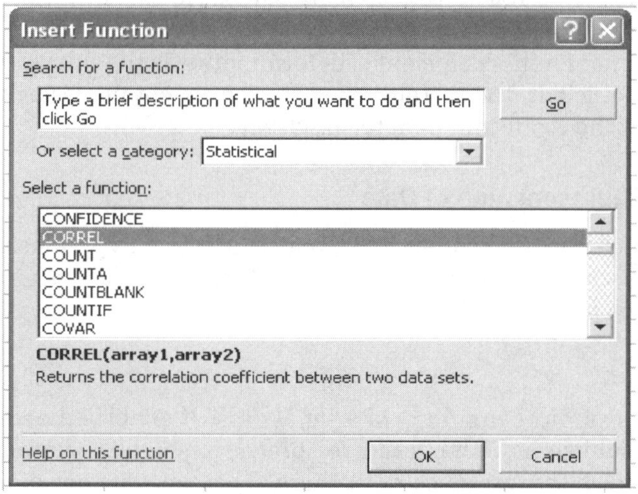

In the "Function Arguments" dialog box, insert the ranges of the variables in *Array1* and *Array2*, and the results appear.

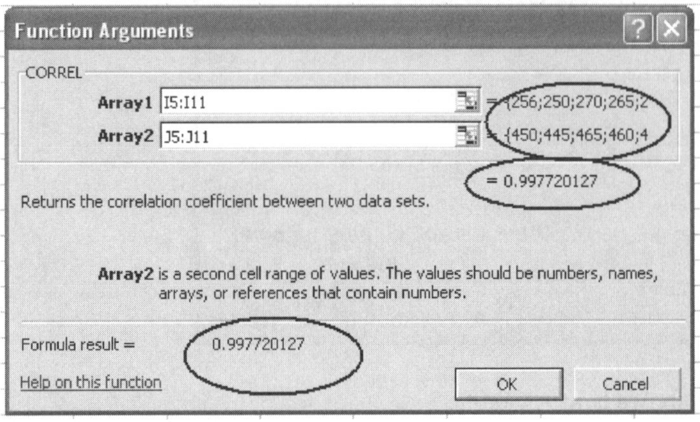

Function Arguments
CORREL
Array1 I5:I11 = {256;250;270;265;2
Array2 J5:J11 = {450;445;465;460;4
= 0.997720127
Returns the correlation coefficient between two data sets.
Array2 is a second cell range of values. The values should be numbers, names, arrays, or references that contain numbers.
Formula result = 0.997720127
Help on this function
OK Cancel

3.3.3 Coefficient of determination

The *coefficient of determination* (r^2) measures the proportion of changes of the dependent variable y that are explained by the independent variable x. It is the square of the correlation coefficient r and for that reason, it is always positive and ranges between zero and one. When the coefficient of determination is zero, the variations of y are not explained by the variations of x. When r^2 equals one, the changes in y are explained fully by the changes in x. Any other value of r^2 must be interpreted according to how close it is to zero or one.

For the previous example, r was equal to 0.998, therefore $r^2 = 0.998 \times 0.998 = 0.996004$. In other words, 99.6004 percent of the variations of y are explained by the variations in x.

Note that even though the coefficient of determination is the square of the correlation coefficient, the correlation coefficient is not necessarily the square root of the coefficient of determination.

3.4 Graphical Representation of Data

Graphical representations can make data easy to interpret by just looking at graphs. Histograms, stem-and-leaf, and box plots are types of graphs commonly used in statistics.

3.4.1 Histograms

A *histogram* is a graphical summary of a set of data. It enables the experimenter to visualize how the data are spread, to see how skewed they are, and detect the presence of outliers. The construction of a histogram starts with the division of a frequency distribution into equal classes, and then each class is represented by a vertical bar.

Using Minitab, we can construct the histogram for the data in Table 3.11. Go to "Graph→ Histogram" and then, in the "Histogram" dialog box, select "With Fit" and select "OK." The "Histogram—With Fit" dialog box pops up and insert the variables into the "Graph variables:" textbox, then select "Multiple Graphs . . . "

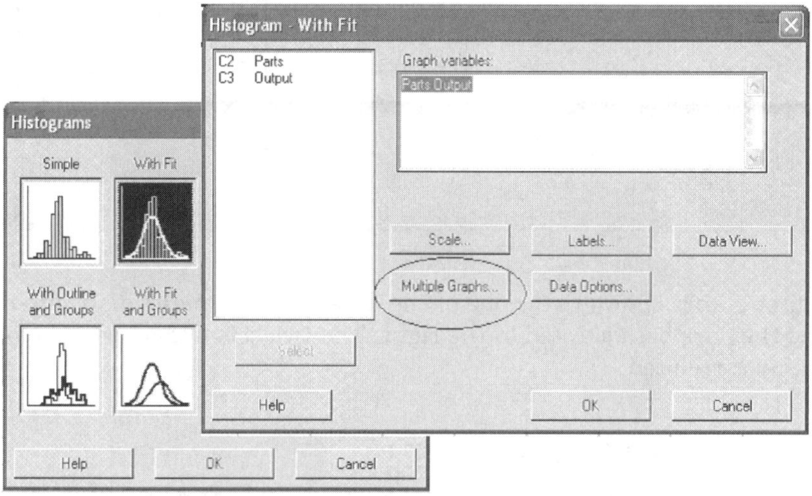

The "Histogram—Multiple Graphs" dialog box pops up, and select the option *In separate panels of the same graph*.

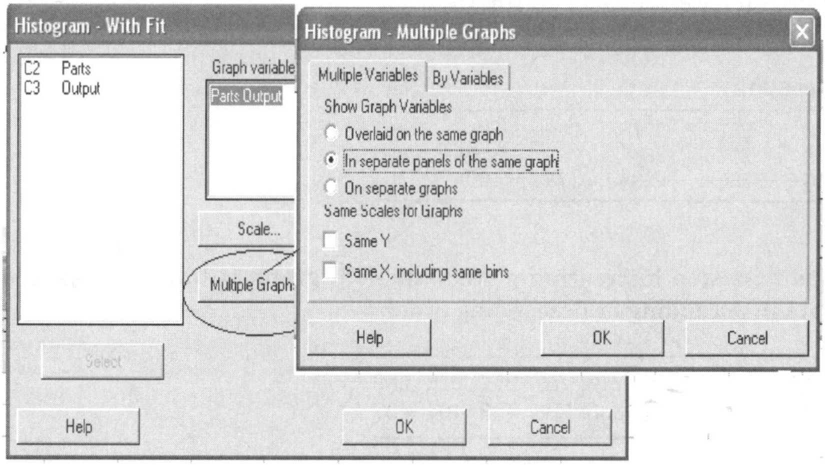

Select "OK" and then "OK" again.

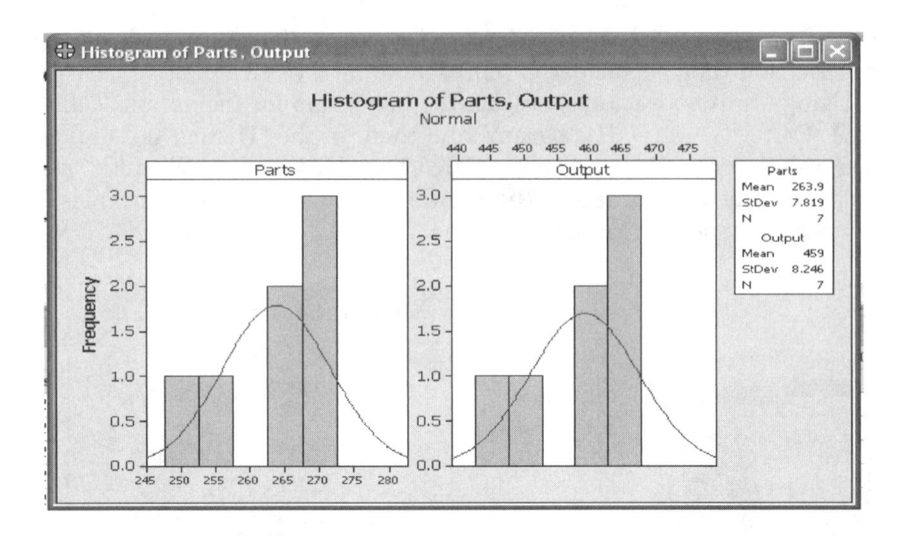

A first glance shows that none of the data set is normally distributed, and they are both skewed to the right. The two sets of data seem to be highly correlated.

3.4.2 Stem-and-leaf graphs

A *stem-and-leaf* graph resembles a histogram and like a histogram, it is used to visualize the spread of a distribution and indicate around what values the data are mainly concentrated. The stem-and-leaf graph is essentially composed of two parts: the stem, which is on the left side of the graph, and the leaf on the right.

Consider the data in Table 3.12.

TABLE 3.12

302	287	277	355	197	403
278	257	286	388	189	407
313	288	213	178	188	404

The first step in creating a stem-and-leaf graph is to reorganize the data in ascending or descending order.

178	188	189	197	213	257	277	278	286
287	288	302	313	355	388	403	404	407

The stem will be composed of the first digits of all numbers, and the leaf will be the second digit. The numbers that start with 1 have 7, 8,

8 again, and 9 as the second digits. There are three numbers starting with 4 and all of them have 0 as a second digit.

1	7889
2	1577888
3	158
4	000

The stem-and-leaf graph shows that most of the data are clustered between 210 and 288. Excel does not provide a function for stem-and-leaf graphs without using macros.

Generating a stem-and-leaf graph using Minitab Open the *Stem and Leaf Graph.mpj* document on the included CD. On the Menu bar, select "Graph" and then select "Stem and Leaf." In the "Stem and Leaf" dialog box, select "Stem and leaf test" and then select "OK." The output should look like that in Figure 3.2.

Note that there is a slight difference in presentation with the graph we first obtained. This is because the Minitab program subdivides the first digits.

Stem-and Leaf Display: Stem and leaf

```
Steam-and-leaf of Stem and leaf N = 18
Leaf Unit = 10


  1   1  7
  4   1  889
  5   2  1
  5   2
  6   2  5
  8   2  77
 (3)  2  888
  7   3  01
  5   3
  5   3  5
  4   3
  4   3  8
  3   4  000
```

Figure 3.2

3.4.3 Box plots

The *box plot*, otherwise known as a box-and-whisker plot or "five number summary," is a graphical representation of data that shows how the data are spread. It has five points of interest, which are the quartiles, the median, and the highest and lowest values. The plot shows how the data are scattered within those ranges. The advantage of using the box plot is that when it is used for multiple variables, not only does it graphically show the variation between the variables but it also shows the variations within the ranges.

To build a box plot we need to find the median first.

39	51	54	61	73	78	87	87	92	93
95	97	97	102	102	107	109	111	113	

The median is the value in the middle of a distribution. In this case, we have an uneven distribution: the median is 93, the observation in the middle. The first quartile will be the median of the observations on the left of 93—in this case, 73. The upper quartile will be the median of the observations on the right of 93—therefore, 102.

The first step in drawing the graph will consist in drawing a graded line and plotting on it the values that have just been determined. The last step will consist in drawing a rectangle, the corners of which will be the quartiles. The interquartile range will be the difference between the upper and lower quartiles, e.g., 102 – 73 = 29.

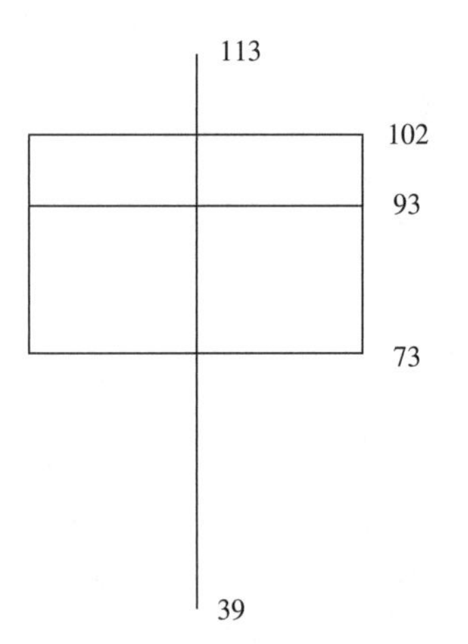

The purpose of a box plot is not only to show how the data are spread but also to make obvious the presence of outliers. To determine the presence of outliers, we first need to find the interquartile range (IQR). The IQR measures the vertical distance of the box; it is the difference between the upper quartile and the lower quartile values. In this case, it will be the difference between 102 and 73, and therefore equal to 29.

An *outlier* is defined as any observation away from the closest quartile by more than 1.5 IQR. An outlier is considered extreme when it is away from the closest quartile by more than 3 IQR.

$$1.5\,\text{IQR} = 1.5 \times 29 = 43.5$$

$$3\,\text{IQR} = 3 \times 29 = 87$$

So any observation smaller than 73 – 43.5 = 29.5 or greater than 102 + 43.5 = 145.5 is considered as an outlier. We do not have any outliers in this distribution but the graph shows that the data are unevenly distributed with more observations concentrated below the median.

Using Minitab, open the worksheet *Box Whiskers.mpj* on the included CD, then from the Graph menu, select "Box plot," then select the "Simple" option, and then select "OK."

Select "Parts" in the "Graph variable" textbox and select "OK." You should obtain the graph shown in Figure 3.3.

Clicking on any line on the graph would give you the measurements of the lines of interest.

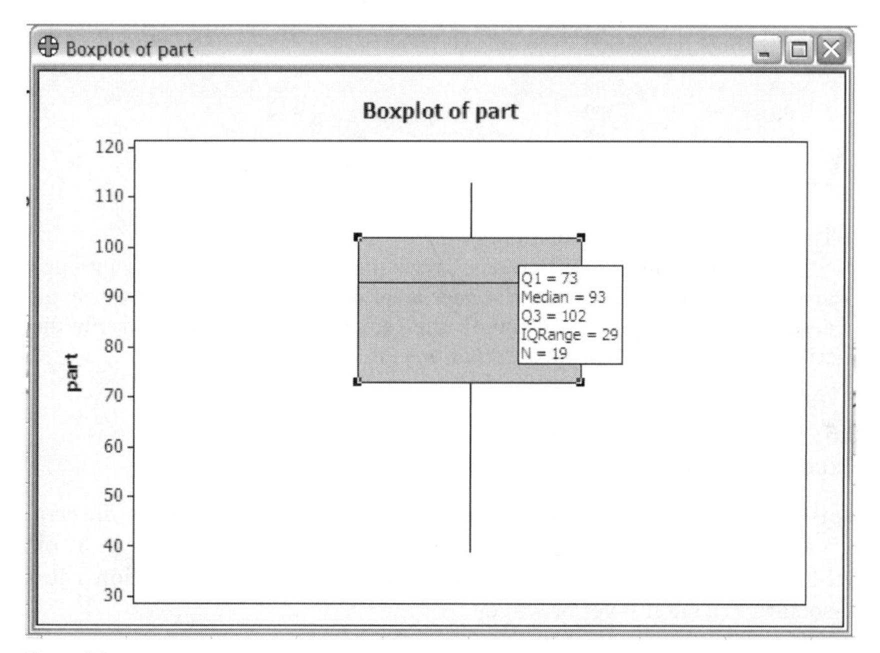

Figure 3.3

Using box plots to compare the variability of several distributions. Not only do box plots show how data are spread within a distribution, they also help compare variability within and between distributions and even visualize the existence of correlations between distributions.

Example The data in Table 3.13 represents the temperature of two rooms: one was built with metal door and window frames and the other one without any metal. We want to determine if there is a difference between the two rooms and the level of temperature variability within the two rooms.

Open *heat.mpj* on the included CD, then from the Graph menu, select "Boxplots." In the "Boxplots" dialog box, select "Multiple Y's Simple" and then select "OK." In the "Multiple Y's Simple" dialog box, insert the "With Metal" and "Without Metal" values in the *Variables* textbox and then select "OK."

TABLE 3.13

With metal	Without metal
52	58
81	62
83	65
79	71
89	59
89	60
98	99
96	60
98	96
99	93
95	87
99	89
99	92
99	85
101	81

The graph of Figure 3.4 should appear.

The graphs show that there is a large disparity between the two groups, and for the room with metal the heat level is predominantly below the median. For the room without metal, the temperatures are more evenly distributed, albeit most of the observations are below the median.

3.5 Descriptive Statistics—Minitab and Excel Summaries

Both Excel and Minitab offer ways to summarize most of the descriptive statistics measurements in one table. The two columns in Table 3.14 represent the wages paid to employees and the retention rates associated to each level of wages.

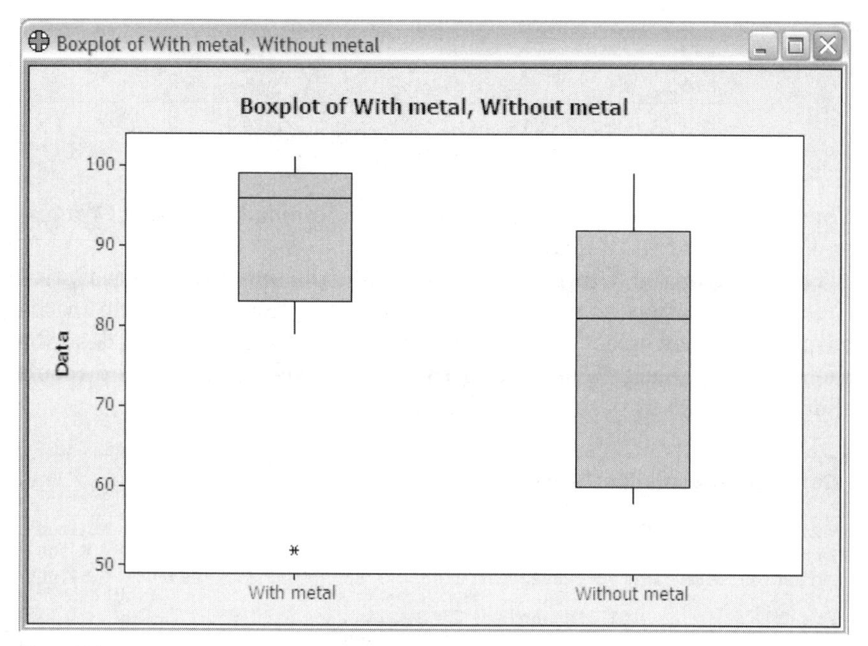

Figure 3.4

Using Minitab and Excel, we want to find the mean, the variance, the standard deviation, the median, and the mode for each column.

TABLE 3.14

Wages ($)	Retention rate (%)
8.70	72
8.90	72
7.70	65
7.60	66
7.50	64
7.80	67
8.70	72
8.90	73
8.70	72
6.70	59
7.50	67
7.60	67
7.80	68
7.90	69
8.00	71

Using Minitab. Open the file *Wage Retention.mpj* on the included CD and from the Stat menu, select "Basic Statistics" and then "Display Descriptive Statistics..."

Stat Graph Editor Tools Window Help
Basic Statistics ▸ Display Descriptive Statistics
Regression ▸ Store Descriptive Statistics..

Then in the "Display Descriptive Statistics" dialog box, insert "Wages" and "Retention" in the *Variables* textbox.

Then click on the "Statistics" command button and in the "Descriptive Statistics—Statistics" dialog box, check the options to include in the output, and then select "OK." If we want to see the graphs that describe the columns, select "Graph" and then make the selection. The result should look like that shown in Figure 3.5.

Descriptive Statistics: Wages, Retention rate

Variable	N	N*	Mean	SE Mean	StDev	Variance	Sum	Minimum
Wages	15	0	8.000	0.166	0.643	0.413	120.000	6.700
Retention rate	15	0	68.27	1.00	3.88	15.07	1024.00	59.00

Variable	Q1	Median	Q3	Maximum
Wages	7.600	7.800	8.700	8.900
Retention rate	66.00	68.00	72.00	73.00

Figure 3.5

Using Excel. Open the file *RetentionWages.xls* on the included CD, then select "Tool" and then "Data Analysis." In the "Data Analysis" dialog box, select "Descriptive Statistics" and then select "OK." We select the two columns at the same time in the "Input Range," check the options *Label* and *Summary Statistics*, and then select "OK." Excel's output should look like that of Figure 3.6.

	A	B	C	D
1	*Wages*		*Retention rate*	
2				
3	Mean	8	Mean	68.26667
4	Standard Error	0.165903012	Standard Error	1.00222
5	Median	7.8	Median	68
6	Mode	8.7	Mode	72
7	Standard Deviation	0.642539604	Standard Deviat	3.88158
8	Sample Variance	0.412857143	Sample Varianc	15.06667
9	Kurtosis	-0.496594592	Kurtosis	0.678743
10	Skewness	-0.041010409	Skewness	-0.83292
11	Range	2.2	Range	14
12	Minimum	6.7	Minimum	59
13	Maximum	8.9	Maximum	73
14	Sum	120	Sum	1024
15	Count	15	Count	15

Figure 3.6

Exercises Bakel Distribution Center maintains a proportion of 0.5 CXS parts to the overall inventory. In other words, the number of CXS parts available in the warehouse must always be 50 percent of the inventory at any given time. But the management has noticed a great deal of increase in back orders, and sometimes CXS parts are overstocked.

Based on the samples taken in Table 3.15, using Minitab and Excel:

TABLE 3.15

Total inventory	Part CXS
1235	125
1234	123
1564	145
1597	156
1456	125
1568	148
1548	195
1890	165
1478	145
1236	123
1456	147
1493	186

a. Show the means and the standard deviations for the two frequencies.
b. Determine if there is a perfect correlation between the inventory and part CXS.

c. Find the coefficient of variation for the inventory and part CXS.
d. Find the correlation coefficient for the inventory and part CXS.
e. Determine the portion of variation in the inventory that is explained by the changes in the volume of part CXS.
f. Using Minitab, draw box plots for the two frequencies on separate graphs.
g. Determine the stem-and-leaf graphs for the two frequencies.
h. Using Minitab and then Excel, show the descriptive statistics summaries.

The table can be found on file *InventoryParts.xls* and *InventoryParts.mpj* on the accompanying CD.

Chapter

4

Introduction to Basic Probability

Learning Objectives:

- Understand the meaning of probability
- Be able to distinguish between discrete and continuous data
- Know how to use basic probability distributions
- Understand when to use a particular probability distribution
- Understand the concept of Rolled Throughput Yield and DPMO and be able to use a probability distribution to find the RTY

In management, knowing with certitude the effects of every decision on Operations is extremely important, yet uncertainty is a constant in any endeavor. No matter how well-calibrated a machine is, it is impossible to predict with absolute certainty how much part-to-part variation it will generate. Based on statistical analysis, an estimation can be made to have an approximate idea about the results. The area of statistics that deals with uncertainty is called *probability*.

We all deal with the concept of probability on a daily basis, sometimes without even realizing it. What are the chances that 10 percent of your workforce will come to work late? What is the likelihood that the shipment sent to the customers yesterday will reach them on time? What are the chances that the circuit boards received from the suppliers are defect free?

So what is probability? It is the chance, or the likelihood, that something will happen. In statistics, the words "chance" and "likelihood" are seldom used to describe the possibilities for an event to take place; instead, the word "probability" is used along with some other basic concepts whose meanings differ from our everyday use. *Probability* is the measure of the possibility for an event to take place. It is a number between zero and one. If there is a 100 percent chance that the event will take place, the probability will be one, and if it is impossible for it to happen, the probability will be zero.

An *experiment* is the process by which one observation is obtained. An example of an experiment would be the sorting out of defective parts from a production line. An *event* is the outcome of an experiment. Determining the number of employees who come to work late twice a month is an experiment, and there are many possible events; the possible outcomes can be anywhere between zero and the number of employees in the company. A *sample space* is the set of all possible outcomes in an experiment.

4.1 Discrete Probability Distributions

A *probability distribution* shows the possible events and the associated probability for each of these events to occur. Table 4.1 is a distribution that shows the weight of a part produced by a machine and the probability of the part meeting quality requirements.

TABLE 4.1

Weight (g)	Probability
5.00	0.99
5.05	0.97
5.10	0.95
5.15	0.94
5.20	0.92
5.25	0.90
5.30	0.88
5.35	0.85

A distribution is said to be *discrete* if it is built on discrete random variables. All the possible outcomes when pulling a card from a stack are finite because we know in advance how many cards are in the stack and how many are being pulled. A random variable is said to be discrete when all the possible outcomes are countable.

The four most used discrete probability distributions in business operations are the binomial, the Poisson, the geometric, and the hypergeometric distributions.

4.1.1 Binomial distribution

The *binomial distribution* assumes an experiment with n identical trials, each trial having only two possible outcomes considered as success or failure and each trial independent of the previous ones. For the remainder of this section, p will be considered as the probability for a success and q as the probability for a failure.

$$q = (1 - p)$$

The formula for a binomial distribution is as follows:

$$P(x) = {}_nC_x\,(p)^x\,(q)^{n-x}$$

where $P(x)$ is the probability for the event x to happen. The variable x may take any value from zero to n and ${}_nC_x$ represents the number of possible outcomes that can be obtained.

$${}_nC_x = \frac{n!}{x!\,(n-x)!}$$

The mean, variance, and standard deviation for a binomial distribution are

$$\mu = np$$

$$\sigma^2 = npq$$

$$\sigma = \sqrt{\sigma^2} = \sqrt{npq}$$

Example A machine produces soda bottles, and 98.5 percent of all bottles produced pass an audit. What is the probability of having only 2 bottles that pass audit in a randomly selected sample of 7 bottles?

$$98.5\% = 0.985$$

$$p = 0.985$$

$$q = 1 - 0.985 = 0.015$$

$${}_7C_2\,(0.985)^2\,(0.015)^5 = 0$$

In other words, the probability of having only two good bottles out of 7 is zero. This result can also be found using the binomial table found in Appendix section.

Using Minitab. Minitab has the capabilities to calculate the probabilities for more than just one event to take place. So in Column C1, we want the probabilities of finding 0 to 10 bottles that pass audit out of the 7 bottles that we selected. Fill in the selected column C1 as shown in Figure 4.1. From the Calc menu, select "Probability Distributions," then select "Binomial," and the "Binomial Distribution" dialog box appears.

We are looking for the probability of an event to take place, not for the cumulative probabilities or their inverse. The number of trials is 7,

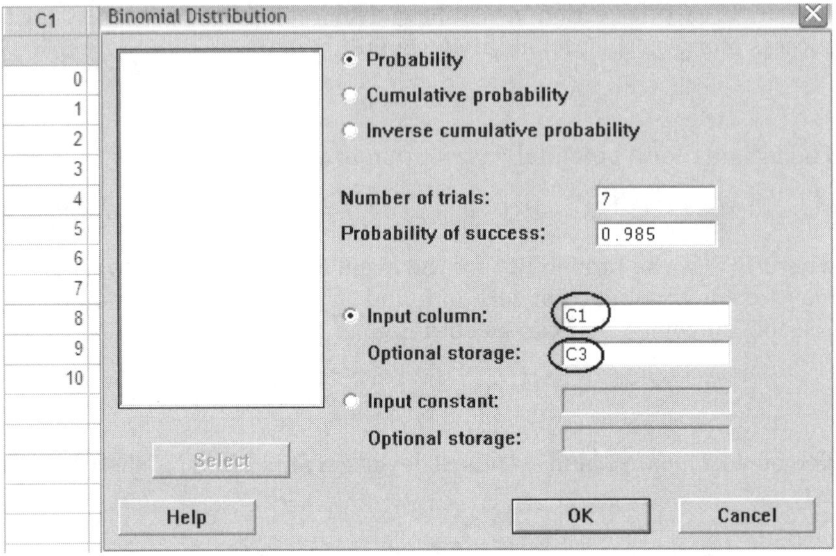

C1
0
1
2
3
4
5
6
7
8
9
10

Binomial Distribution

- Probability
- Cumulative probability
- Inverse cumulative probability

Number of trials: 7

Probability of success: 0.985

- Input column: C1

Optional storage: C3

- Input constant:

Optional storage:

Select

Help OK Cancel

Figure 4.1

and the probability for a success is 0.985. The *Input column* is the one that contains the data that we are looking for and the *Output column* is the column where we want to store the results; in this case, it is C3. After making these entries, click "OK."

The results obtained are shown in Figure 4.2.

C1	C2	C3
0		0.000000
1		0.000000
2		0.000000
3		0.000002
4		0.000111
5		0.004381
6		0.095897
7		0.899609
8		0.000000
9		0.000000
10		0.000000

Figure 4.2

If we want to know the probability of having between 3 and 6 bottles that pass audit, all we would need to do is add the probabilities of having 3, 4, 5, and 6, and we would obtain 0.100391.

Using Excel. After having inserted the values of p, n, and x in selected cells, select the cell where the result should be output and click the "Insert Function" (f_x) shortcut button.

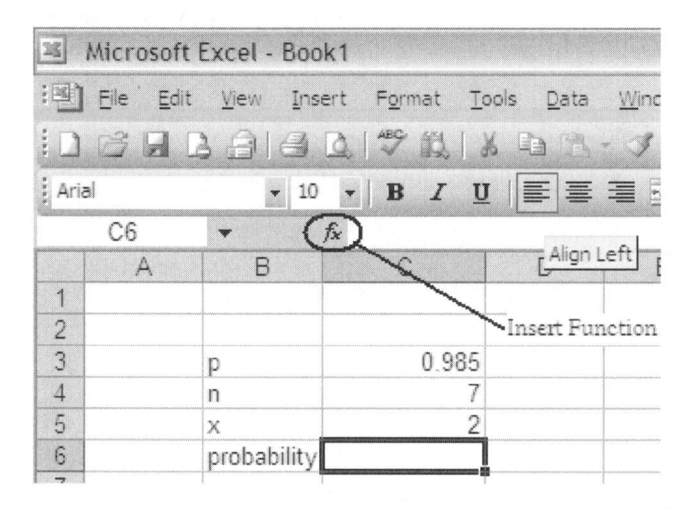

When the "Insert Function" dialog box appears, select "Statistical" in the *Or select a category:* textbox and select "BINOMDIST" from the *Select a function:* list. Then, select "OK."

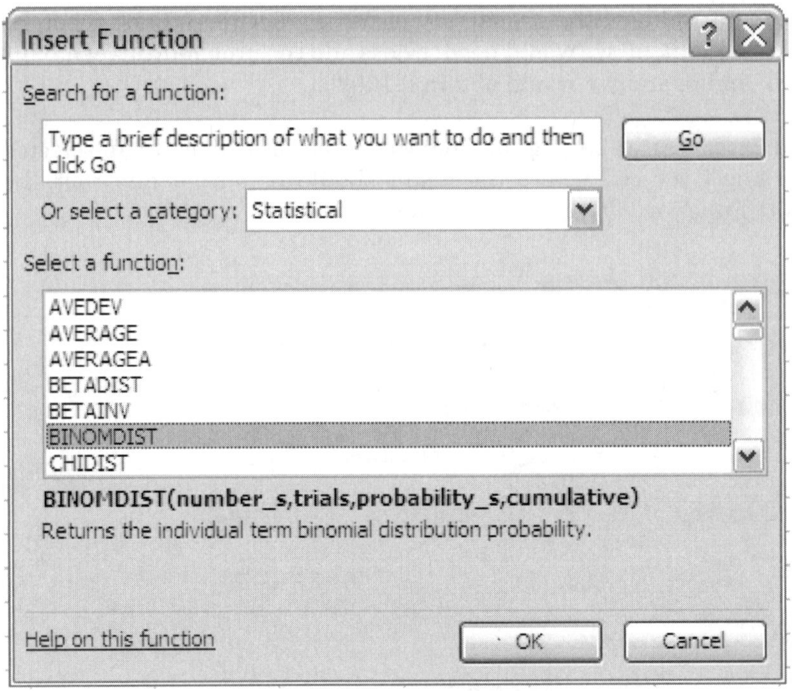

Insert Function

Search for a function:

Type a brief description of what you want to do and then click Go

Go

Or select a category: Statistical

Select a function:

AVEDEV
AVERAGE
AVERAGEA
BETADIST
BETAINV
BINOMDIST
CHIDIST

BINOMDIST(number_s,trials,probability_s,cumulative)

Returns the individual term binomial distribution probability.

Help on this function

OK Cancel

In the "Function Arguments" dialog box, fill the fields as shown in Figure 4.3.

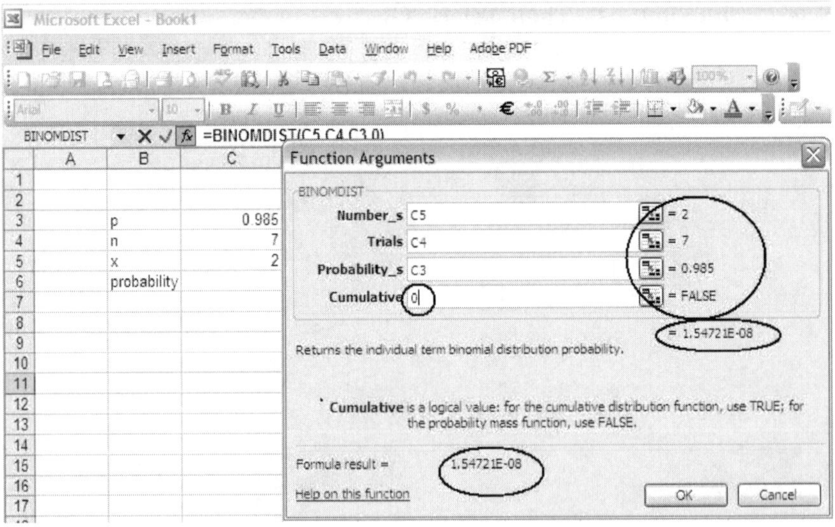

Microsoft Excel - Book1

File Edit View Insert Format Tools Data Window Help Adobe PDF

BINOMDIST =BINOMDIST(C5,C4,C3,0)

	A	B	C
1			
2			
3		p	0.985
4		n	7
5		x	2
6		probability	
7			
8			
9			
10			
11			
12			
13			
14			
15			
16			
17			

Function Arguments

BINOMDIST

Number_s C5 = 2

Trials C4 = 7

Probability_s C3 = 0.985

Cumulative 0 = FALSE

= 1.54721E-08

Returns the individual term binomial distribution probability.

Cumulative is a logical value: for the cumulative distribution function, use TRUE; for the probability mass function, use FALSE.

Formula result = 1.54721E-08

Help on this function

OK Cancel

Figure 4.3

The result that appears in *Formula result* shows that it is infinitesimal. We use zero in the *Cumulative* field because we are not looking for a cumulative result.

Exercise. A machine produces ceramic pots, and 68.9 percent of all pots weigh 5 pounds. What is the probability of selecting 3 pots that weigh 5 pounds in a randomly selected sample of 8 pots?

4.1.2 Poisson distribution

The *Poisson distribution* focuses on the probability for a number of events occurring over some interval or continuum where μ, the average of such an event occurring, is known. For instance, a Quality Control manager may want to know the probability of finding a defective part on a manufactured circuit board.

The following example deals with an event that occurs over a continuum, and therefore can be solved using the Poisson distribution. The formula for the Poisson distribution is

$$P(x) = \frac{\mu^x e^{-\mu}}{x!}$$

where $P(x)$ is the probability of the event x to occur, μ is the arithmetic mean number of occurrences in a particular interval, and e is the constant 2.718282. The mean and the variance of the Poisson distribution are the same, and the standard deviation is the square root of the mean,

$$\mu = \sigma^2$$
$$\sigma = \sqrt{\mu} = \sqrt{\sigma^2}$$

Binomial problems can be approximated by the Poisson distribution when the sample sizes are large ($n > 20$) and p is small ($np \leq 7$). In this case, $\mu = np$.

Example A product failure has historically averaged 3.84 occurrences per day. What is the probability of 5 failures in a randomly selected day?

$$\mu = 3.84$$
$$x = 5$$
$$P(5) = \frac{3.84^5 e^{-3.84}}{5!} = 0.149549$$

The same result can be found in the Poisson table on Appendix 2.

Using Minitab. The process of finding the probability for a Poisson distribution in Minitab is the same as that for the Binomial distribution. The output is shown in Figure 4.4.

↓	C1	C2	C3
1	0		0.021494
2	1		0.082535
3	2		0.158468
4	3		0.202839
5	4		0.194726
6	5		0.149549
7	6		0.095711
8	7		0.052505
9	8		0.025202
10	9		0.010753

Figure 4.4

Using Excel. The same process can be found for Excel. The Excel output is shown in Figure 4.5.

Exercise. A machine has averaged a 97 percent pass rate per day. What is the probability of having more than 7 defective products in one day?

4.1.3 Poisson distribution, rolled throughput yield, and DPMO

Because it seeks to improve business performance, Six Sigma, whenever possible should use traditional business metrics. But because it is

Function Arguments

POISSON

X 5 = 5

Mean 3.84 = 3.84

Cumulative 0 = FALSE

= 0.14954919

Returns the Poisson distribution.

Cumulative is a logical value: for the cumulative Poisson probability, use TRUE; for the Poisson probability mass function, use FALSE.

Formula result = 0.14954919

Help on this function

OK Cancel

Figure 4.5

deeply rooted in statistical analysis, some of the techniques it uses are not commonly applied in business. The Defect Per Million Opportunity (DPMO) and the rolled throughput yield (RTY) are just a few examples.

Defects per unit (DPU) and yield. Consider a company that manufactures circuit boards. A circuit board is composed of multiple elements such as switches, resistors, capacitors, computer chips, and so on. Having every part of a board within specification is critical to the quality of each manufactured unit. Any time an element of a unit (a switch, for this example) is outside its specified limits, it is considered as a defect —in other words, a *defect* is a nonconforming element on a defective unit.

To measure the quality of his throughput, the manufacturer will want to know how many defects are found per unit. Because there are multiple parts per unit, it is conceivable to have more than one defect on one unit. If we call the number of defects D and the number of units U, then the *defects per unit* (DPU) is

$$DPU = \frac{D}{U}$$

Consider 15 units with defects spread as shown in Table 4.2:

TABLE 4.2

Units	2	3	9	1	
Defects	3	2	0	1	
Total number of defects	2 × 3 = 6	3 × 2 = 6	9 × 0 = 0	1	6 + 6 + 1 = 13

$$DPU = \frac{13}{15} = 0.86667$$

In this case, the probability of finding defects on a unit follows a Poisson distribution because the defects can occur randomly throughout an interval that can be subdivided into independent subintervals.

$$P(x) = \frac{\mu^x e^{-\mu}}{x!}$$

where $P(x)$ is the probability for a unit to contain x defects, and μ is the mean defect per unit. This equation can be rewritten if the DPU is known,

$$P(x) = \frac{DPU^x e^{-DPU}}{x!}$$

Example If the DPU is known to be 0.5, what is the probability of having two defects on a unit?

$$P(2) = \frac{(0.5)^2 e^{-0.5}}{2!} = \frac{0.25e^{-0.5}}{2} = \frac{0.606531 \times 0.25}{2}$$
$$= \frac{0.151633}{2} = 0.0758165$$

The probability of having two defects is 0.0758165. What is the probability of having one defect?

$$P(1) = \frac{0.5^1 e^{-0.5}}{1!} = \frac{0.5e^{-0.5}}{1} = 0.303265$$

The probability of having one defect on a unit will be 0.303265.

The objective of a manufacturer is to produce defect-free products. The probability to produce defect-free (zero defect) units will be

$$P(0) = \frac{DPU^0 e^{-DPU}}{0!}$$

Because $DPU^0 = 1$ and $0! = 1$,

$$P(0) = e^{-DPU}$$

Manufacturing processes are made up of several operations with several linked steps. The probability for a unit to pass a step defect-free will be

$$P(0) = e^{-DPU}$$

If we call the *yield* (y) the probability of a unit passing a step the first time defect-free, then

$$y = e^{-DPU}$$

If y is known, DPU can be found by simply rearranging the previous formula,

$$\ln(y) = -DPU \ln e$$

Because $\ln e = 1$,

$$DPU = -\ln(y)$$

Example If a process has a first pass yield of 0.759,

$$DPU = -\ln(0.759) = 0.27575$$

Rolled throughput yield (RTY). A yield measures the probability of a unit passing a step defect-free, and the *rolled throughput yield* (RTY) measures the probability of a unit passing a set of processes defect-free.

TABLE 4.3

Process 1	Process 2	Process 3	Process 4
0.78	0.86	0.88	0.83

The RTY is obtained by multiplying the individual yields of the different processes,

$$RTY = \prod_{i=1}^{n} y_i$$

What is the RTY for a product that goes through four processes with the respective yields shown in Table 4.3 for each process? What is the DPU?

$$RTY = 0.78 \times 0.86 \times 0.88 \times 0.83 = 0.489952$$

The probability of a unit passing all the processes defect-free is 0.489952. The probability of a defect will be 1 – 0.489952 = 0.510048.

$$DPU = -\ln(y)$$

$$DPU = -\ln(0.489952) = 0.71345$$

An opportunity is defined as any step in the production process where a defect can occur. The Defect Per Opportunity –DPO would be the ratio of the defects actually found to the number of opportunities.

$$DPO = \frac{D}{O}$$

Example The products at a Memphis distribution center have to go through a cycle made up of 12 processes. The products have to be pulled from the trucks by transfer associates, they are systemically received by another associate before being transferred again from the receiving dock to the setup stations, and from there they are taken to the packaging area where they are packaged before being transferred to their stocking locations.

On the outbound side, orders are dropped into the system and allocated to the different areas of the warehouse, and then they are assigned to the picking associates who pick the products and take them to the packing stations where another associate packs them. After the products are packed, they are taken to the shipping dock where they are processed and moved to the trucks that will ship them to the customers.

Each one of the processes presents an average of five opportunities for making a mistake and causing defects. So the number of opportunities for a part that goes through the 12 processes to be defective will be 60. Each part has 60 opportunities to be defective. A total of 15 Parts XY1122AB have been audited and two defects have been found. What is the DPMO?

Solution Find the Defects Per Opportunity (DPO) first:

$$DPMO = DPO \times 10^6$$

Total opportunity for defects $= 15 \times 60 = 900$

$$DPO = \frac{2}{900}$$

$$DPMO = 10^6 \times \frac{2}{900} = 2222.222$$

4.1.4 Geometric distribution

When we studied the binomial distribution, we were only interested in the probability of a success or a failure to occur and the outcomes had an equal opportunity to occur because the trials were independent. The *geometric distribution* addresses the number of trials necessary before the first success. If the trials are repeated k times until the first success, we would have $k-1$ failures. If p is the probability for a success and q the probability for a failure, the probability of the first success to occur at the kth trial will be

$$P(k, p) = pq^{k-1}$$

The probability that more than n trials are needed before the first success will be

$$P(k > n) = q^n$$

The mean and standard deviation for the geometric distribution are

$$\mu = \frac{1}{p}$$

$$\sigma = \frac{\sqrt{q}}{p}$$

Example The probability for finding an error by an auditor in a production line is 0.01. What is the probability that the first error is found at the 70th part audited?

Solution

$$P(k, p) = pq^{k-1}$$

$$P(70, 0.01) = 0.01 \times 0.99^{70-1} = 0.004998$$

The probability that the first error is found at the 70th part audited will be 0.004998.

Example What is the probability that more than 50 parts must be audited before the first error is found?

Solution

$$P(k > n) = q^n$$
$$P(k > 50) = 0.99^{50} = 0.605$$

4.1.5 Hyper-geometric distribution

One of the conditions of a binomial distribution was the independence of the trials; the probability of a success is the same for every trial. If successive trials are performed without replacement and the sample size or population is small, the probability for each observation will vary.

If a sample has 10 stones, the probability of taking a particular stone out of the 10 will be 1/10. If that stone is not replaced into the sample, the probability of taking another one will be 1/9. But if the stones are replaced each time, the probability of taking a particular one will remain the same, 1/10.

When the sampling is finite (relatively small and known) and the outcome changes from trial to trial, the *hyper-geometric distribution* is used instead of the binomial distribution. The formula for the hyper-geometric distribution is as follows,

$$P(x) = \frac{C_x^k C_{n-x}^{N-k}}{C_n^N}$$

where x is an integer whose value is between zero and n.

$$x \leq k$$
$$\mu = n\left(\frac{k}{N}\right)$$
$$\sigma^2 = n\left(\frac{k}{N}\right)\left(1-\frac{k}{N}\right)\left(\frac{N-n}{N-1}\right)$$

Example A total of 75 parts are received from the suppliers. We are informed that 8 defective parts were shipped by mistake, and 5 parts have already been installed on machines. What is the probability that exactly 1 defective part was installed on a machine? What is the probability of finding less than 2 defective parts?

Solution The probability that exactly 1 defective was installed on a machine is

$$P(1) = \frac{C_1^8 C_4^{67}}{C_5^{75}} = \frac{6131840}{17259390} = 0.355275592$$

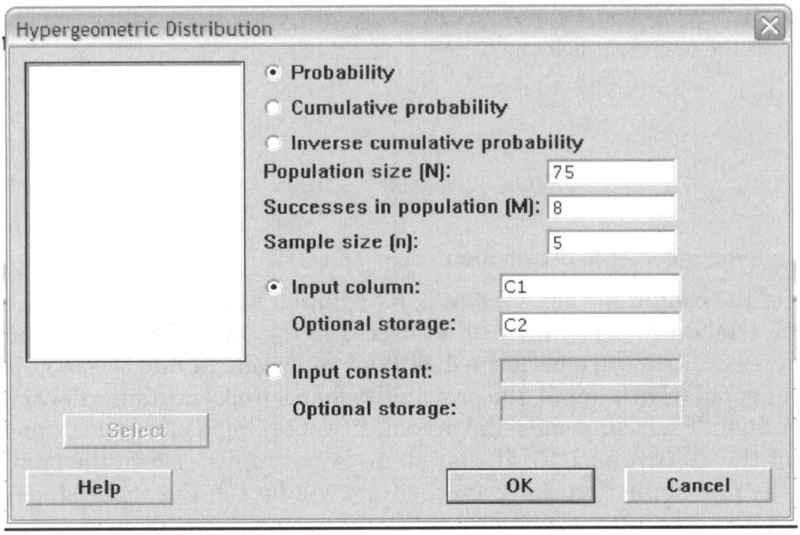

Hypergeometric Distribution

- Probability
- Cumulative probability
- Inverse cumulative probability

Population size (N): 75

Successes in population (M): 8

Sample size (n): 5

- Input column: C1

  Optional storage: C2

- Input constant:

  Optional storage:

Select

Help | OK | Cancel

Figure 4.6

Using Minitab. Here the process of the hyper-geometric distribution is the same as for the Poisson and binomial distributions. From the Calc menu, select "Probability Distributions" and then select "Hypergeometric." In the "Hypergeometric Distribution" dialog box, enter the data as shown in Figure 4.6.

The results appear as shown in Figure 4.7.

↓	C1	C2
1	0	0.559559
2	1	0.355276
3	2	0.077717
4	3	0.007174
5	4	0.000272
6	5	0.000003
7	6	0.000000
8	7	0.000000
9	8	0.000000
10	9	0.000000

Figure 4.7

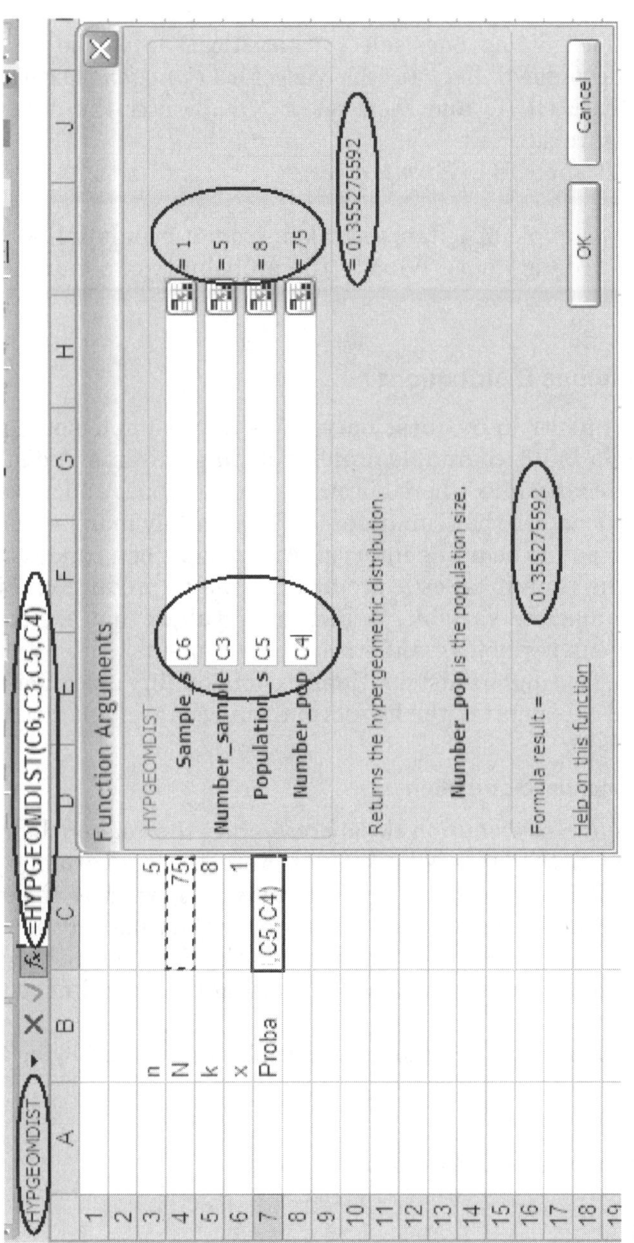
HYPGEOMDIST
=HYPGEOMDIST(C6,C3,C5,C4)
n 5
N 75
k 8
x 1
Proba
,C5,C4)
Function Arguments
HYPGEOMDIST
Sample_s C6 = 1
Number_sample C3 = 5
Population_s C5 = 8
Number_pop C4 = 75
= 0.355275592
Returns the hypergeometric distribution.
Number_pop is the population size.
Formula result = 0.355275592
Help on this function
OK
Cancel

Figure 4.8

Using Excel. The process of finding the probability for the hypergeometric distribution is the same as for the previous distributions. Click on the "Insert function" (f_x) shortcut button, then in the "Insert Function" dialog box, select "Statistical" from the *Or select a category:* drop-down list. In the *Select a Function* textbox, select "HYPERGEOMDIST" and then select "OK." Enter the data as indicated in Figure 4.8.

The result appears in *Formula result.*

Exercise. A sample of 7 items is taken from a population of 19 items containing 11 blue items. What is the probability of obtaining exactly 3 blue items?

4.2 Continuous Distributions

Most experiments in business operations have sample spaces that do not contain a finite, countable number of simple events. A distribution is said to be *continuous* when it is built on continuous random variables, which are variables that can assume the infinitely many values corresponding to points on a line interval. An example of a random variable would be the time it takes a production line to produce one item. In contrast to discrete variables, which have values that are countable, the continuous variables' values are measurements.

The main continuous distributions used in quality operations are the normal, the exponential, the log-normal, and the Weibull distributions.

4.2.1 Exponential distribution

The *exponential distribution* closely resembles the Poisson distribution. The Poisson distribution is built on discrete random variables and describes random occurrences over some intervals, whereas the exponential distribution is continuous and describes the time between random occurrences. Examples of an exponential distribution are the time between machine breakdowns and the waiting time in a line at a supermarket.

The exponential distribution is determined by the following formula,

$$P(x) = \lambda e^{-\lambda x}$$

The mean and the standard deviation are

$$\mu = \frac{1}{\lambda}$$

$$\sigma = \frac{1}{\lambda}$$

The shape of the exponential distribution is determined by only one parameter, λ. Each value of λ determines a different shape of the curve. Figure 4.9 shows the graph of the exponential distribution:

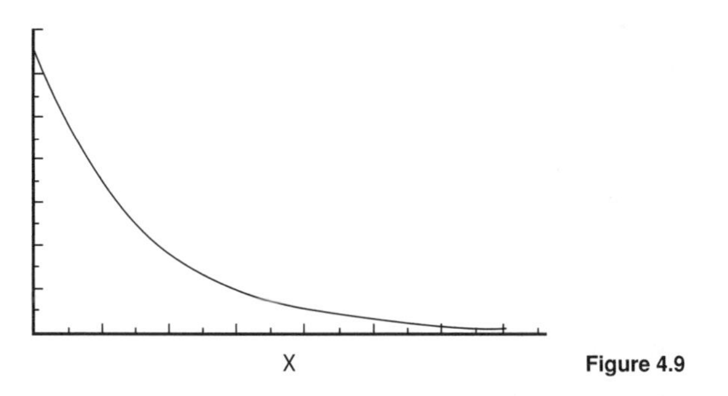

Figure 4.9

The area under the curve between any two points determines the probabilities for the exponential distribution. The formula used to calculate that probability is

$$P(x \geq a) = e^{-\lambda a}$$

with $a \geq 0$. If the number of events taking place in a unit time has a Poisson distribution with a mean λ, then the interval between these events are exponentially distributed with the mean interval time equal to $1/\lambda$.

Example If the number of items arriving at inspection at the end of a production line follows a Poisson distribution with a mean of 10 items an hour, then the time between arrivals follows an exponential distribution with a mean between arrival times of $\mu = 6$ minutes because

$$1/\lambda = 1/10 = 0.1$$

$$0.1 \times 60 \text{ mn} = 6 \text{ minutes}$$

Example Suppose that the time in months between line stoppages on a production line follows an exponential distribution with $\lambda = 0.5$

a. What is the probability that the time until the line stops again will be more than 15 months?

b. What is the probability that the time until the line stops again will be less than 20 months?

c. What is the probability that the time until the line stops again will be between 10 and 15 months?

d. Find μ and σ. Find the probability that the time until the line stops will be between $(\mu - 3\sigma)$ and $(\mu + 3\sigma)$.

Solution

a.

$$P(x > 15) = e^{-15\lambda} = e^{(-15 \cdot 0.5)} = e^{-7.5} = 0.000553$$

The probability that the time until the line stops again will be more than 15 months is 0.000553.

b.

$$P(x < 20) = 1 - P(x > 20) = 1 - e^{-(20 \cdot 0.5)} = 1 - e^{-10}$$
$$= 1 - 0.0000454 = 0.9999$$

The probability that the time until the line stops again will be less than 20 months is 0.9999.

c. We have already found that

$$P(x > 15) = 0.000553$$

We need to find the probability that the time until the line stops again will be more than 10 months,

$$P(x > 10) = 0.006738$$

The probability that the time until the line stops again will be between 10 and 15 months is the difference between 0.13533 and 0.000553,

$$P(10 \leq x \leq 15) = 0.006738 - 0.000553 = 0.006185$$

d. The mean and the standard deviation are given by $\mu = \sigma = 1/\lambda$, therefore:

$$\mu = \sigma = \frac{1}{0.5} = 2$$
$$(\mu - 3\sigma) = 2 - 6 = -4$$
$$(\mu + 3\sigma) = 2 + 6 = 8$$

So we need to find $P(-4 \leq x \leq 8)$ which is equal to $P\left(0 \leq x \leq 8\right)$

$$P(0 \leq x \leq 8) = 1 - P\left(x \geq 8\right)$$

Therefore:

$$P(-4 \leq x \leq 8) = 1 - e^{-8 \cdot 0.5} = 1 - e^{-4} = 1 - 0.018315638 = 0.9817$$

The probability that the time until the line stops again will be between $(\mu - 3\sigma)$ and $(\mu + 3\sigma)$ is 0.9817.

4.2.2 Normal distribution

The *normal distribution* is certainly one of the most widely used probability distributions. Most of nature and human characteristics are normally distributed, and so are most production outputs for well-calibrated machines. Six Sigma derives its statistical definition from it. When a population is normally distributed, most of the observations are clustered around the mean. The mean, the mode, and the median become good measures of estimates.

The average height of an adult male is 5 feet and 8 inches. This does not mean all adult males are of that height but more than 80 percent of them are very close. The weight and shape of apples are very close to their mean.

The normal probability is given by

$$f(x) = \frac{1}{\sigma\sqrt{2\pi}} e^{-\frac{(x-\mu)^2}{2\sigma^2}}$$

Where

$$e = 2.7182828$$
$$\pi = 3.1416$$

The equation of the distribution depends on μ and σ. The curve associated with that function is bell-shaped and has an apex at the center. It is symmetrical about the mean, and the two tails of the curve extend indefinitely without ever touching the horizontal axis. The area between the curve and the horizontal line is estimated to be equal to one.

$$\int f(x)\,dx = \frac{1}{\sigma\sqrt{2\pi}} \int e^{-\frac{(x-\mu)^2}{2\sigma^2}}\,dx = 1$$

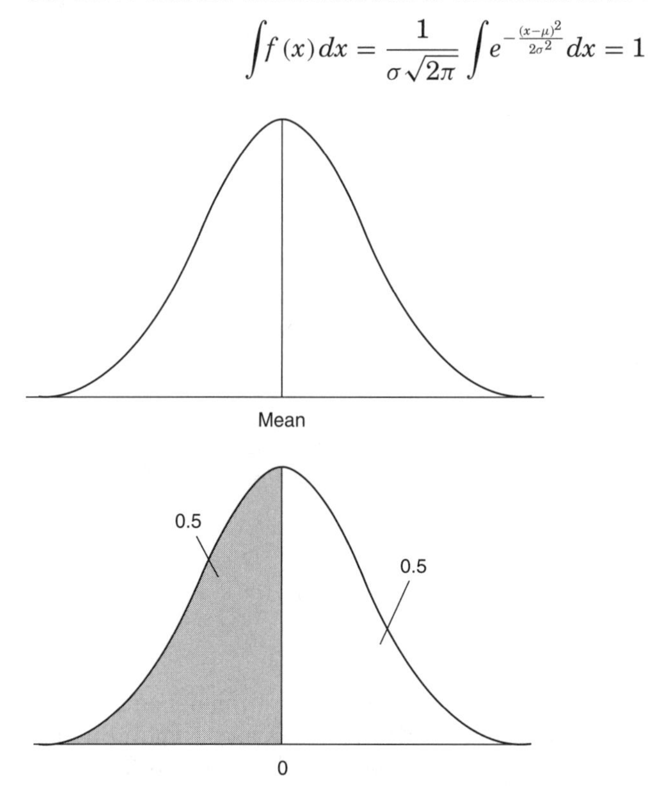

Remember that the total area under the curve is equal to 1, and half of that area is equal to 0.5. The area on the left side of any point on the horizontal represents the probability of an event being "less than" that point of estimate, and the area on the right represents the probability of an event being "more than" the point of estimate, and the point itself represents the probability of an event being "equal to" the point of estimate.

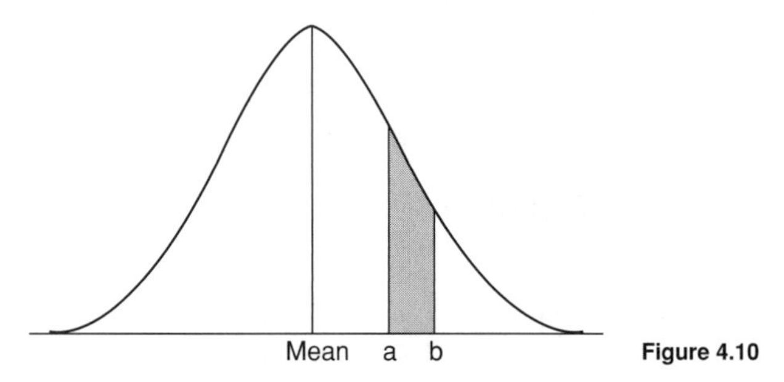

Figure 4.10

In Figure 4.10, the shaded area under the curve between a and b represents the probability that a random variable assumes a certain value in that interval.

For a sigma-scaled normal distribution, the area under the curve has been determined. Approximately 68.26 percent of the area lies between $\mu - \sigma$ and $\mu + \sigma$.

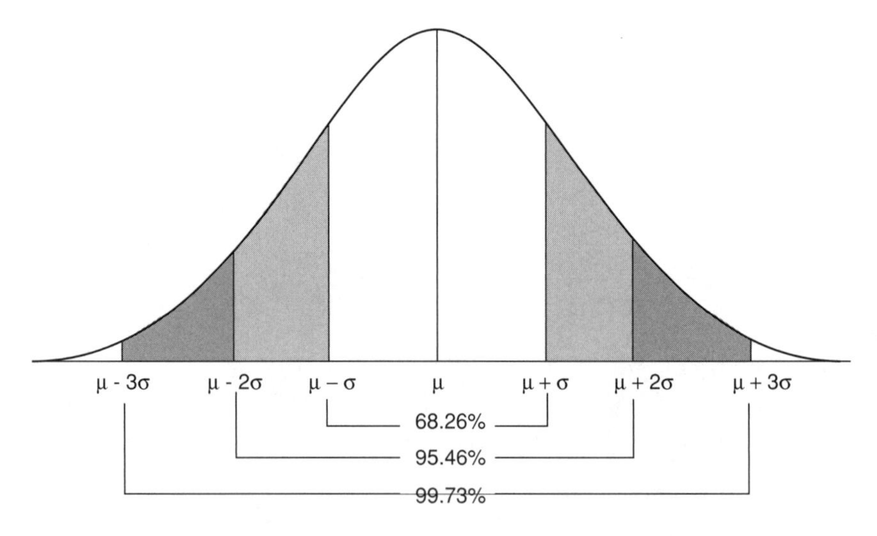

Z-transformation. The shape of the normal distribution depends on two factors, the mean and the standard deviation. Every combination of μ and σ represent a unique shape of a normal distribution. Based on the mean and the standard deviation, the complexity involved in the normal distribution can be simplified and it can be converted into the simpler z-distribution. This process leads to the *standardized normal distribution*,

$$Z = \frac{X - \mu}{\sigma}$$

Because of the complexity of the normal distribution, the standardized normal distribution is often used instead.

Consider the following example. The weekly profits of a large group of stores are normally distributed with a mean of $\mu = 1200$ and a standard deviation of $\sigma = 200$. What is the Z value for a profit for $x = 1300$? For $x = 1400$?

For $x = 1300$	For $x = 1400$
$Z = \frac{1300 - 1200}{200} = 0.5$	$Z = \frac{1400 - 1200}{200} = 1$

Example In the example above, what is the percentage of the stores that make \$1500 or more a week?

Solution

$$Z = \frac{1500 - 1200}{200} = \frac{300}{200} = 1.5$$

On the Z score table (Appendix 3), 1.5 corresponds to 0.4332. This represents the area between \$1200 and \$1500. The area beyond \$1500 is found by deducting 0.4332 from 0.5 (0.5 is half of the area under the curve). This area is 0.0668; in other words, 6.68 percent of the stores make more than \$1500 week.

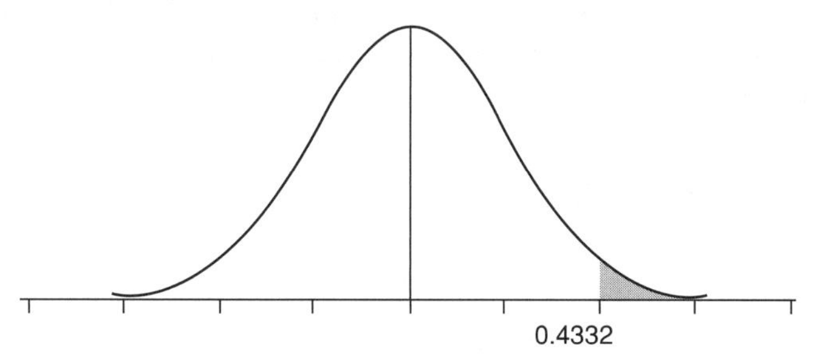

Using Minitab. Open a Minitab worksheet and enter “1500” in the first cell of column C1. Then from the Calc menu, select the “Probability distributions” option and then select “Normal.” Fill in the fields in the “Normal Distribution” dialog box as indicated in Figure 4.11 and then select “OK.”

Normal Distribution

- Probability density
- (•) Cumulative probability
- Inverse cumulative probability

Mean: 1200
Standard deviation: 200

(•) Input column: C1
Optional storage: C2
Input constant:
Optional storage:

Select
Help OK Cancel

Worksheet

↓	C1
1	1500
2	
3	
4	

Figure 4.11

The result appears in C2, the column we chose to store the result.

↓	C1	C2
1	1500	0.933192

The question that was asked was, "What is the percentage of the stores that make $1500 or more a week?" A total of 0.933192, or 93.3192 percent, is the percentage of the stores that make less than $1500. The percentage of stores that make more than $1500 will be 100 − 93.3192 = 0.066807, or 6.6807 percent.

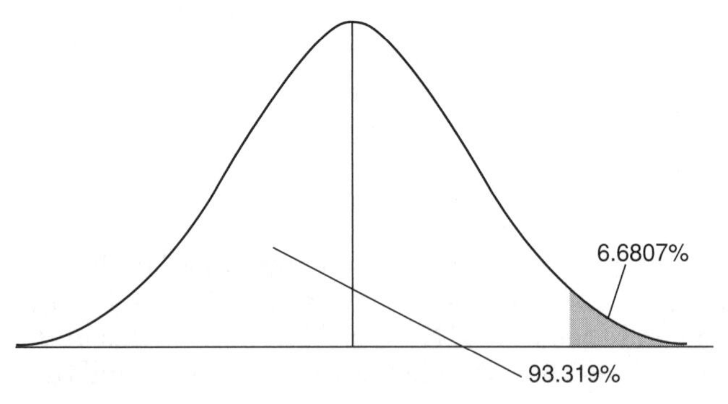

Figure 4.12

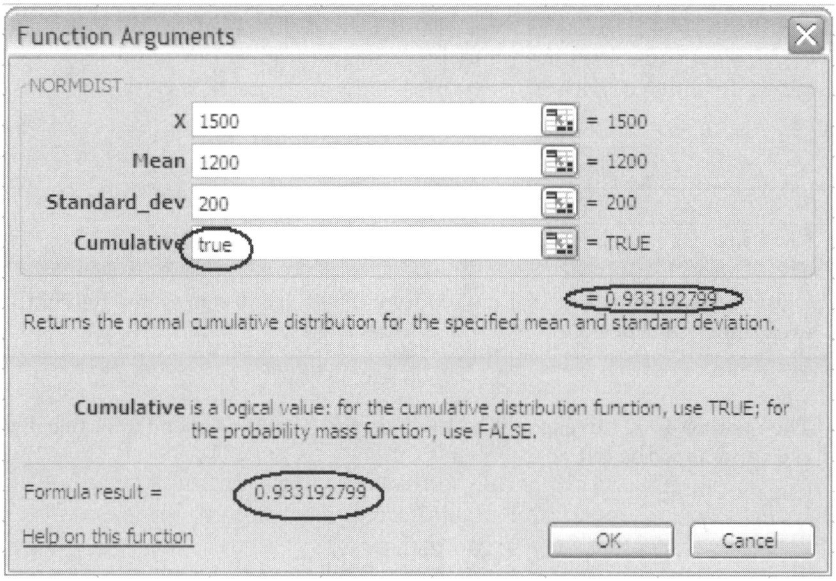

Function Arguments

NORMDIST

X 1500 = 1500

Mean 1200 = 1200

Standard_dev 200 = 200

Cumulative true = TRUE

= 0.933192799

Returns the normal cumulative distribution for the specified mean and standard deviation.

Cumulative is a logical value: for the cumulative distribution function, use TRUE; for the probability mass function, use FALSE.

Formula result = 0.933192799

Help on this function OK Cancel

Figure 4.13

The darkened tail of the area under the curve in Figure 4.12 represents the stores that make more than $1500, and the area on the left of this area represents the stores that make less than that amount.

Using Excel. We can use Excel to come to the same result. Click on the "Insert Function" (f_x) button, then select "Statistical" from the *Or select a category:* drop-down list, select "NORMDIST" from the *Select a function* list, and the "Function Argument" dialog box appears. Fill in the fields as indicated in Figure 4.13. Notice that for *Cumulative* we entered "true" — this is because the question was asking for the stores that make more. Had the question been asked for the stores that make exactly $1500, then we would have entered "false."

Example A manufacturer wants to set a minimum life expectancy on a newly manufactured light bulb. A test has revealed a mean of $\mu = 250$ hours and a standard deviation of $\sigma = 15$. The production of light bulbs is normally distributed. The manufacturer wants to set the minimum life expectancy of the light bulbs so that less than 5 percent of the bulbs will have to be replaced. What minimum life expectancy should be put on the light bulb labels?

Solution

The shaded area in Figure 4.14 under the curve between x and the end of the tail represents the 5 percent (or 0.0500) of the light bulbs that might need to be replaced. The area between x and μ (250) represents the 95 percent of good light bulbs.

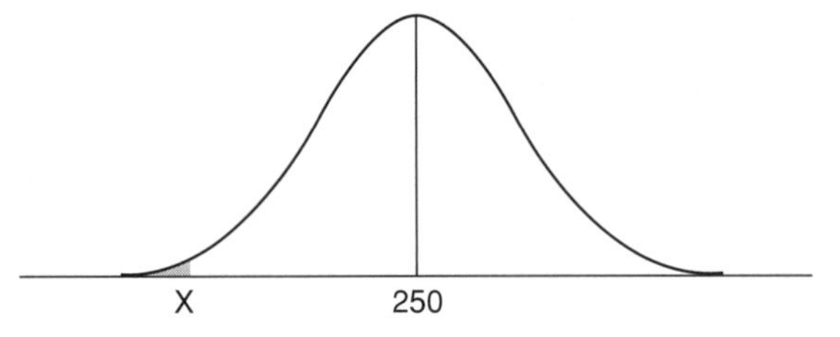

Figure 4.14

To find Z, we must deduct 0.0500 from 0.500 (0.500 represents half of the area under the curve)

$$0.5 - 0.05 = 0.45$$

The result 0.4500 corresponds to 1.645 on the Z table (Appendix 3). Because the value is to the left of μ,

$$Z = -1.645$$

$$Z = \frac{X - 250}{15} = -1.645$$

$$x = 225.325$$

The minimum life expectancy for the light bulb will be 225.325 hours.

Example The mean number of defective parts that come from a production line is $\mu = 10.5$ with a standard deviation of $\sigma = 2.5$. What is the probability that the number of defective parts for a randomly selected sample will be less than 15?

Solution

$$Z = \frac{15 - 10.5}{2.5} = 1.8$$

The result 1.8 corresponds to 0.4641 on the Z table (Appendix 3). So the probability that the number of defective parts will be less than 15 is 0.9641 (0.5 + 0.4641).

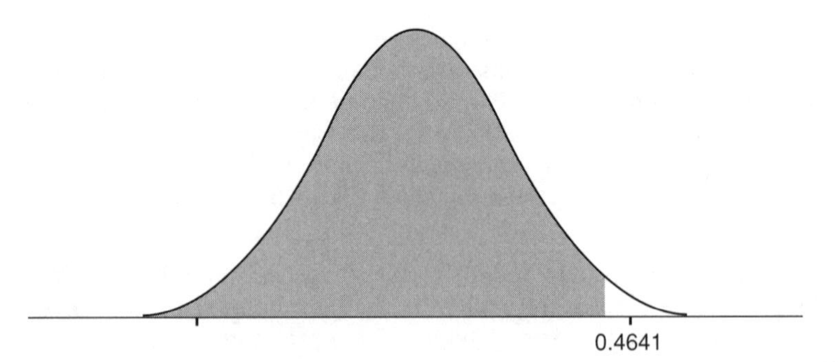

4.2.3 The log-normal distribution

Along with the Weibull distribution, the *log-normal distribution* is frequently used in risk assessment, in reliability, and in material strength and fatigue analysis. A random variable is said to be log-normally distributed if its logarithm is normally distributed. Because the lognormal distribution is derived from the normal distribution, the two share most of the same properties. The formula of the log-normal distribution is

$$f(x) = \frac{1}{x\sigma\sqrt{2\pi}}e^{-\frac{1}{2}\left(\frac{\ln z-\mu}{\sigma}\right)}, x > 0$$

where μ represents the log of the mean and σ, the scale parameter, represents the log of the standard deviation.

The log-normal cumulative distribution is

$$F(x) = \theta\left(\frac{\ln x - \mu}{\sigma}\right)$$

and the reliability function is

$$R(x) = 1 - F(x)$$

Reliability is defined as the probability that the products will be functional throughout their engineered specified life-time. where $\theta(x)$ represents the standard cumulative distribution function.

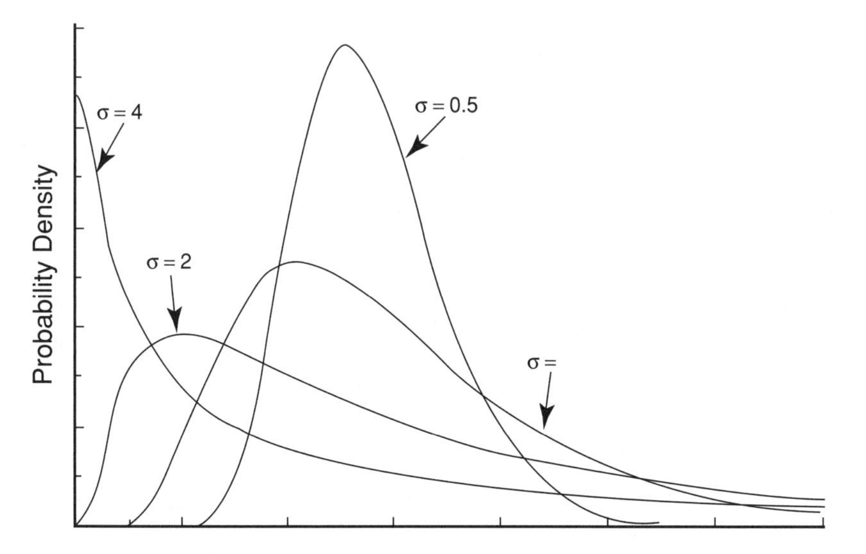

The shape of the log-normal distribution depends on the scale parameter, σ

Chapter

5

How to Determine, Analyze, and Interpret Your Samples

Learning Objectives:

- Understand the importance of sampling in manufacturing and services
- Understand how sample sizes are determined
- Understand the Central Limit Theorem
- Estimate population parameters based on sample statistics

Sampling consists of taking a subset of a population for analysis to make an inference about the population from which the samples were taken. It is a method very often used in quality control. In a large scale production environment, testing every single product is not cost-effective because it would require a plethora of manpower and a great deal of time and space.

Consider a company that produces 100,000 tires a day. If the company is open 16 hours a day (two shifts) and it takes an employee 10 minutes to test a tire, the testing of all the tires would require one million minutes, or 16,667 hours, and the company would need at least 1042 employees in the quality control department to test every single tire that comes out of production, as well as a tremendous amount of space for the QA department and the outbound inventory.

Machine productions are generally normally distributed when the machines are well-calibrated. For a normally distributed production output, taking a sample of the output and testing it can help determine the quality level of the whole production. Under some specific conditions, even if the production output is not normally distributed, sampling can be used as a method for estimating population parameters.

5.1 How to Collect a Sample

Sampling consists of testing a subset of the population to derive a conclusion for the whole population. Depending on the type of data being analyzed and the purpose of the analysis, several methods can be used to collect samples. The way the samples are collected and their sizes are crucial for the statistics derived from their analysis to be reflective of the population parameters.

First of all, it is necessary to distinguish between random and nonrandom sampling. In a *random* sampling, all the items in the population are presumed identical and they have the same probability of being selected for testing. For instance, the products that come from a manufacturing line are presumed identical and the auditor can select any one of them for testing. Albeit sampling is more often random, *nonrandom* sampling is also used in production. For example, if the production occurs over 24 hours and the auditor only works 4 hours a day, the samples he or she takes cannot be considered random because they can only have been produced by the people who work on the same shift as the auditor.

5.1.1 Stratified sampling

Stratified sampling begins with subdividing the population being studied into groups and selecting the samples from each group. In so doing, the opportunities for errors are significantly reduced. For instance, if we are testing the performance of a machine based on its output and it produces several different products, when sampling the products it would be more effective to subgroup the products by similarities.

5.1.2 Cluster sampling

In stratified sampling, the groupings are homogeneous—all the items in a group are identical. In *cluster sampling*, every grouping is representative of the population; the items it contains are diverse.

5.1.3 Systematic sampling

In a *systematic sampling*, the auditor examines the pattern of the population and determines the size of the sample he or she wants to take, decides the appropriate intervals between the items he or she would select, and then takes every ith item from the population.

5.2 Sampling Distribution of Means

If the means of all possible samples are obtained and organized, we could derive the *sampling distribution of the means*.

Consider the following example. We have five items labeled 5, 6, 7, 8 and 9 and we want to create a sampling distribution of the means for all the items. The size of the samples is two, so the number of samples will be

$$_5C_2 = \frac{5!}{2!(5-2)!} = \frac{5 \times 4 \times 3 \times 2 \times 1}{(2 \times 1)(3 \times 2 \times 1)} = 10$$

Because the number of samples is 10, the number of means will also be 10. The samples and their means will be distributed as shown in Table 5.1.

TABLE 5.1

Combinations	Means
(6, 5)	5.5
(6, 7)	6.5
(6, 8)	7.0
(6, 9)	7.5
(5, 7)	6.0
(5, 8)	6.5
(5, 9)	7.0
(7, 8)	7.5
(7, 9)	8.0
(8, 9)	6.5

Exercise. How many samples of five items can we obtain from a population of 30?

Exercise. Based on the data in Table 5.2, build a distribution of the means for samples of two items.

TABLE 5.2

9	12	14	13	12	16	15

5.3 Sampling Error

The sample statistics may not always be exactly the same as their corresponding population parameters. The difference is known as the *sampling error*.

Suppose a population of 10 bolts has diameter measurements of 9, 11, 12, 12, 14, 10, 9, 8, 7, and 9 mm. The mean μ for that population would be 10.1 mm. If a sample of only three measurements—9, 14, and 10 mm—is taken from the population, the mean of the sample would be $(9 + 14 + 10)/3 = 11$ mm and the sampling error (E) would be

$$E = \overline{X} - \mu = 11 - 10.1 = 0.9$$

Take another sample of three measurements — 7, 12, and 11 mm. This time, the mean will be 10 mm and the sampling error will be

$$E = \overline{X} - \mu = 10 - 10.1 = -0.1$$

If another sample is taken and estimated, its sampling error might be different. These differences are said to be due to chance.

We have seen in the example of the bolt diameters that the mean of the first sample was 11 mm and the mean of the second was 10 mm. In that example, we had 10 bolts, and if all possible samples of three were computed, there would have been 120 samples and means.

$${}_N C_n = \frac{N!}{n!\,(N-n)!} = \frac{10!}{3!(10-3)!} = 120$$

Exercise. Based on the population given in Table 5.3, what is the sampling error for the following samples: (9, 15), (12, 16), (14, 15), and (13, 15).

TABLE 5.3

9	12	14	13	12	16	15

So if it is possible to make mistakes while estimating the population's parameters from a sample, how can we be sure that sampling can help get a good estimate? Why use sampling as a means of estimating the population parameters?

The Central Limit Theorem can help us answer these questions.

5.4 Central Limit Theorem

The *Central Limit Theorem* states that for sufficiently large sample sizes ($n \geq 30$), regardless of the shape of the population distribution, if samples of size n are randomly drawn from a population that has a mean μ and a standard deviation σ, the samples' means $\overline{X}$ are approximately normally distributed. If the populations are normally distributed, the samples' means are normally distributed regardless of the sample sizes. The implication of this theorem is that for sufficiently large populations, the normal distribution can be used to analyze samples drawn from populations that are not normally distributed, or whose distribution characteristics are unknown.

When means are used as estimators to make inferences about a population's parameters and $n \geq 30$, the estimator will be approximately normally distributed in repeated sampling. The mean and standard deviation of that sampling distribution are given as

$$\mu_{\overline{x}} = \mu$$

$$\sigma_{\overline{x}} = \frac{\sigma}{\sqrt{n}}$$

where $\mu_{\overline{X}}$ is the mean of the samples and $\sigma_{\overline{X}}$ is the standard deviation of the samples. If we know the mean and the standard deviation for the population, we can easily derive the mean and the standard deviation for the sample distribution,

$$\mu = \mu_{\overline{X}}$$

$$\sigma = \sigma_{\overline{X}}\sqrt{n}$$

Example Gajaga Electronics is a company that manufactures circuit boards. The average imperfection on a board is $\mu = 5$ with a standard deviation of $\sigma = 2.34$ when the production process is under statistical control. A random sample of $n = 36$ circuit boards has been taken for inspection and a mean of $\overline{x} = 6$ defects per board was found. What is the probability of getting a value of $\overline{x} \leq 6$ if the process is under control?

Solution Because the sample size is greater than 30, the Central Limit Theorem can be used in this case even though the number of defects per board follows a Poisson distribution. Therefore, the distribution of the sample mean $\overline{x}$ is approximately normal with the standard deviation

$$\sigma_{\overline{x}} = \frac{\sigma}{\sqrt{n}} = \frac{2.34}{\sqrt{36}} = 0.39$$

$$z = \frac{\overline{x} - \mu}{\sigma/\sqrt{n}} = \frac{6 - 5}{0.39} = \frac{1}{0.39} = 2.56$$

The result $Z = 2.56$ corresponds to 0.4948 on the table of normal curve areas (Appendix 3).

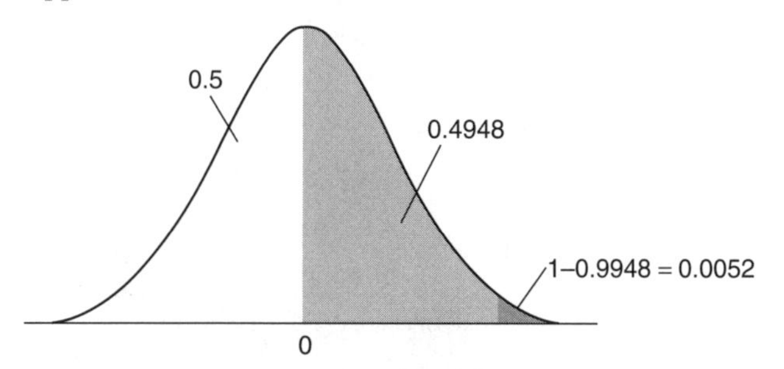

Remember from our normal distribution discussion that the total area under the curve is equal to one and half of that area is equal to 0.5. The area on the left side of any point on the horizontal line represents the probability of an event being "less than" that point of estimate, the area on the right represents the probability of an event being "more than" that point of estimate, and that the point itself represents the probability of an event being "equal to" that point of estimate. Therefore, the probability of getting a value of $\overline{x} \leq 6$ is $0.5 + 0.4948 = 0.9948$.

$$P\left(\overline{x} \leq 6\right) = 0.9948$$

We can use Minitab to come to the same result. From the Stat menu, select "Basic Statistics" and then select "1-Sample Z. . . "

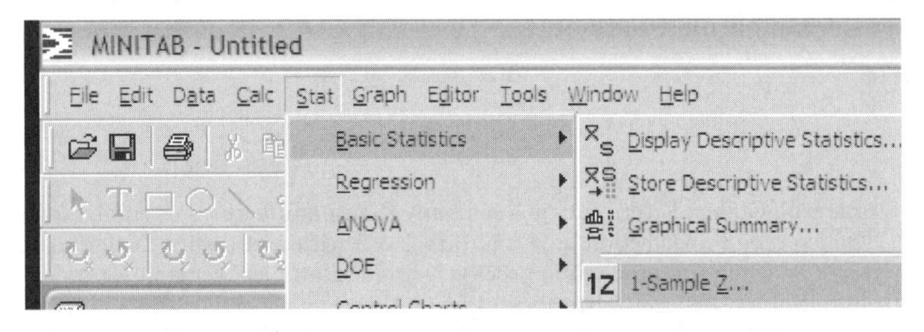

In the "1-Sample Z" dialog box, fill in the fields as indicated in Figure 5.1, then select "OK."

1-Sample Z (Test and Confidence Interval)

Samples in columns:

Summarized data

Sample size: 36

Mean: 6

Standard deviation: 2.34

Perform hypothesis test

Hypothesized mean: 5

Select | Graphs... | Options...

Help | OK | Cancel

Figure 5.1

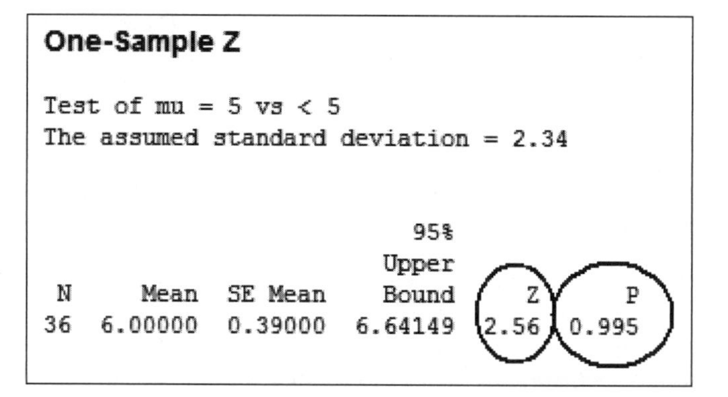

One-Sample Z

Test of mu = 5 vs < 5
The assumed standard deviation = 2.34

N	Mean	SE Mean	95% Upper Bound	Z	P
36	6.00000	0.39000	6.64149	2.56	0.995

Example The average number of parts that reach the end of a production line defect-free at any given hour of the first shift is 372 parts with a standard deviation of 7. What is the probability that a random sample of 34 different productions' first-shift hours would yield a sample mean between 369 and 371 parts that reach the end of the line defect-free?

Solution In this case, $\mu = 372$, $\sigma = 7$, and $n = 34$. We must determine the probability of having the mean between 369 and 371. We will first find the probability that the mean would be equal to 369 and then for it to be equal to 371.

$$z = \frac{369 - 372}{\frac{7}{\sqrt{34}}} = \frac{-3}{1.2} = -2.5$$

In the Z score table in Appendix 3, a value of 2.5 corresponds to 0.4938.

$$z = \frac{371 - 372}{\frac{7}{\sqrt{34}}} = \frac{-1}{1.2} = -0.8333$$

In the Z score table in Appendix 3, a value of 0.833 corresponds to 0.2967.

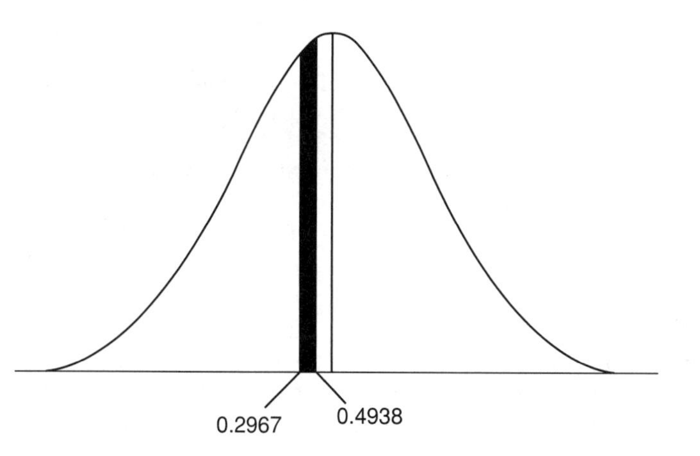

The probability for the mean to be within the interval [369, 371] will be the difference between 0.4938 and 0.2967, which is equal to 0.1971.

5.5 Sampling from a Finite Population

The previous example is valid for an extremely large population. Sampling from a finite population will require some adjustment called the *finite correction factor*:

$$fc = \sqrt{\frac{N-n}{N-1}}$$

The variable Z will therefore become

$$Z = \frac{\overline{x} - \mu}{\frac{\sigma}{\sqrt{n}}\sqrt{\frac{N-n}{N-1}}}$$

Example A city's 450 restaurant employees average \$35 in tips per day with a standard deviation of \$9. If a sample of 50 employees is taken, what is the probability that the sample will have an average of less than \$37 tips a day?

Solution

$$N = 450$$
$$n = 50$$
$$\mu = 35$$
$$\sigma = 9$$
$$\overline{x} = 37$$
$$z = \frac{37 - 35}{\frac{9}{\sqrt{50}}\sqrt{\frac{400}{449}}} = \frac{2}{1.27 * 0.89} = \frac{2}{1.13} = 1.77$$

On the Z score table (Appendix 3), a value of 1.77 corresponds to 0.4616; therefore the probability of getting an average daily tip of less than \$37 will be $0.4616 + 0.5 = 0.9616$.

If the finite correction factor were not taken into account, Z would have been 1.57, which corresponds to 0.4418 on the Z score table, and therefore the probability of having a daily tip of less than \$37 would have been 0.9418.

5.6 Sampling Distribution of $\overline{p}$

When the data being analyzed are measurable, as is the case of the two previous examples or in the case of distance or income, the sample mean is often privileged. However, when the data are countable—as in the case of people in a group, or defective items on a production line—the sample proportion is the statistic of choice.

The *sample proportion* $\overline{p}$ applies to situations that would have required a binomial distribution where p is the probability for a success and q the probability for a failure, with $q = 1 - p$. When a random sample of n trials is selected from a binomial population (that is, an experiment with n identical trials with each trial having only two possible outcomes considered as success or failure) with parameter p, the sampling distribution $\overline{p}$ of the sample proportion will be

$$\overline{p} = \frac{x}{n}$$

where x is the number of success. The mean and standard deviation will be

$$\mu_{\overline{p}} = p$$

$$\sigma_{\overline{p}} = \sqrt{\frac{pq}{n}}$$

If $0 \le \mu_{\overline{p}} \pm 2\sigma_{\overline{p}} \le 1$, then the sampling distribution of $\overline{p}$ can be approximated using the normal distribution.

Example In a sample of 100 workers, 25 might be coming late once a week. The sample proportion $\overline{p}$ of the latecomers will be $25/100 = 0.25$. In this example,

$$\mu_{\overline{p}} = 0.25$$

$$\sigma_{\overline{p}} = \sqrt{\frac{0.25 \times 0.75}{100}} = 0.0433$$

If $np > 5$ and $nq > 5$, the Central limit Theorem applies to the sample proportion. The Z formula for the sample proportion is given as

$$Z = \frac{\overline{p} - p}{\sigma_{\overline{p}}} = \frac{\overline{p} - p}{\sqrt{\frac{pq}{n}}}$$

where $\overline{p}$ is the sample proportion, p is the population proportion, n is the sample size, and $q = 1 - p$.

Example If 40 percent of the parts that come off a production line are defective, what is the probability of taking a random sample of size 75 from the line and finding that 70 percent or less are defective?

Solution

$$p = 0.4$$
$$\overline{p} = 0.7$$
$$n = 75$$
$$Z = \frac{0.7 - 0.4}{\sqrt{\frac{0.4 \times 0.6}{75}}} = \frac{0.2}{0.057} = 3.54$$

In the standard normal distribution table (Appendix 3), a value of 3.54 correspond to 0.4998. So the probability of finding 70 percent or less defective parts is $0.5 + 0.4998 = 0.9998$.

Example Forty percent of all the employees have signed up for the stock option plan. An HR specialist believes that this ratio is too high. She takes a sample of 450 employees and finds that 200 have signed up. What is the probability of getting a sample proportion larger than this if the population proportion is really 0.4?

Solution

$$p = 0.4$$
$$q = 0.6$$
$$n = 450$$
$$\overline{p} = 0.44$$
$$Z = \frac{0.44 - 0.4}{\sqrt{\frac{0.4 \times 0.6}{450}}} = \frac{0.04}{0.0231} = 1.73$$

This corresponds to 0.4582 on the standard normal distribution table. The probability of getting a sample proportion larger than 0.4 will be

$$0.5 - 0.4582 = 0.0418.$$

5.7 Estimating the Population Mean with Large Sample Sizes

Suppose a company has just developed a new process for prolonging the life of a light bulb. The engineers want to be able to date each light bulb to determine its longevity, yet it is not possible to test each light bulb in a production process that generates hundreds of thousands of light bulbs a day. But they can take a random sample and determine its average longevity, and from there they can estimate the longevity of the whole population.

Using the Central Limit Theorem, we have determined that the Z value for sample means can be used for large samples.

$$z = \frac{\overline{X} - \mu}{\sigma/\sqrt{n}}$$

By rearranging this formula, we can derive the value of μ,

$$\mu = \overline{X} - Z\frac{\sigma}{\sqrt{n}}$$

Because Z can be positive or negative, a more accurate formula would be

$$\mu = \overline{X} \pm Z\frac{\sigma}{\sqrt{n}}$$

In other words, μ will be within the following interval:

$$\overline{X} - Z\frac{\sigma}{\sqrt{n}} \leq \mu \leq \overline{X} + Z\frac{\sigma}{\sqrt{n}}$$

where

$$\overline{X} - Z\frac{\sigma}{\sqrt{n}}$$

is the *lower confidence limit* (LCL) and

$$\overline{X} + Z\frac{\sigma}{\sqrt{n}}$$

is the *upper confidence limit* (UCL).

But a confidence interval presented as such does not take into account α, the area under the normal curve that is outside the confidence interval. α measures the confidence level. We estimate with some confidence that the mean μ is within the interval

$$\overline{X} - Z\frac{\sigma}{\sqrt{n}} \leq \mu \leq \overline{X} + Z\frac{\sigma}{\sqrt{n}}$$

But in this case, we cannot be absolutely certain that it is within this interval unless the confidence level is 100 percent. For a two-tailed normal curve, if we want to be 95 percent sure that μ is within that interval, the confidence level will be equal to 0.95, $(1 - \alpha)$ or $(1 - 0.05)$, and the areas under the tails will be

$$\alpha/2 = 0.05/2 = 0.025$$

Therefore

$$Z_{\alpha/2} = Z_{0.025}$$

TABLE 5.4

Confidence interval $(1 - \alpha)$	α	$Z_{\alpha/2}$
0.90	0.10	1.645
0.95	0.05	1.960
0.99	0.01	2.580

which corresponds to a value of 1.96 on the Z score table (Appendix 3). The confidence interval should be rewritten as

$$\overline{X} - Z_{\alpha/2}\frac{\sigma}{\sqrt{n}} \leq \mu \leq \overline{X} + Z_{\alpha/2}\frac{\sigma}{\sqrt{n}}$$

or

$$\overline{X} - Z_{.025}\frac{\sigma}{\sqrt{n}} \leq \mu \leq \overline{X} + Z_{.025}\frac{\sigma}{\sqrt{n}}$$

Table 5.4 shows the most commonly used confidence coefficients and their Z-score values.

Example A survey was conducted of companies that use solar panels as a primary source of electricity. The question that was asked was this: How much of the electricity used in your company comes from the solar panels? A random sample of 55 responses produced a mean of 45 megawatts. Suppose the population standard deviation for this question is 15.5 megawatts. Find the 95 percent confidence interval for the mean.

Solution

$$n = 55$$

$$\overline{X} = 45$$

$$\sigma = 15.5$$

$$Z_{\alpha/2} = 1.96$$

$$45 - 1.96\,\frac{15.5}{\sqrt{55}} \leq \mu \leq 45 + 1.96\,\frac{15.5}{\sqrt{55}}$$

$$40.90 \leq \mu \leq 49.1$$

We can be 95 percent sure that the mean will be between 40.9 and 49.1 megawatts. In other words, the probability for the mean to be between 40.9 and 49.1 will be 0.95.

$$P(40.9 \leq \mu \leq 49.1) = 0.95$$

Using Minitab. From the Stat menu, select "Basic Statistics" and then select "1-Sample Z. . . " In the "1-Sample Z" dialog box, fill in the fields as indicated in Figure 5.2 and then select "OK."

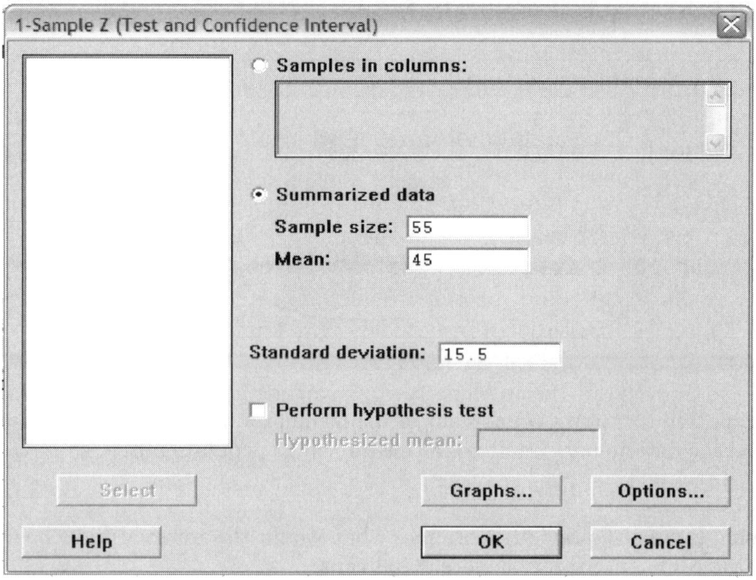

1-Sample Z (Test and Confidence Interval)

Samples in columns:

Summarized data

Sample size: 55

Mean: 45

Standard deviation: 15.5

Perform hypothesis test

Hypothesized mean:

Select | Graphs... | Options...

Help | OK | Cancel

Figure 5.2

The results appear as shown in Figure 5.3.

One-Sample Z

```
The assumed standard deviation = 15.5


 N     Mean   SE Mean          95% CI
55  45.0000    2.0900  (40.9036, 49.0964)
```

Figure 5.3

Example A sample of 200 circuit boards was taken from a production line, and it revealed the number of average defects to be 7 with a standard deviation of 2. What is the 95 percent confidence interval for the population mean μ?

Solution When the sample size is large ($n \geq 30$), the sample standard deviation can be used as an estimate of the population standard deviation.

$$\overline{X} = 7$$

$$s = 2$$

$$7 - 1.96\frac{2}{\sqrt{200}} \leq \mu \leq 7 + 1.96\frac{2}{\sqrt{200}}$$

$$6.723 \leq \mu \leq 7.277$$

The Minitab output is shown in Figure 5.4.

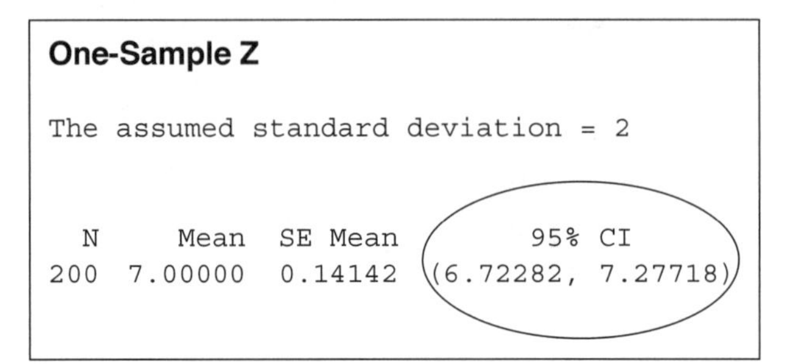

Figure 5.4

In repeated sampling, 95 percent of the confidence intervals will enclose the average defects per circuit board for the whole population μ.

Example From the previous example, what would the interval have been like if the confidence interval were 90 percent?

Solution

$$7 - 1.645\frac{2}{\sqrt{200}} \leq \mu \leq 7 + 1.645\frac{2}{\sqrt{200}}$$

$$6.77 \leq \mu \leq 7.233$$

The Minitab output is shown in Figure 5.5.

One-Sample Z

The assumed standard deviation = 2

N	Mean	SE Mean	90% CI
200	7.00000	0.14142	(6.76738, 7.23262)

Figure 5.5

Exercise. The mean number of phone calls received at a call center per day is 189 calls with a standard deviation of 12. A sample of 35 days has been taken for analysis, what would be the probability for the mean to be between 180 and 193 at a confidence level of 99 percent?

In repeated sampling, 90 percent of the confidence intervals will enclose the average defects per circuit board for the whole population μ.

5.8 Estimating the Population Mean with Small Sample Sizes and σ Unknown: *t*-Distribution

We have seen that when the population is normally distributed and the standard deviation is known, μ can be estimated to be within the interval $\overline{X} \pm Z_{\alpha/2}\frac{\sigma}{\sqrt{n}}$. But as in the case of the previous example, σ is not known; in these cases, it can be replaced by S, the sample's standard deviation, and μ is found within the interval $\overline{X} \pm Z_{\alpha/2}\frac{s}{\sqrt{n}}$. Replacing σ with S can only be a good approximation if the sample sizes are large ($n > 30$). In fact, the Z formula has been determined not to always generate normal distributions for small sizes if the population is not normally distributed. So in the case of small samples and when σ is not known, the t-*distribution* is used instead.

The formula for the t-distribution is given as

$$t = \frac{\overline{X} - \mu}{s/\sqrt{n}}$$

This equation is identical to the one for the Z formula but the tables used to determine the values are different from the ones used for the Z values.

Just as in the case of the Z formula, the t formula can also be manipulated to estimate μ, but because the sample sizes are small, to not produce a biased result we must convert of them to degrees of freedom (df),

$$df = n - 1$$

So the mean μ will be found within the interval

$$t_{\alpha/2,n-1} \pm \frac{\overline{X} - \mu}{s/\sqrt{n}}$$

Therefore,

$$\overline{X} \pm t_{\alpha/2,n-1}\frac{s}{\sqrt{n}}$$

or

$$\overline{X} - t_{\alpha/2,n-1}\frac{S}{\sqrt{n}} \leq \mu \leq \overline{X} + t_{\alpha/2,n-1}\frac{S}{\sqrt{n}}$$

Example A manager of a car rental company wants to estimate the average number of times luxury cars would be rented a month. She takes a random sample of 19 cars that produces the following number of times the cars are rented in a month. result:

$$3, 7, 12, 5, 9, 13, 2, 8, 6, 14, 6, 1, 2, 3, 2, 5, 11, 13, 5$$

She wants to use these data to construct a 95 percent confidence interval to estimate the average.

Solution

$$3 + 7 + 12 + 5 + 9 + 13 + 2 + 8 + 6 + 14 + 6 + 1 + 2 + 3 + 2 + 5 + 11 + 13 + 5 = 127$$

$$\overline{X} = \frac{127}{19} = 6.68$$

$$s = 4.23$$

$$n = 19$$

$$df = n - 1 = 18$$

From the t table in Appendix 4,

$$t_{0.05,0.25} = 2.101$$

$$6.68 - 2.101\frac{4.23}{\sqrt{19}} \leq \mu \leq 6.68 + 2.101\frac{4.23}{\sqrt{19}}$$

$$4.641 \leq \mu \leq 8.72$$

$$P(4.641 \leq \mu \leq 8.72) = 0.95$$

The probability for μ to be between 4.64 and 8.72 is 0.95.

Using Minitab. Open the file *Car rental.mpj* on the included CD and from the Stat menu, select "Basic Statistics" and then select "1-Sample t. . ." Select "Samples in Columns" and insert the column title (C1) into the textbox. Select "Options," and the default for the confidence interval should be 95.0 percent. Select the *Alternative* from the drop-down list and select *Not equal*, then select "OK" and "OK" once again.

The Minitab output will be

One-Sample T: C1

```
Variable   N      Mean    StDev  SE Mean          95% CI
C1        19  6.68421  4.23022  0.97048  (4.64531, 8.72311)
```

5.9 Chi Square (χ^2) Distribution

In quality control, in most cases the objective of the auditor is not to find the mean of a population but rather to determine the level of variation of the output. For instance, they would want to know how much variation the production process exhibits about the target to see what adjustments are needed to reach a defect-free process.

We have seen that if the means of all possible samples are obtained and organized we can derive the sampling distribution of the means. The same principle applies to the variances, and we would obtain the sampling distribution of the variances. Whereas the distribution of the means follows a normal distribution when the population is normally distributed or when the samples are greater than 30, the distribution of the variance follows a *chi square (χ^2) distribution.*

We have already seen that the sample variance is determined as

$$S^2 = \frac{\sum(X - \overline{X})^2}{n - 1}$$

The χ^2 formula for single variance is given as

$$\chi^2 = \frac{(n-1)S^2}{\sigma^2}$$

The shape of the χ^2 distribution resembles the normal curve but it is not symmetrical, and its shape depends on the degrees of freedom.

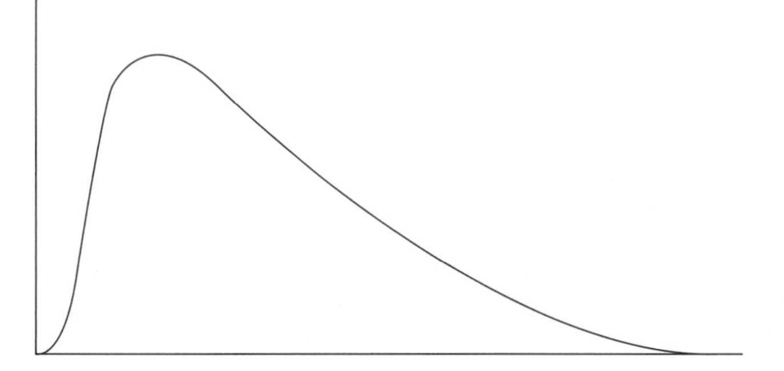

The χ^2 formula can be rearranged to find σ^2. The value σ^2 will be within the interval

$$\frac{(n-1)S^2}{\chi^2_{\alpha/2}} \leq \sigma^2 \leq \frac{(n-1)S^2}{\chi^2_{1-\alpha/2}}$$

with a degree of freedom of $n-1$.

Example A sample of 9 screws was taken out of a production line and the sizes of the diameters are shown in Table 5.5.

TABLE 5.5

13.00 mm
13.00 mm
12.00 mm
12.55 mm
12.99 mm
12.89 mm
12.88 mm
12.97 mm
12.99 mm

Estimate the population variance σ^2 with 95 percent confidence.

Solution We need to determine the *point of estimate*, which is the sample's variance.

$$S^2 = 0.1122$$

with a degree of freedom (df) of $n - 1 = 8$. Because we want to estimate σ with a confidence level of 95 percent,

$$\alpha = 1 - 0.95 = 0.05$$
$$\alpha/2 = .025$$
$$1 - \alpha/2 = 0.975$$

So σ^2 will be within the interval

$$\frac{8 \times 0.1122}{\chi^2_{0.025}} \leq \sigma^2 \leq \frac{8 \times 0.1122}{\chi^2_{0.975}}$$

From the χ^2 table in Appendix 5, the values of $\chi^2_{0.025}$ and $\chi^2_{0.975}$ for a degree of freedom of 8 are 17.5346 and 2.17973, respectively.

So the confidence interval becomes

$$\frac{0.8976}{17.5346} \leq \sigma^2 \leq \frac{0.8976}{2.17973}$$
$$0.0512 \leq \sigma^2 \leq 0.412$$

and

$$P(0.0512 \leq \sigma^2 \leq 0.412) = 0.95$$

The probability for σ^2 to be between 0.0512 and 0.412 is 0.95.

Exercise. From the data in Table 5.6, find the population's variance at a confidence level of 99 percent.

TABLE 5.6

23	25	26	24	28	39	31	38	37	36

5.10 Estimating Sample Sizes

In most cases, sampling is used in quality control to make an inference for a whole population because of the cost associated in actually studying every individual part of that population. But again, the question of the sample size arises. What size of a sample best reflects the condition of the whole population being estimated? Should we consider a sample of 150 or 1000 of products from a production line to determine the quality level of the output?

5.10.1 Sample size when estimating the mean

At the beginning of this chapter, we defined the sampling error E as being the difference between the sampling mean $\overline{X}$ and the population mean μ.

$$E = \overline{X} - \mu$$

We also have seen when studying the sampling distribution of $\overline{X}$ that when μ is being determined, we can use the Z formula for sampling means,

$$Z_{\alpha/2} = \frac{\overline{X} - \mu}{\sigma/\sqrt{n}}$$

We can clearly see that the numerator is nothing but the sampling error, E. We can therefore replace $\overline{X} - \mu$ by E in the Z formula and come up with

$$Z_{\alpha/2} = \frac{E}{\sigma/\sqrt{n}}$$

We can determine n from this equation,

$$\sqrt{n} = \frac{Z_{\alpha/2}\sigma}{E}$$

$$n = \frac{Z_{\alpha/2}^{2}\sigma^{2}}{E^{2}} = \left(\frac{Z_{\alpha/2}\sigma}{E}\right)^{2}$$

Example A production manager at a call center wants to know how much time on average an employee should spend on the phone with a customer. She wants to be within two minutes of the actual length of time, and the standard deviation of the average time spent is known to be three minutes. What sample size of calls should she consider if she wants to be 95 percent confident of her result?

Solution

$$Z = 1.96$$
$$E = 2$$
$$\sigma = 3$$
$$n = \frac{(1.96 \times 3)^2}{2^2} = \frac{34.5744}{4} = 8.6436$$

Because we cannot have 8.6436 calls, we can round up the result to 9 calls. The manager can be 95 percent confident that with a sample of 9 calls, she can determine the average length of time an employee must spend on the phone with a customer.

5.10.2 Sample size when estimating the population proportion

To determine the sample size needed when estimating p, we can use the same procedure as the one we used when determining the sample size for μ.

We have already seen that the Z formula for the sample proportion is given as

$$Z = \frac{\overline{p} - p}{\sigma_{\overline{p}}} = \frac{\overline{p} - p}{\sqrt{\frac{pq}{n}}}$$

The error of estimation (or sampling error) in this case will be $E = \overline{p} - p$. We can replace the numerator $\overline{p} - p$ by its value in the Z formula and obtain

$$Z_\alpha = \frac{E}{\sqrt{\frac{pq}{n}}}$$

We can then derive n from this equation,

$$n = \frac{Z_\alpha pq}{E^2}$$

Example A study is conducted to determine the extent to which companies promote Open Book Management. The question asked to employees is, "Do your managers provide you with enough information about the company?" It was previously estimated that only 30 percent of companies did actually provide the information needed to their employees. If the researcher wants to be 95 percent confident in the results and be within 0.05 of the true population proportion, what size of sample should be taken?

$$E = 0.05$$
$$p = 0.3$$
$$q = 0.7$$
$$Z_{0.05} = 1.96$$
$$n = \frac{(1.96)^2(0.3)(0.7)}{(0.05)^2} = 322.69$$

The sample must include 323 companies.

Chapter

6

Hypothesis Testing

Learning Objectives:

- Understand how to use samples to make an inference about a population
- Understand what sample size should be used to make an inference about a population
- How to test the normality of data

The confidence interval can help estimate the range within which we can, with a certain level of confidence, estimate the values of a population mean or the population variance after analyzing a sample. Another method of determining the significance or the characteristics of a magnitude is the hypothesis testing. The hypothesis testing is about assessing the validity of a hypothesis made about a population.

A *hypothesis* is a value judgment, a statement based on an opinion about a population. It is developed to make an inference about that population. Based on experience, a design engineer can make a hypothesis about the performance or qualities of the products she is about to produce, but the validity of that hypothesis must be ascertained to confirm that the products are produced to the customer's specifications. A test must be conducted to determine if the empirical evidence does support the hypothesis. Some examples of hypotheses are:

1. The average number of defects per circuit board produced on a given line is 3.
2. The lifetime of a given light bulb is 350 hours.
3. It will take less than 10 minutes for a given drug to start taking effect.

Most of the time, the population being studied is so large that examining every single item would not be cost effective. Therefore, a sample will be taken and an inference will be made for the whole population.

6.1 How to Conduct a Hypothesis Testing

Suppose that Sikasso, a company that produces computer circuit boards, wants to test a hypothesis made by an engineer that exactly 20 percent of the defects found on the boards are traceable to the CPU socket. Because the company produces thousands of boards a day, it would not be cost effective to test every single board to validate or reject that statement, so a sample of boards is analyzed and statistics computed. Based on the results found and some decision rules, the hypothesis is or is not rejected. (Note that we did not say that the hypothesis is accepted, because not finding enough evidence to reject the hypothesis does not necessarily mean that it must be accepted.)

If exactly 10 percent or 29 percent of the defects on the sample taken are actually traced to the CPU socket, the hypothesis will certainly be rejected, but what if 19.95 percent or 20.05 percent of the defects are actually traced to the CPU socket? Should the 0.05 percent difference be attributed to a sampling error? Should we reject the statement in this case? To answer these questions, we must understand how a hypothesis testing is conducted. There are six steps in the process of testing a hypothesis to determine if it is to be rejected or not beyond a reasonable doubt. The following six steps are usually followed to test a hypothesis.

6.1.1 Null hypothesis

The first step consists in stating the hypothesis. In the case of the circuit boards at Sikasso, the hypothesis would be: "On average, exactly 20 percent of the defects on the circuit board are traceable to the CPU socket." This statement is called the *null hypothesis*, denoted H_0, and is read "H sub zero." The statement will be written as:

$$H_0 : \mu = 20\%$$

6.1.2 Alternate hypothesis

If the hypothesis is not rejected, exactly 20 percent of the defects will actually be traced to the CPU socket. But if enough evidence is statistically provided that the null hypothesis is untrue, an *alternate hypothesis* should be assumed to be true. That alternate hypothesis, denoted H_1, tells what should be concluded if H_0 is rejected.

$$H_1 : \mu \neq 20\%$$

6.1.3 Test statistic

The decision made on whether to reject H_0 or fail to reject it depends on the information provided by the sample taken from the population being studied. The objective here is to generate a single number that will be compared to H_0 for rejection. That number is called the *test statistic.*

To test the mean μ, the Z formula is used when the sample sizes are greater than 30,

$$Z = \frac{\overline{X} - \mu}{\sigma/\sqrt{n}}$$

and the t formula is used when the samples are smaller,

$$t = \frac{\overline{X} - \mu}{s/\sqrt{n}}$$

These two equations look alike but, remember that the tables that are used to compute the Z-statistic and t-statistic are different.

6.1.4 Level of significance or level of risk

The *level of risk* addresses the risk of failing to reject a hypothesis when it is actually false, or rejecting a hypothesis when it is actually true. Suppose that in the case of the defects on the circuit boards, a sample of 40 boards was randomly taken for analysis and 45 percent of the defects were actually found to be traceable to the CPU sockets. In that case, we would have rejected the null hypothesis as false. But what if the sample were taken from a substandard population? We would have rejected a null hypothesis that might be true. We therefore would have committed what is called the *Type I* or *Alpha error*.

However, if we actually find that 20 percent of the defects are traceable to the CPU socket from a sample and only the boards on that sample out of the whole population happened to have those defects, we would have made the *Type II* or *Beta error.* We would have assumed the null hypothesis to be true when it actually is false.

The probability of making a Type I error is referred to as α, and the probability of making a Type II error is referred to as β. There is an inverse relationship between α and β.

6.1.5 Decision rule determination

The *decision rule* determines the conditions under which the null hypothesis is rejected or not. The one-tailed (right-tailed) graph in Figure 6.1 shows the region of rejection, the location of all the values for which the probability of the null hypothesis being true is infinitesimal.

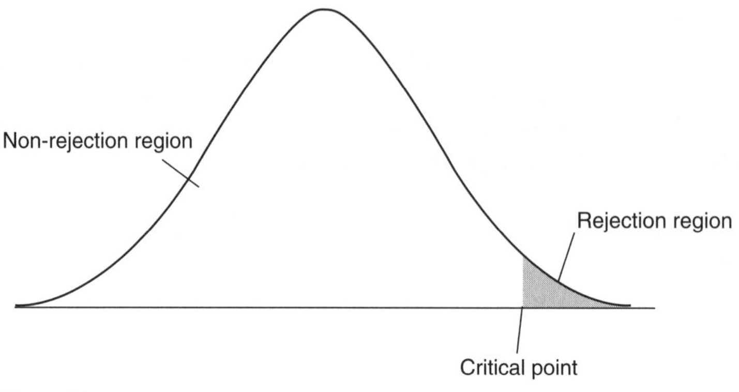

Figure 6.1

The critical value is the dividing point between the area where H_0 is rejected and the area where it is assumed to be true.

6.1.6 Decision making

Only two decisions are considered, either the null hypothesis is rejected or it is not. The decision to reject a null hypothesis or not depends on the level of significance. This level often varies between 0.01 and 0.10. Even when we fail to reject the null hypothesis, we never say "we accept the null hypothesis" because failing to reject the null hypothesis that was assumed true does not equate proving its validity.

6.2 Testing for a Population Mean

6.2.1 Large sample with known σ

When the sample size is greater than 30 and σ is known, the Z formula can be used to test a null hypothesis about the mean.

Example An old survey had found that the average income of operations managers for Fortune 500 companies was $80,000 a year. A pollster wants to test that figure to determine if it is still valid. She takes a random sample of 150 operations managers to determine if their average income is $80,000. The mean of the sample is found to be $78,000 with a standard deviation assumed to be $15,000. The level of significance is set at 5 percent. Should she reject $80,000 as the average income or not?

Solution The null hypothesis will be $80,000 and the alternate hypothesis will be anything other than $80,000,

$$H_0 : \mu = \$80,000$$
$$H_1 : \mu \neq \$80,000$$

Because the sample size n is larger than 30, we can use the Z formula to test the hypothesis. Because the significance level is set at 5 percent (in other words, $\alpha = 0.05$) and we are dealing with a two-tailed test, the area under each tail of the distribution will be $\alpha/2 = 0.025$. The area between the mean μ and the critical value on each side will be 0.4750 (or 0.05 − 0.025). The critical Z-value is obtained from the Z score table by using the 0.4750 area under the curve. A value of $Z_{\alpha/2} = \pm 1.96$ corresponds to 0.4750. The null hypothesis will not be rejected if $-1.96 \leq Z \leq +1.96$ and rejected otherwise.

$$Z = \frac{\overline{X} - \mu}{\sigma/\sqrt{n}} = \frac{78000 - 80000}{\dfrac{15000}{\sqrt{150}}} = \frac{-2000}{1224.745} = -1.633$$

Because Z is within the interval ± 1.96, the statistical decision should be to not reject the null hypothesis. A salary of \$78,000 is just the sample mean; if a confidence interval were determined, \$80,000 would have been the estimate point.

Another way to solve it. Because we already know that

$$Z_{\alpha/2} = \frac{\overline{X} - \mu}{\sigma/\sqrt{n}}$$

we can transform this equation to find the interval within which μ is located,

$$\overline{X} - Z_{\alpha/2}\frac{\sigma}{\sqrt{n}} \leq \mu \leq \overline{X} + Z_{\alpha/2}\frac{\sigma}{\sqrt{n}}$$
$$\overline{X} = 78,000$$
$$\sigma = 15,000$$
$$n = 150$$
$$Z_{\alpha/2} = 1.96$$

Therefore,

$$78,000 - 1.96\frac{15,000}{\sqrt{150}} \leq \mu \leq 78,000 + 1.96\frac{15,000}{\sqrt{150}}$$
$$75,599.4998 \leq \mu \leq 80,400.5002$$

Because \$78,000 is within that interval, we cannot reject the null hypothesis.

Using Minitab. Open Minitab and from the Stat menu, select "Basic Statistics" and then select "1-Sample Z..." The "1-Sample Z" dialog box appears, and values are entered as shown in Figure 6.2.

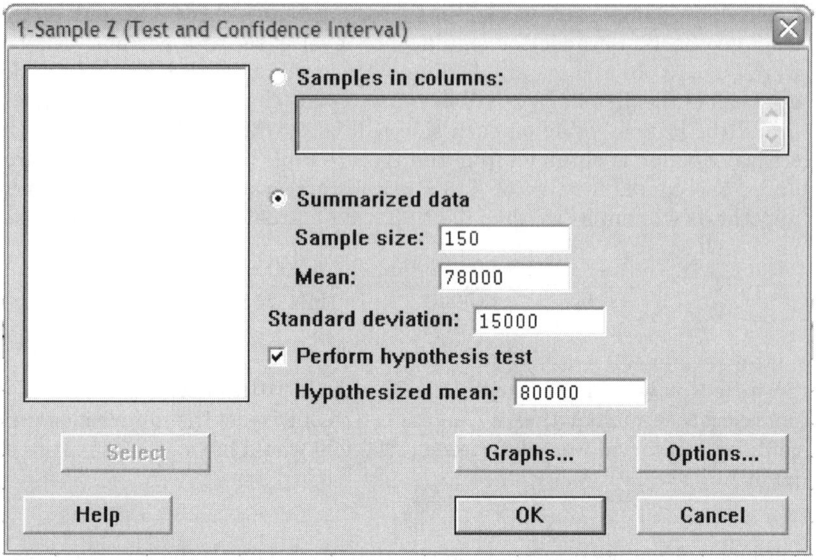

1-Sample Z (Test and Confidence Interval)

Samples in columns:

Summarized data
Sample size: 150
Mean: 78000
Standard deviation: 15000
Perform hypothesis test
Hypothesized mean: 80000

Select | Graphs... | Options...
Help | OK | Cancel

Figure 6.2

After selecting "OK," the Minitab output should show as shown in Figure 6.3.

The Minitab output suggests that for a 95 percent confidence level, the mean is expected to fall within the interval 75,599.5 and 80,400.5. Because the mean obtained from the sample is 78,000, we cannot reject the null hypothesis.

One-Sample Z

```
Test of mu = 80000 vs not = 80000
The assumed standard deviation = 15000


  N     Mean  SE Mean          95% CI              Z      P
150  78000.0   1224.7  (75599.5, 80400.5)  -1.63  0.102
```

Figure 6.3

6.2.2 What is the *p*-value and how is it interpreted?

In the previous example, we did not reject the null hypothesis because the value of the test statistic Z was within the interval $[-1.96, +1.96]$. Had it been outside that interval, we would have rejected the null hypothesis and concluded that \$80,000 is not the average income for the managers. The reason why ± 1.96 was chosen is that the confidence level α was set at 95 percent. If α were set at another level, the interval would have been different. The results obtained do not allow a comparison with a single value to make an assessment; any value of $\overline{X}$ that falls within that interval would lead to a non-rejection of the null hypothesis.

The use of the p-value method enables the value of α not to be preset. The null hypothesis is assumed to be true, and the p-value sets the smallest value of α for which the null hypothesis must be rejected. For instance, in the example above the p-value is 0.102 and $\alpha = 0.05$; therefore, α is smaller than the p-value and 0.102 is the smallest value for which the null hypothesis must be rejected. We cannot reject the null hypothesis in this case.

Example The diameter of the shafts produced by a machine has historically been 5.02 mm with a standard deviation of 0.08 mm. The old machine has been discarded and replaced with a new one. The reliability engineer wants to make sure that the new machine performs as well as the old one. He takes a sample of 35 shafts just produced by the new machine and measures their diameter, and obtains the results in file *Diameter.mpj* on the included CD.

We want to test the validity of the null hypothesis,

$$H_0 : \mu = 5.02$$
$$H_1 : \mu \neq 5.02$$

Solution Open the file *Diameter.mpj* on the included CD. From the Stat menu, select "Basic Statistics." From the drop-down list, select "1-Sample Z." Select "Diameter" for the *Samples in Columns* option. Enter "0.08" into the *Standard Deviation* field. Check the option *Perform hypothesis test.* Enter "5.02" into the *Hypothesized mean* field. Select "Options" and make sure that "Not equal" is selected from the *Alternative* drop-down list and that the *Confidence level* is 95 percent. Select "OK" to get the output shown in Figure 6.4.

One-Sample Z: Diameter

```
Test of mu = 5.02 vs not = 5.02
The assumed standard deviation = 0.08

Variable   N   Mean     StDev    SE Mean   95% CI                Z     P
Diameter  35   5.04609  0.06949  0.01352  (5.01958, 5.07259)  1.93  0.054
```

Figure 6.4

Interpretation of the results. The mean of the sample is 5.04609, and the sample size is 35 with a standard deviation of 0.06949. The confidence interval is therefore [5.01958, 5.07259]. If the value of the mean falls within this interval, we cannot reject the null hypothesis. The value of the sample mean (5.04609) is indeed within that interval. The p-value, 0.054, is greater than α, which is 0.05; therefore, we cannot reject the null hypothesis.

6.2.3 Small samples with unknown σ

The Z test statistic is used when the population is normally distributed or when the sample sizes are greater than 30. This is because when the population is normally distributed and σ is known, the sample means will be normally distributed, and when the sample sizes are greater than 30, the sample means will be normally distributed based on the Central Limit Theorem.

If the sample being analyzed is small ($n \leq 30$), the Z test statistic would not be appropriate; the t test should be used instead. The formula for the t test resembles the one for the Z test but the tables used to compute the values for Z and t are different. Because σ is unknown, it will be replaced by s, the sample standard deviation.

$$t = \frac{\overline{X} - \mu}{s/\sqrt{n}}$$

$$df = n - 1$$

Example A machine used to produce gaskets has been stable and operating under control for many years, but lately the thickness of the gaskets seems to be smaller than they once were. The mean thickness was historically 0.070 inches. A Quality Assurance manager wants to determine if the age of the machine is causing it to produce poorer quality gaskets. He takes a sample of 10 gaskets for testing and finds a mean of 0.074 inches and a standard deviation of 0.008 inches. Test the hypothesis that the machine is working properly with a significance level of 0.05.

Solution The null hypothesis should state that the population mean is still 0.070 inches—in other words, the machine is still working properly—and the alternate hypothesis should state that the mean is different from 0.070.

$$H_0 : \mu = 0.070$$
$$H_1 : \mu \neq 0.070$$

We have an equality, therefore we are faced with a two-tailed test and we will have $\alpha/2 = 0.025$ on each side. The degree of freedom $(n - 1)$ is equal to 9. The value of t that we will be looking for is $t_{0.025,9} = 2.26$. If the computed value t falls within the interval $[-2.26, +2.26]$, we will not reject the null hypothesis; otherwise, we will.

$$t = \frac{\overline{X} - \mu}{s/\sqrt{n}} = \frac{0.074 - 0.07}{0.008/\sqrt{10}} = 1.012$$

The computed value of t is 1.012, therefore it falls within the interval $[-2.262, +2.262]$. We conclude that we cannot reject the null hypothesis.

Another way to solve it. We already know that

$$t_{\alpha/2} = \frac{\overline{X} - \mu}{s/\sqrt{n}}$$

We can rearrange this equation to find the interval within which μ resides,

$$\overline{X} - t_{\alpha/2}\frac{s}{\sqrt{n}} \leq \mu \leq \overline{X} + t_{\alpha/2}\frac{s}{\sqrt{n}}$$

We can now plug in the numbers:

$$0.074 - 2.26\frac{0.008}{\sqrt{10}} \leq \mu \leq 0.074 + 2.26\frac{0.008}{\sqrt{10}}$$
$$0.074 - 0.00571739 \leq \mu \leq 0.074 + 0.00571739$$
$$0.06828 \leq \mu \leq 0.07972$$

Using Minitab. From the Stat menu, select "Basic Statistics" and from the drop-down list, select "1-Sample t. . . "

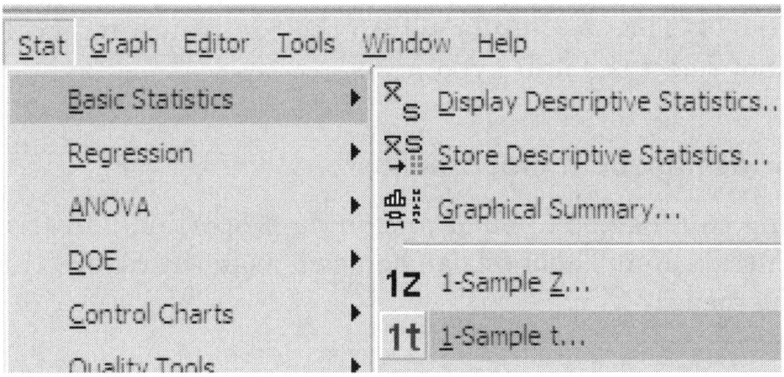

Fill out the "1-Sample t" dialog box as shown in Figure 6.5.

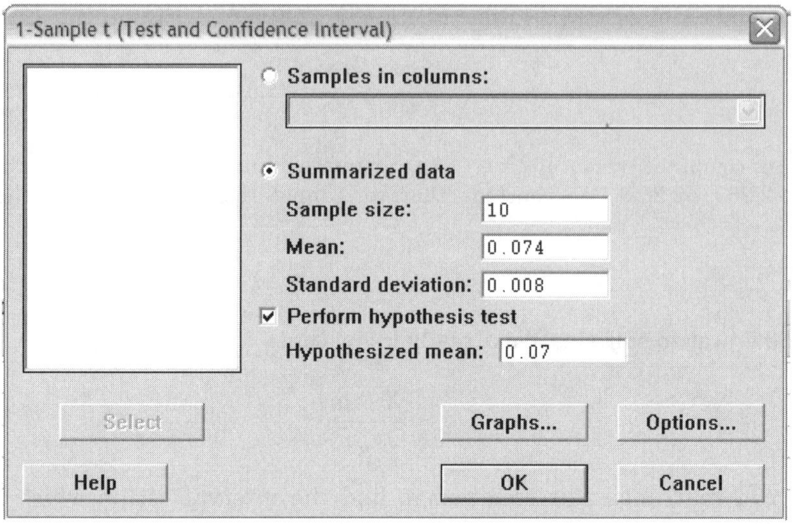

1-Sample t (Test and Confidence Interval)

Samples in columns:

Summarized data

Sample size: 10

Mean: 0.074

Standard deviation: 0.008

Perform hypothesis test

Hypothesized mean: 0.07

Select | Graphs... | Options...

Help | OK | Cancel

Figure 6.5

Select "OK" to get the result shown in Figure 6.6.

One-Sample T

```
Test of mu = 0.07 vs not = 0.07

 N      Mean     StDev   SE Mean          95% CI             T      P
10  0.074000  0.008000  0.002530  (0.068277, 0.079723)  1.58  0.148
```

Figure 6.6

Interpretation of the results. The p-value of 0.148 is greater than the value $\alpha = 0.05$. The confidence interval is [0.068277, 0.079723] and the sample mean is 0.074. The mean falls within the confidence interval, therefore we cannot reject the null hypothesis.

6.3 Hypothesis Testing about Proportions

Hypothesis testing can also be applied to sample proportions. In this situation, the Central Limit Theorem can be used, as in the case of the distribution of the mean:

$$Z = \frac{\hat{p} - p}{\sqrt{\dfrac{pq}{n}}}$$

where $\hat{p}$ is the sample proportion, p is the population proportion, n is the sample size, and $q = 1 - p$.

Example A design engineer claims that 90 percent of the of alloy bars he created become 120 PSI (pound per square inch) strong 12 hours after they are produced. In a sample of 10 bars, 8 were 120 PSI strong after 12 hours. Determine whether the engineer's claim is legitimate at a confidence level of 95 percent.

Solution In this case, the null and alternate hypotheses will be

$$H_0 : p = 0.90$$
$$H_1 : p \neq 0.90$$

The sample proportion is

$$\hat{p} = \frac{8}{10} = 0.8$$
$$q = 1 - 0.90 = 0.10$$

Therefore

$$Z = \frac{0.8 - 0.9}{\sqrt{\dfrac{0.1 \times 0.9}{10}}} = \frac{-0.1}{0.09487} = -1.054$$

For a confidence level of 95 percent, the rejection area would be anywhere outside the interval [−1.96, +1.96]. The value −1.054 is within that interval, and therefore we cannot reject the null hypothesis.

6.4 Hypothesis Testing about the Variance

We saw in Chapter 5 that the distribution of the variance follows a chi-square distribution, with the χ^2 formula for single variance being

$$\chi^2 = \frac{(n-1)s^2}{\sigma^2}$$

where σ^2 is the population variance, s^2 is the sample variance, and n is the sample size.

Example Kanel Incorporated's Days Sales Outstanding (DSO) have historically had a standard deviation of 2.78 days. The last 17 days, the standard deviation has been 3.01 days. At an α level of 0.05, test the hypothesis that the variance has increased.

Solution The null and alternate hypotheses will be:

$$H_0 : \sigma > 2.78$$
$$H_1 : \sigma < 2.78$$

with

$$n = 17$$
$$s^2 = 3.01 \times 3.01 = 9.0601$$
$$\sigma^2 = 2.78 \times 2.78 = 7.7284$$
$$\chi^2 = \frac{(n-1)s^2}{\sigma^2} = \frac{(17-1)(9.0601)}{7.7284} = 18.757$$

We are faced with a one-tailed graph with a degree of freedom of 16 and $\alpha = 0.05$. From the chi-square table, this corresponds to $\chi^2_{0.05,16} = 26.2962$. The calculated $\chi^2(18.757)$ is lower than the critical value $\chi^2_{0.05,16}$, therefore the decision should not be rejected.

Doing it another way. The same result can be obtained another way. Instead of looking for the calculated χ^2, we can look for the critical S value,

$$S_c^2 = \frac{\sigma^2 \chi^2_{0.05,16}}{17-1} = \frac{7.7284 \times 26.2962}{16} = 12.702$$

In this case again, the S_c^2 critical value is greater than the sample variance (which was 9.0601); therefore, we do not reject the null hypothesis.

6.5 Statistical Inference about Two Populations

So far, all our discussion has been focused on samples taken from one population. We have learned how to determine sample sizes, how to determine confidence intervals for χ^2, for proportions and for μ, and how to test a hypothesis about these statistics.

Very often, it is not enough to be able to make statistical inference about one population. We sometimes want to compare two populations. A quality controller might want to compare data from a production line to see what effect the aging machines are having on the production process over a certain period of time. A manager might want to know how the productivity of her employees compares to the average productivity in the industry. In this section, we will learn how to test and estimate the difference between two population means, proportions, and variances.

6.5.1 Inference about the difference between two means

Just as in the analysis of a single population, to estimate the difference between two populations the researcher would draw samples from each population. The best estimator for the population mean μ was the sample mean $\overline{X}$, so the best estimator of the difference between the population means $(\mu_1 - \mu_0)$ will be the difference between the samples' means $(\overline{X}_1 - \overline{X}_0)$.

The Central Limit Theorem applies in this case, too. When the two populations are normal, $(\overline{X}_1 - \overline{X}_0)$ will be normally distributed and it will be approximately normal if the samples sizes are large $(n \geq 30)$. The standard deviation for $(\overline{X}_1 - \overline{X}_0)$ will be

$$\sqrt{\frac{\sigma_1^2}{n_1} + \frac{\sigma_0^2}{n_0}}$$

and its expected value

$$E(\overline{X}_1 - \overline{X}_0) = (\mu_1 - \mu_0)$$

Therefore,

$$Z = \frac{(\overline{X}_1 - \overline{X}_0) - (\mu_1 - \mu_0)}{\sqrt{\frac{\sigma_1^2}{n_1} + \frac{\sigma_0^2}{n_0}}}$$

This equation can be transformed to obtain the confidence interval

$$(\overline{X}_1 - \overline{X}_0) - Z_{\alpha/2}\sqrt{\frac{\sigma_1^2}{n_1} + \frac{\sigma_0^2}{n_0}} \leq (\mu_1 - \mu_0)$$

$$\leq (\overline{X}_1 - \overline{X}_0) + Z_{\alpha/2}\sqrt{\frac{\sigma_1^2}{n_1} + \frac{\sigma_0^2}{n_0}}$$

Example In December, the average productivity per employee at Senegal-Electric was 150 machines per hour with a standard deviation of 15 machines. For the same month, the average productivity per employee at Cazamance Electromotive was 135 machines per hour with a standard deviation of 9 machines. If 45 employees at Senegal-Electric and 39 at Cazamance Electromotive were randomly sampled, what is the probability that the difference in sample averages would be greater than 20 machines?

Solution

$$\mu_1 = 150, \quad \sigma_1 = 15 \quad n_1 = 45$$
$$\mu_0 = 135, \quad \sigma_0 = 9 \quad n_0 = 39$$

$$\overline{X}_1 - \overline{X}_0 = 20$$

$$Z = \frac{(\overline{X}_1 - \overline{X}_0) - (\mu_1 - \mu_0)}{\sqrt{\dfrac{\sigma_1^2}{n_1} + \dfrac{\sigma_0^2}{n_0}}} = \frac{20 - (150 - 135)}{\sqrt{\dfrac{15^2}{45} + \dfrac{9^2}{39}}} = \frac{5}{2.66} = 1.88$$

From the Z score table, the probability of getting a value between zero and 1.88 is 0.4699, and the probability for Z to be larger than 1.88 will be 0.5 − 0.4699 = 0.0301. Therefore, the probability that the difference in the sample averages will be greater than 20 machines is 0.0301. In other words, there exists a 3.01 percent chance that the difference would be at least 20 machines.

Because the populations' standard deviations are seldom known, the previous formula is rarely used; therefore, the standard error of the sampling distribution must be estimated. At least two conditions must be considered—the approach we take when making an inference about the two means depends on whether their variances are equal or not.

6.5.2 Small independent samples with equal variances

In the previous example, the sample sizes were both greater than 30, so the Z test was used to determine the confidence interval. If one or both samples are smaller than 30, the t statistic must be used.

If the population variances σ_1^2 and σ_0^2 are unknown and we assume that they are equal, they can be estimated using the sample variances S_1^2 and S_0^2. The estimate S_p^2 based on the two sample variances is called the *pooled sample variance*. Its formula is given as

$$S_p^2 = \frac{(n_1 - 1)S_1^2 + (n_0 - 1)S_0^2}{n_1 + n_0 - 2}$$

where $(n_1 - 1)$ is the degree of freedom for Sample 1, and $(n_0 - 1)$ is the degree of freedom for Sample 0. The denominator is just the sum of the two degrees of freedom,

$$n_1 + n_0 - 2 = (n_1 - 1) + (n_0 - 1)$$

Example The variances of two populations are assumed to be equal. A sample of 15 items was taken from Population 1 with a standard deviation of 3, and a sample of 19 items was taken from Population 0 with a standard deviation of 2. Find the pooled sample variance.

Solution

$$S_p^2 = \frac{(n_1 - 1)S_1^2 + (n_0 - 1)S_0^2}{n_1 + n_0 - 2}$$

$$S_p^2 = \frac{(15-1)3 + (19-1)2}{19 + 15 - 2} = \frac{78}{32} = 2.4375$$

Note that

$$S_0^2 \leq S_p^2 \leq S_1^2$$

If

$$n_1 = n_0 = n$$

then S_p^2 can be simplified:

$$S_p^2 = \frac{nS_1^2 - S_1^2 + nS_0^2 - S_0^2}{n + n - 2}$$

$$S_p^2 = \frac{n(S_1^2 + S_0^2) - (S_1^2 + S_0^2)}{2(n-1)}$$

$$S_p^2 = \frac{(S_1^2 + S_0^2)(n-1)}{2(n-1)}$$

Therefore,

$$S_p^2 = \frac{S_1^2 + S_0^2}{2}$$

For sample sizes smaller than 30, the t statistic will be used,

$$t = \frac{(\overline{x}_1 - \overline{x}_0) - (\mu_1 - \mu_0)}{\sqrt{\dfrac{\sigma_1^2}{n_1} + \dfrac{\sigma_0^2}{n_0}}}$$

But because σ_1^2 and σ_0^2 are unknown and they can be estimated based on the samples' standard deviation, the denominator will be changed:

$$\sqrt{\frac{\sigma_1^2}{n_1} + \frac{\sigma_0^2}{n_0}} \approx \sqrt{\frac{S_p^2}{n_1} + \frac{S_p^2}{n_0}} = \sqrt{S_p^2\left(\frac{1}{n_1} + \frac{1}{n_0}\right)}$$

and therefore,

$$t = \frac{\left(\overline{x}_1 - \overline{x}_0\right) - \left(\mu_1 - \mu_0\right)}{\sqrt{S_p^2\left(\frac{1}{n_1} + \frac{1}{n_0}\right)}}$$

This equation can be transformed to obtain the confidence interval for the populations' means,

$$(\overline{x}_1 - \overline{x}_0) \pm t_{\alpha/2}\sqrt{S_p^2\left(\frac{1}{n_1} + \frac{1}{n_0}\right)}$$

Example The general manager of Jolof-Semiconductors oversees two production plants and has decided to raise the customer satisfaction index (CSI) to at least 98. To determine if there is a difference in the mean of the CSI in the two plants, random samples are taken over several weeks. For the Kayor plant, a sample of 17 weeks has yielded a mean of 96 CSI and a standard deviation of 3, and for the Matam plant, a sample of 19 weeks has generated a mean of 98 CSI and a standard deviation of 4. At the 0.05 level, determine if a difference exists in the mean level of CSI for the two plants, assuming that the CSIs are normal and have the same variance.

Solution

$$\begin{aligned} \alpha &= 0.05 \\ H_0 &= (\mu_1 - \mu_0) = 0 \\ H_1 &= (\mu_1 - \mu_0) \neq 0 \end{aligned}$$

$$\begin{aligned} n_1 &= 17 \\ \overline{x}_1 &= 96 \\ S_1 &= 3 \\ n_0 &= 19 \\ \overline{x}_0 &= 98 \\ s_0 &= 4 \end{aligned}$$

Estimate the common variance with the pooled sample variance, S_p^2.

$$S_p^2 = \frac{(n_1 - 1)S_1^2 + (n_0 - 1)S_0^2}{n_1 + n_0 - 2}$$

$$S_p^2 = \frac{16(3)^2 + 18(4)^2}{34} = \frac{432}{34} = 12.71$$

The value of the test statistic is

$$t = \frac{(\overline{x}_1 - \overline{x}_0) - (\mu_1 - \mu_0)}{\sqrt{S_p^2\left(\dfrac{1}{n_1} + \dfrac{1}{n_0}\right)}}$$

Therefore

$$t = \frac{(96 - 98) - 0}{\sqrt{12.71\left(\dfrac{1}{17} + \dfrac{1}{19}\right)}} = -\frac{2}{1.19} = -1.68$$

Because the alternate hypothesis does not involve "greater than" or "less than" but rather "is different from," we are faced with a two-tailed rejection region with

$$\alpha/2 = 0.05/2 = 0.025$$

at the end of each tail with a degree of freedom of 34. From the t table, we obtain $t_{0.025} = 2.03$ and H_0 is not rejected when $-2.03 < t < +2.03$.
$t = -1.68$ is well within the interval, we therefore cannot reject the null hypothesis.

Using Minitab. From the Stat menu, select "Basic Statistics" and from the drop-down list, select "2-Sample t. . . " When the "2-Sample t" dialog box pops up, select the *Summarized data* option and fill out the fields as shown in Figure 6.7.

Then select "OK" to get the output shown in Figure 6.8.

Because we are faced with a two-tailed graph, the graph that illustrates the results obtained in the previous example should look like that in Figure 6.9.

Because $t_{0.025} = -1.68$ is not in the rejected region, we cannot reject the null hypothesis. There is not enough evidence at a significance level

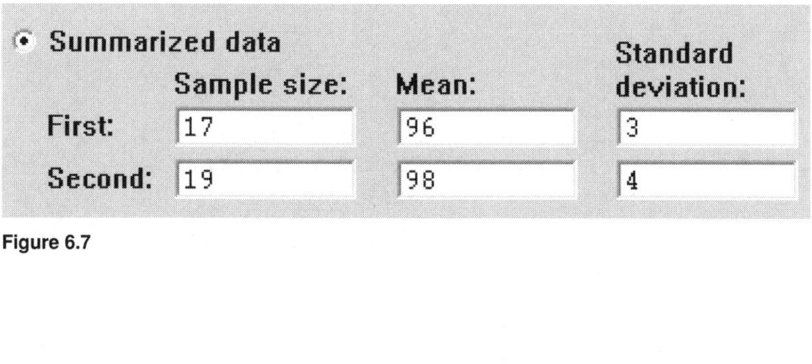

Summarized data

	Sample size:	Mean:	Standard deviation:
First:	17	96	3
Second:	19	98	4

Figure 6.7

Two-Sample T-Test and CI

```
Sample   N    Mean   StDev  SE Mean
1       17   96.00    3.00     0.73
2       19   98.00    4.00     0.92

Difference = mu (1) - mu (2)
Estimate for difference: -2.0000
95% CI for difference: (-4.41840, 0.41840)
T-Test of difference = 0 (vs not=): T-Value = -1.68 T-Value = 0.102 DF = 34
Both use Pooled stDev = 3.5645
```

Figure 6.8

of 0.05 to conclude that there is a difference in the mean level of the CSIs for the two plants.

Example The amount of water that flows through between two points of equal distance of pipes X and Y are given in Table 6.1 in liters per minute (found on the included CD under files *Waterflow.xl* and *Waterflow.mpj*). An engineer wants to determine if there is a statistical significance between the speed of the water flow through the two pipes at a significance level of 95 percent.

Solution The null and alternate hypotheses in this case would be:

H_0 : The speed of the water flow inside the two pipes is the same.
H_1 : There is a difference in the speed of the water flow inside the two pipes.

Using Minitab. Open the file *Waterflow.mpj* on the included CD. From the Stat menu, select "Basic Statistics." From the drop-down list, select "2-t 2-Sample t. . . " Select the second option, "Samples in different columns." Select "X" for *First* and "Y" for *Second*. Select the *Assume*

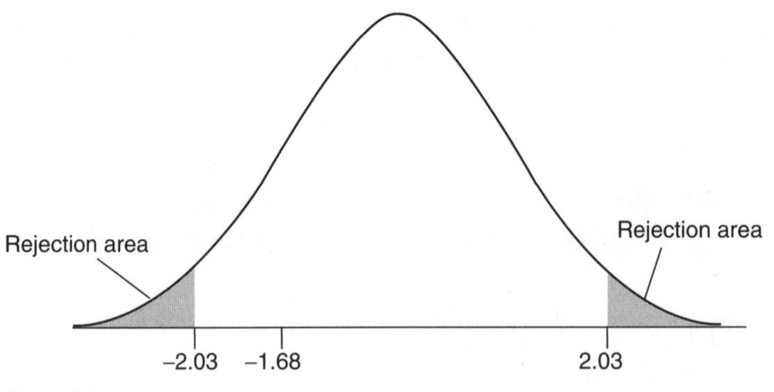

Figure 6.9

TABLE 6.1

X	Y
163	167
150	157
171	149
155	145
186	135
145	157
154	135
173	167
152	154
150	165
143	170
138	165
166	154
193	176
158	155
175	157
167	134
150	156
158	147

equal variances option. Select "OK" to get the output shown in Figure 6.10.

Using Excel. Open the file *Waterflow.xl* from the included CD. From the Tools menu, select "Data Analysis." The "Data Analysis" dialog box appears and select "t-test: two samples assuming equal variances," then select "OK." Select the range for X for the field of *Variable 1 Range*. Select the range for Y for the field of *Variable 2 Range*. For *Hypothesized mean difference*, insert "0." If the titles "X" and "Y" were selected with their respective ranges, select the *Labels* option; otherwise, do not. *Alpha* should be "0.05."

Two-Sample T-Test and CI: X, Y

```
Two-sample T for X vs Y

    N   Mean  StDev  SE Mean
X  19  160.4   14.5      3.3
Y  19  155.0   12.0      2.8


Difference = mu (X) - mu (Y)
Estimate for difference: 5.36842
95% CI for difference: (-3.41343, 14.15027)
T-Test of dfference = 0 (vs not =): T-Value = 1.24  P-Value = 0.223  DF = 36
Both use Pooled StDev = 13.3463
```

Figure 6.10

Select "OK" to get the output shown in Figure 6.11.

t-Test: Two-Sample Assuming Equal Variances		
	X	Y
Mean	160.3684	155
Variance	211.2456	145
Observations	19	19
Pooled Variance	178.1228	
Hypothesized Mean Difference	0	
df	36	
t Stat	1.239791	
P(T<=t) one-tail	0.111537	
t Critical one-tail	1.688298	
P(T<=t) two-tail	0.223074	
t Critical two-tail	2.028094	

Figure 6.11

The *p*-value of 0.223 suggests that at a confidence level of 95 percent, there is not a statistically significant difference between the speed of the water flow in the two pipes.

6.5.3 Testing the hypothesis about two variances

Very often in quality management, the statistic of interest is the standard deviation or the variance instead of the measures of central tendency. In those cases, the tests of hypothesis are more often about the variance. Most statistical tests for the mean require the equality of the variances for the populations. The hypothesis test for the variance can help assess the equality of the populations' variances.

The hypothesis testing of two population variances is done using samples taken from those populations. The F distribution is used in this case. The calculated F statistic is given as

$$F = \frac{S_1^2}{S_2^2}$$

The values of interest besides the sample sizes are the degrees of freedom. The graph for an F distribution is shown in Figure 6.12.

Here again, what must be compared are the calculated F and the critical F obtained from the table. Two values are of interest: the critical value of F at the lower tail and the value at the upper tail. The critical

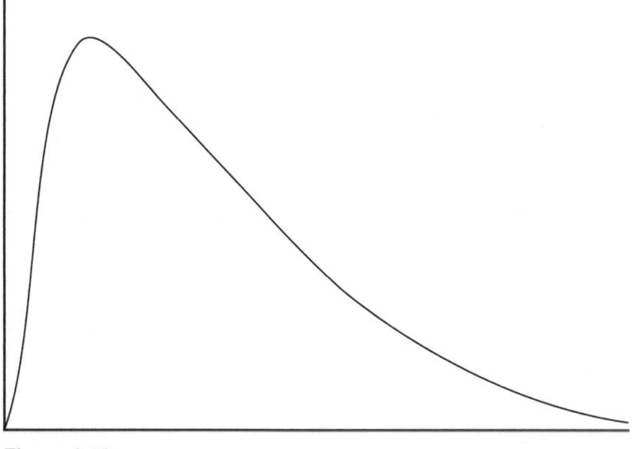

Figure 6.12

value of the upper tail is $F_{1-\alpha,n_1,n_2}$. The critical value for the lower tail is given as the inverse of the value for the upper tail,

$$F_{1-\alpha,n_1,n_2} = \frac{1}{F_{\alpha,n_1,n_2}}$$

For the null hypothesis not to be rejected, the value of the calculated F should be between the value for the upper tail and the value for the lower tail.

Example Kolda Automotive receives gaskets for its engines from two suppliers. The QA manager wants to compare the variance in thickness of the gaskets with $\alpha = 0.05$. He takes a sample of 10 gaskets from supplier A and 12 from supplier B and obtains a standard deviation of 0.087 from A and 0.092 from B.

Solution The null and alternate hypotheses will be:

$$H_0 : \sigma_A^2 = \sigma_B^2$$

$$H_1 : \sigma_A^2 \neq \sigma_B^2$$

Therefore,

$$F = \frac{S_A^2}{S_B^2} = \frac{0.087 \times 0.087}{0.092 \times 0.092} = \frac{0.007569}{0.008464} = 0.894$$

Now we must find the critical F from the table (Appendix 6). Because α = 0.05 and we are faced with a two-tailed test, the critical F must be found at $\alpha/2 = 0.025$ with degrees of freedom of 9 and 11. The critical F for the upper tail is $F_{0.025,9,11} = 3.59$. The critical F for the lower tail is the inverse of this value,

$$F_{0.975,11,9} = \frac{1}{3.59} = 0.279$$

The value of the calculated F (0.894) is well within the interval [0.279, 3.59]; therefore, we cannot reject the null hypothesis.

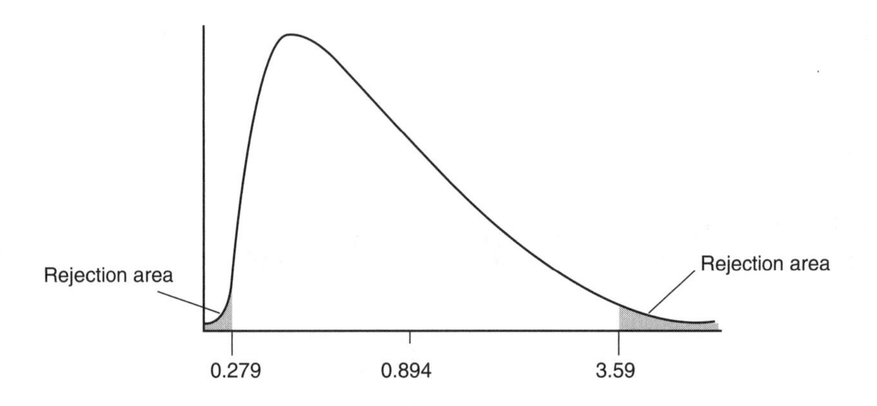

6.6 Testing for Normality of Data

The normality or non-normality of data is extremely important in quality control and Six Sigma, as we will see in the coming chapters. Several options are given by Minitab to test the normality of data. The data contained in the Minitab worksheet of the file *O Ring.mpj* (reproduced in Table 6.2) represents the diameters in inches of rings produced by a machine, and we want to know if the diameters are normally distributed. If the data are normally distributed, they all should be close to the mean and when we plot them on a graph, they should cluster closely about each other.

The null hypothesis for normality will be

$$H_0 : \text{The data are normally distributed.}$$

and the alternate hypothesis will be

$$H_1 : \text{The data are not normally distributed.}$$

To run the normality test, Minitab offers several options. We can run the Anderson-Darling, the Kolmogorov-Smirnov, or the Ryan-Joyner test. For this example, we will run the Anderson-Darling Test.

TABLE 6.2

O-Ring	
1.69977	1.69969
1.70001	1.69973
1.69956	1.69990
1.70007	1.70004
1.70021	1.69999

From the Minitab's Stat menu, select "Basic Statistics" and then "Normality Test." The "Normality Test" dialog box pops up and it should be filled out as shown in Figure 6.13.

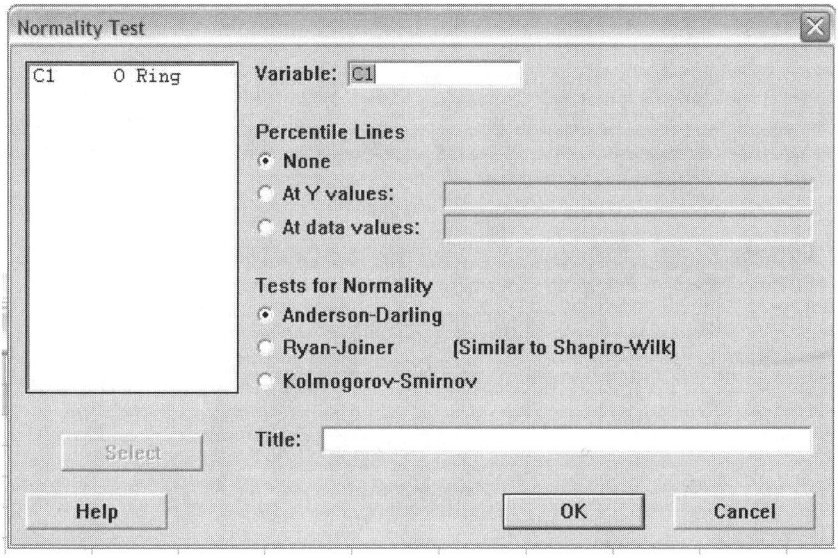

Figure 6.13

Then, select "OK."

Notice that all the dots in Figure 6.14 are closely clustered about the regression line. The p-value of 0.705 suggests that at a confidence level of 95 percent, we should not reject the null hypothesis; therefore, we must conclude that the data are normally distributed.

Example Open the file *Circuit boards.mpj* and run a normality test using the Anderson-Darling method. The output we obtain should look like Figure 6.15.

It is clear that the dots are not all closely clustered about the regression line, and they follow a certain pattern that does not suggest normality. The p-value of 0.023 indicates that the null hypothesis of normality should be rejected at a confidence level of 95 percent.

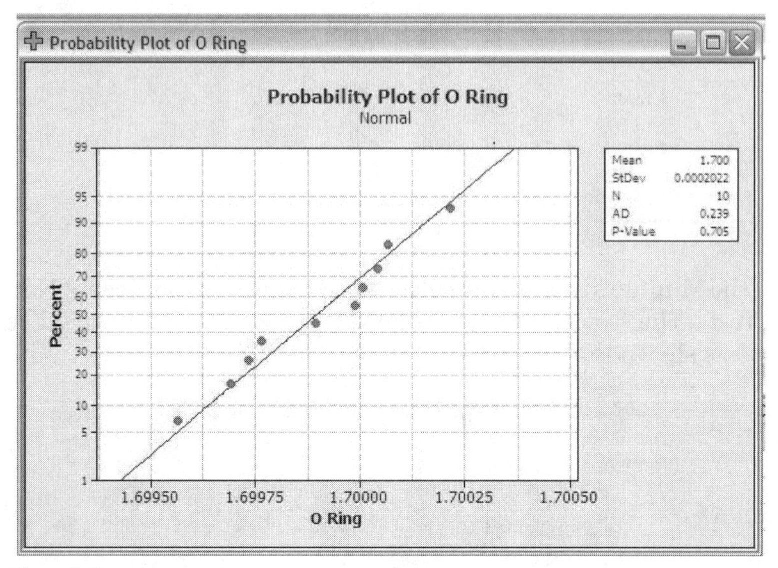

Figure 6.14

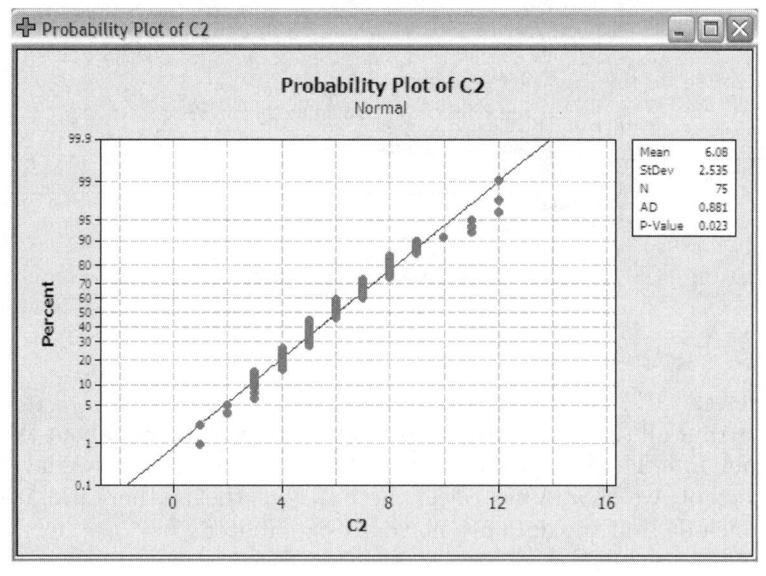

Figure 6.15

Chapter

7

Statistical Process Control

Learning Objectives:

- Know what a control chart is
- Understand how control charts are used to monitor a production process
- Be able to differentiate between variable and attribute control charts
- Understand the WECO rules

The ultimate objective of quality improvement is not just to provide good quality products to customers; it is also to improve productivity while improving customers' satisfaction. In fact, improving productivity and enhancing customer satisfaction must go together because productivity improvement enables companies to lower the cost of quality improvement. One way of improving productivity is through the reduction of defects and rework. The reduction of rework and defects is not achieved through inspection at the end of production lines; it is done by instilling quality in the production processes themselves and by inspecting and monitoring the processes in progress before defective products or services are generated.

The prerequisites for improving customer satisfaction while improving productivity address two aspects of operations: the definition of the optimal level of the quality of the products delivered to customers and the stability and predictability of the processes that generate the products. Once those optimal levels (that will be referred to as *targets*) are defined, tolerances are set around them to address the inevitable variations in the quality of the product and in the production processes.

Variations are nothing but deviations from the preset targets, and no matter how well controlled a process is, variations will always be present.

For instance, if a manufacturer of gaskets sets the length of the products to 15 inches, chances are that when a sample of 10 gaskets is randomly taken from the end of the production line under normal production conditions, there would still be differences in length between them.

The causes of the variations are divided into two categories:

- They are said to be *common* (E. Deming) or *random* (W. Shewhart) when they are inherent to the production process. Machine tune-ups are an example of common causes of variation.
- They are said to be *special* (E. Deming) or *assignable* (W. Shewhart) when they can be traced to a source that is not part of the production process. A sleepy machine operator would be an example of an assignable cause of variation

To be able to predict the quality level of the products or services, the processes used to generate them must be stable. The *stability* refers to the absence of special causes of variation.

Statistical Process Control (SPC) is a technique that enables the quality controller to monitor, analyze, predict, control, and improve a production process through control charts. *Control charts* were developed as a monitoring tool for SPC by Shewhart; they are among the most important tools in the analysis of production process variations.

A typical control chart plots sample statistics and is made up of at least four lines: a vertical line that measures the levels of the samples' means; the two outmost horizontal lines that represent the UCL and the LCL; and the center line, which represents the mean of the process. If all the points plot between the UCL and the LCL in a random manner, the process is considered to be "in control."

What is meant by an "in control" process is not a total absence of variation but instead, when the variations are present, they exhibit a random pattern. They are not outside the control limits and based on their pattern, the process trends can be predicted because the variations are strictly due to common causes. The control chart shown in Figure 7.1 exhibits variability around the mean but all the observations are

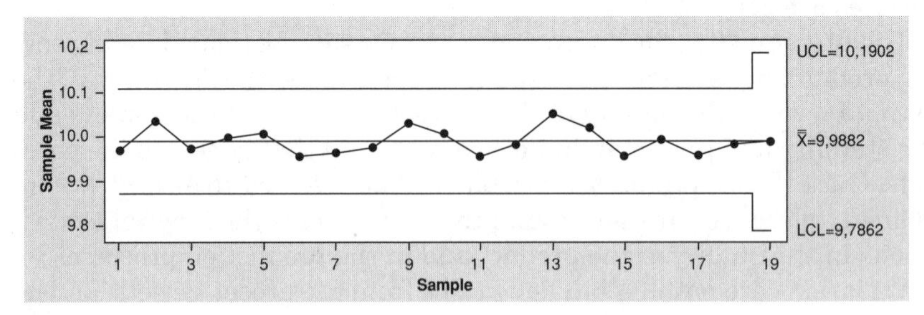

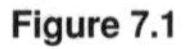
Figure 7.1

within the control limits and close to the center line. The process can be said to be "stable" because the variations follow a pattern that is fairly predictable.

The purpose of using control charts is:

- To help prevent the process from going out of control. The control charts help detect the assignable causes of variation in time so that appropriate actions can be taken to bring the process back in control.
- To keep from making adjustments when they are not needed. Most production processes allow operators a certain level of leeway to make adjustments on the machines that they are using when it is necessary. Yet over-adjusting machines can have a negative impact on the output. Control charts can indicate when the adjustments are necessary and when they are not.
- To determine the natural range (control limits) of a process and to compare this range to its specified limits. If the range of the control limits is wider than the one of the specified limits, the process will be generating defective products and will need to be adjusted.
- To inform about the process capabilities and stability. The process *capability* refers to its ability to constantly deliver products that are within the specified limits, and the *stability* refers to the quality auditor's ability to predict the process trends based on past experience. A long-term analysis of the control charts can help monitor the machine's long-term capabilities. Machine wear-out will reflect on the production output.
- To fulfill the need of a constant process monitoring. If the production process is not monitored, defective products will be produced resulting in extra rework or defects sent to customers.
- To facilitate the planning of production resources allocation. Being able to predict the variation of the quality level of a production process is very important because the variations determine the quantity of defects and the amount of work or rework that might be required to deliver customer orders on time.

7.1 How to Build a Control Chart

The control charts that we are addressing are created for a production process in progress. Samples must be taken at preset intervals and tested to make sure that the quality of the products sent to customers meets their expectations. If the tested samples are within specification, they are put back into production and sent to the customers; otherwise, they are either discarded or sent back for rework. If the products are found to be defective, the reasons for the defects are investigated and adjustments are made to prevent future defects. Making adjustments to the production process does not necessarily lead to a total elimination of variations; in some cases, it may even lead to further defects if done improperly or done when not warranted.

While the production process is in progress, whether adjustments are made or not, the process continues to be monitored, samples continue to be taken, and their statistics plotted and trends observed. Ultimately, what is being monitored using the control charts is not really how much of the production output meets engineered specification but rather how the production process is performing, how much variability it exhibits, and therefore how stable and predictable it is. The expected amount of defects that the process produces is measured by a method called *Process Capability Analysis*, which will be dealt with in the next chapter.

Consider y, a sample statistic that measures a CTQ characteristic of a product (length, color, and thickness), with a mean μ_y and a standard deviation σ_y. The UCL, the center line (CL), and the LCL for the control chart will be given as

$$UCL = \mu_y + k\sigma_y$$

$$CL = \mu_y$$

$$LCL = \mu_y - k\sigma_y$$

where $k\sigma_y$ is the distance between the center line and the control limits, k is a constant, μ_y is the mean of the samples' mean, and σ_y is the standard deviation.

Consider the length as being the critical characteristic of manufactured bolts. The mean length of the bolts is 17 inches with a known standard deviation of 0.01. A sample of five bolts is taken every half hour for testing, and the mean of the sample is computed and plotted on the control chart. That control chart will be called the $\overline{X}$ (read "X bar") control chart because it plots the means of the samples.

Based on the Central Limit Theorem, we can determine the sample standard deviation and the mean,

$$\sigma_{\bar{y}} = \frac{\sigma}{\sqrt{n}} = \frac{0.01}{\sqrt{5}} = 0.0045$$

The mean will still be the same as the population's mean, 17. For 3σ control limits, we will have

$$UCL = 17 + 3(0.0045) = 17.013$$

$$CL = 17$$

$$LCL = 17 - 3(0.0045) = 16.99$$

Control limits on a control chart are readjusted every time a significant shift in the process occurs.

Control charts are an effective tool for detecting the special causes of variation. One of the most visible signs of assignable causes of variation is the presence of an outlier on a control chart. If some points are outside the control limits, this will indicate that the process is out of control and corrective actions must be taken.

The chart in Figure 7.2 plots sample means of a given product at the end of a production line. The process seems to be stable with only common variations until Sample 25 was plotted. That sample is way outside the control limits. Because the process had been stable until that sample was taken, something unique must have happened to cause it to be outside the limits. The causes of that special variation must be investigated so that the process can be brought back under control.

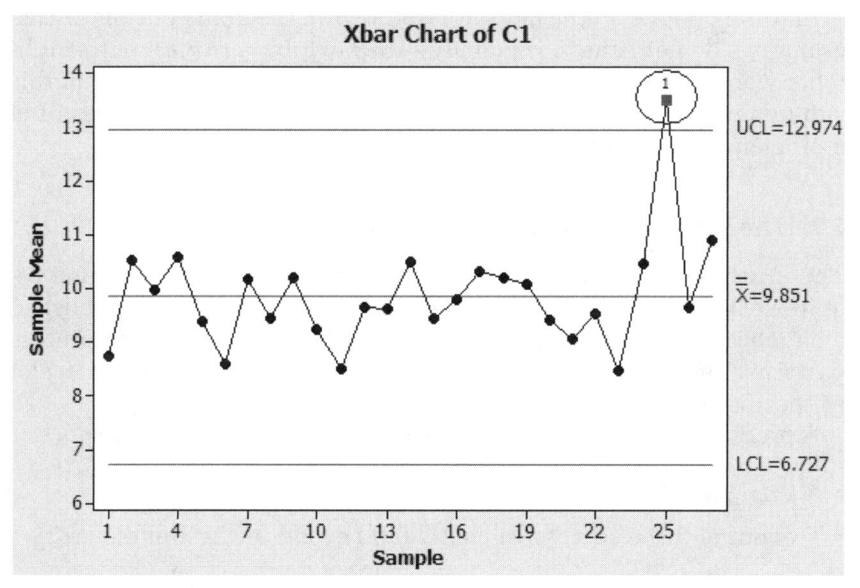

Figure 7.2

The chart in Figure 7.3 depicts a process in control and within the specified limits. The USL and LSL represent the engineered standards, whereas the right side is the control chart. The specification limits determine whether the products meet the customers' expectations and the control limits determine whether the process is under statistical control. These two charts are completely separate entities. There is no statistical relationship between the specification limits and the control limits.

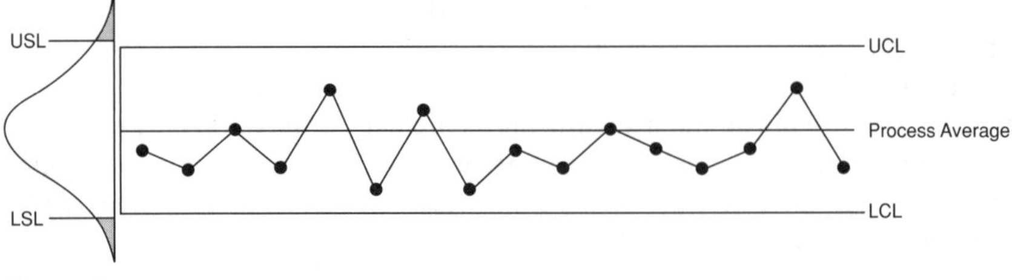

Figure 7.3

Note that a process with all the points between the control limits is not necessarily synonymous with an acceptable process. A process can be within the control limits with a high variability, or too many of the plotted points are too close to one control limit and away from the target.

The chart in Figure 7.4 is a good example of an out-of-control process with all the points plotted within the control limits.

In this example, all the plots are well within the limits but the circled groupings do not behave randomly—they exhibit a *run-up* pattern. In other words, they follow a steady (increasing) trend. The causes of this run-up pattern must be investigated because it might be the result of a problem with the process.

7.2 The Western Electric (WECO) Rules

The interpretation of the control charts patterns is not easy and requires experience and know-how. Western Electric (WECO) published a handbook in 1956 to determine the rules for interpreting the process patterns. These rules are based on the probability for the points to plot at specified areas of the control charts.

A process is said to be out-of-control if one the following occur:

- A single point falls outside the 3σ limit
- Two out of three successive points fall beyond the 2σ limits

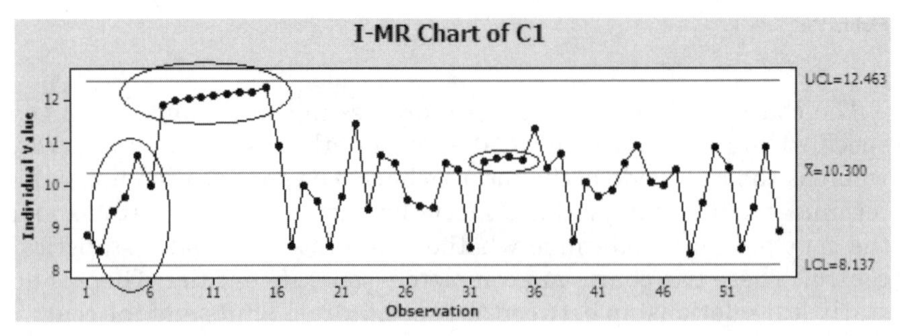

Figure 7.4

- Four out of five successive points fall beyond 1σ from the mean
- Eight successive points fall on one side of the center line

The WECO rules are very good guidelines for interpreting the charts, but they must be used with caution because they add sensitivity to the trends of the mean.

When the process is out-of-control, production is stopped and corrective actions are taken. The corrective actions start with the determination of the category of the variation. The causes of variation can be random or assignable. If the causes of variation are solely due to chance, they are called *chance causes* (Shewhart) or *common causes* (Deming). Not all variations are due to chance; some of them can be traced to specific causes that are not part of the process. In this case, the variations are said to be due to *assignable causes* (Shewhart) or *special causes* (Deming). Finding and correcting special causes of variation are easier than correcting common causes because the common causes are inherent to the process.

7.3 Types of Control Charts

Control charts are generally classified into two groups: they are said to be *univariate* when they monitor a single CTQ characteristic of a product or service, and they are said to be *multivariate* when they monitor more than one CTQ. The univariate control charts are classified according to whether they monitor *attribute* data or *variable* data.

7.3.1 Attribute control charts

Attribute characteristics resemble binary data — they can only take one of two given forms. In quality control, the most common attribute characteristics used are "conforming" or "not conforming," or "good" or "bad." Attribute data must be transformed into discrete data to be meaningful.

The types of charts used for attribute data are:

- The *p*-chart
- The *np*-chart
- The *c*-chart
- The *u*-chart

The *p*-chart. The *p*-chart is used when dealing with ratios, proportions, or percentages of conforming or nonconforming parts in a given sample. A good example for a *p*-chart is the inspection of products on a production line. They are either conforming or nonconforming. The probability distribution used in this context is the binomial distribution with p representing the nonconforming proportion and q (which is equal to $1 - p$) representing the proportion of conforming items. Because the products are only inspected once, the experiments are independent from one another.

The first step when creating a p-chart is to calculate the proportion of nonconformity for each sample.

$$p = \frac{m}{b}$$

where m represents the number of nonconforming items, b is the number of items in the sample, and p is the proportion of nonconformity.

$$\overline{p} = \frac{p_1 + p_2 + \cdots p_k}{k}$$

where $\overline{p}$ is the mean proportion, k is the number of samples audited, and p_k is the kth proportion obtained. The control limits of a p-chart are

$$LCL = \overline{p} - 3\sqrt{\frac{\overline{p}(1-\overline{p})}{n}}$$
$$CL = \overline{p}$$
$$UCL = \overline{p} + 3\sqrt{\frac{\overline{p}(1-\overline{p})}{n}}$$

and $\overline{p}$ represents the center line.

Example Table 7.1 represents defects found on 27 lots taken from a production line over a period of time at Podor Tires. We want to build a control chart that monitors the proportions of defects found on each sample taken.

TABLE 7.1

Defects Found	Lots Inspected	Defects Found	Lots Inspected	Defects Found	Lots Inspected
1	25	1	28	0	24
2	21	2	24	2	29
2	19	1	29	2	20
2	25	2	23	2	17
2	24	3	23	2	20
3	26	3	23		
3	19	1	32		
3	24	2	19		
1	21	3	20		
1	27	3	20		
2	26	2	20		

Open the file *Podor tire.mpj* on the included CD. From the Stat menu, select "Control charts," then select "Attributes charts" and select "P." Fill out the *p*-chart dialog box as indicated in Figure 7.5.

Figure 7.5

Select "OK" to obtain the chart shown in Figure 7.6.

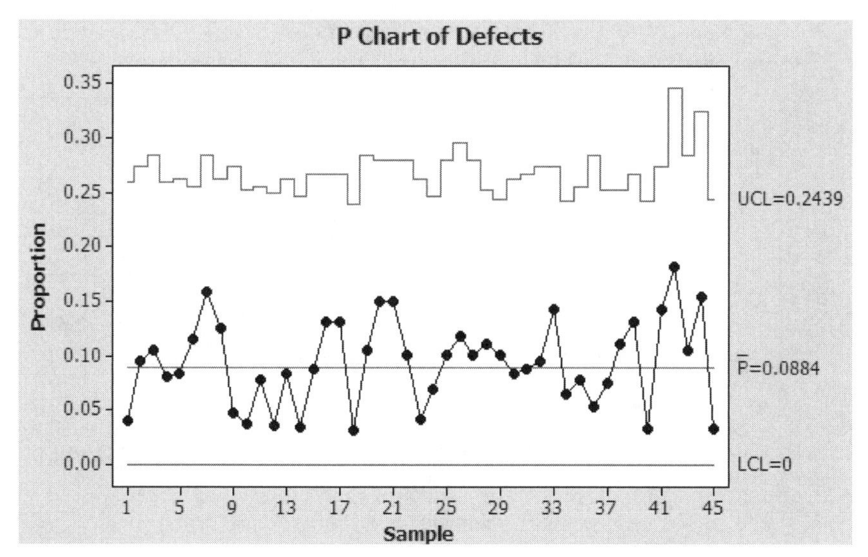

Figure 7.6

What are plotted on the chart are not the defects or the sample sizes but rather the proportions of defects found on the samples taken. In this case, we can say that the process is stable and under control because all the plots are within the control limits and the variation exhibits a random pattern around the mean.

One of the advantages of using the *p*-chart is that the variations of the process change with the sizes of the samples or the defects found on each sample.

The *np*-chart. The *np*-chart is one of the easiest to build. While the *p*-chart tracks the proportion of nonconformities per sample, the *np*-chart plots the number of nonconforming items per sample. The audit process of the samples follows a binomial distribution—in other words, the expected outcome is "good" or "bad," and therefore the mean number of successes is np.

The control limits for an *np*-chart are

$$UCL = n\overline{p} + 3\sqrt{n\overline{p}(1-\overline{p})}$$

$$CL = n\overline{p}$$

$$LCL = n\overline{p} - 3\sqrt{n\overline{p}(1-\overline{p})}$$

Using the same data on the file *Podor Tires.mpj* on the included CD and the same process that was used to build the *p*-chart previously, we can construct the *np*-control chart shown in Figure 7.7.

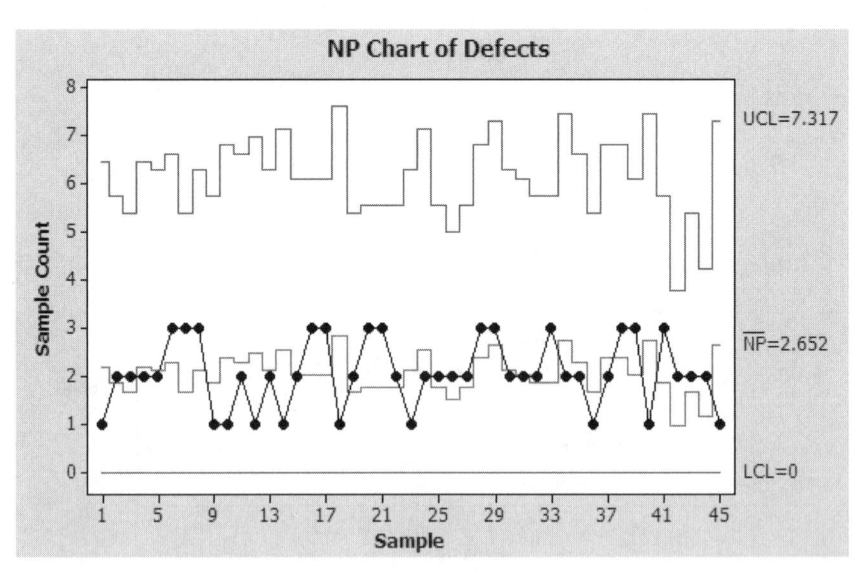

Figure 7.7

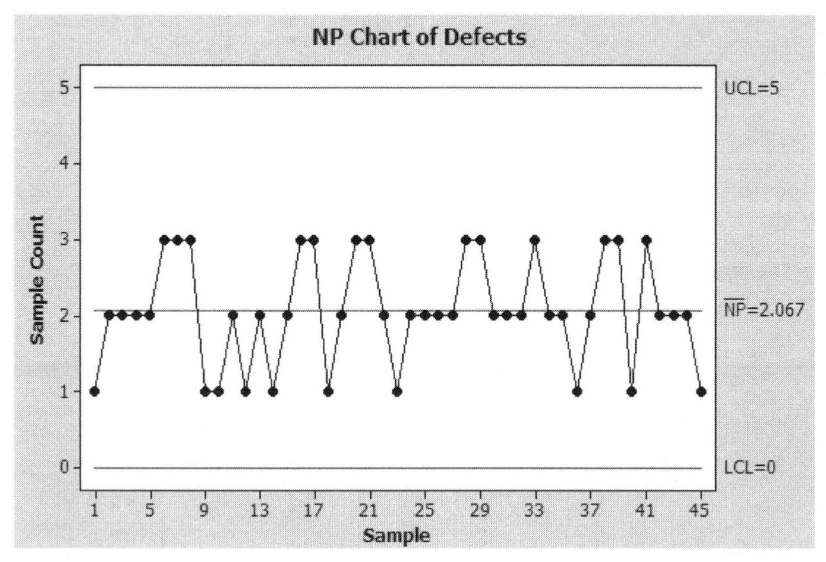

Figure 7.8

Note that the pattern of the chart does not take into account the sample sizes; it just shows how many defects there are on a sample. Sample 2 was of size 21 and had 2 defects, and Sample 34 was of size 31 and had 2 defects, and they are both plotted at the same level on the chart. The chart does not plot the defects relative to the sizes of the samples from which they are taken. For that reason, the *p*-chart has superiority over the *np*-chart.

Consider the same data used to build the chart in Figure 7.7 with all the samples being equal to 5. We obtain the chart shown in Figure 7.8.

These two charts are patterned the same way, with two minor differences being the UCL and the CL. If the sample size for the *p*-chart is a constant, the trends for the *p*-chart and the *np*-chart would be identical but the control limits would be different.

The *p*-chart in Figure 7.9 depicts the same data used previously with all the sample sizes being equal to 5.

The *c*-chart. The *c*-chart monitors the process variations due to the fluctuations of defects per item or group of items. The *c*-chart is useful for the process engineer to know not just how many items are not conforming but how many defects there are per item. Knowing how many defects there are on a given part produced on a line might in some cases be as important as knowing how many parts are defective. Here, nonconformance must be distinguished from defective items because there can be several nonconformities on a single defective item.

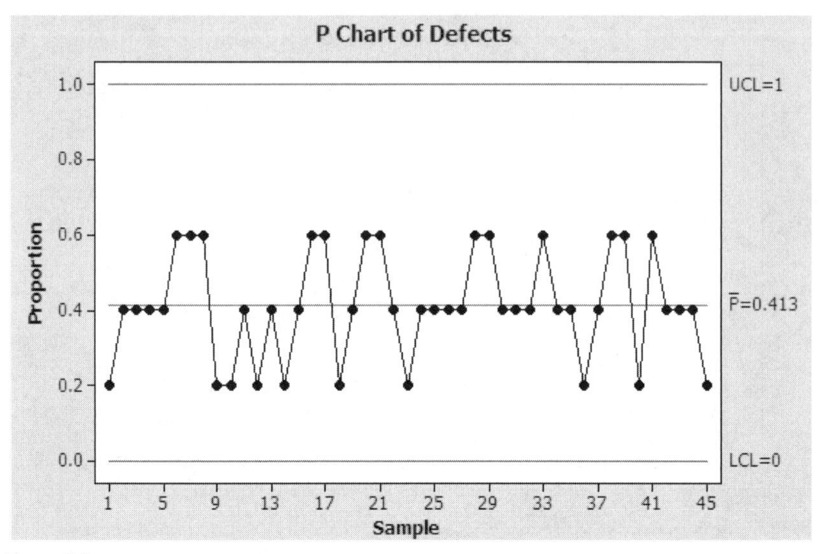

Figure 7.9

The probability for a nonconformity to be found on an item in this case follows a Poisson distribution. If the sample size does not change and the defects on the items are fairly easy to count, the *c*-chart becomes an effective tool to monitor the quality of the production process.

If *c* is the average nonconformity on a sample, the UCL and the LCL limits will be given similar to those for a $k\sigma$ control chart:

$$UCL = \bar{c} + 3\sqrt{\bar{c}}$$

$$CL = \bar{c}$$

$$LCL = \bar{c} - 3\sqrt{\bar{c}}$$

with

$$\bar{c} = \frac{c_1 + c_2 + \cdots c_k}{k}$$

Example Saloum Electrical makes circuit boards for television sets. Each board has 3542 parts, and the engineered specification is to have no more than five cosmetic defects per board. The table on the worksheet on the file *Saloum Electrical.mpj* on the included CD contains samples of boards taken for inspection and the number of defects found on them. We want to build a control chart to monitor the production process and determine if it is stable and under control.

Solution Open the file *Saloum Electrical.mpj* on the included CD. From the Stat menu, select "Control Charts," from the drop-down list, select "Attributes Charts," then select "C." For the variable field, select *Defects* and then select "OK" to obtain the graph shown in Figure 7.10.

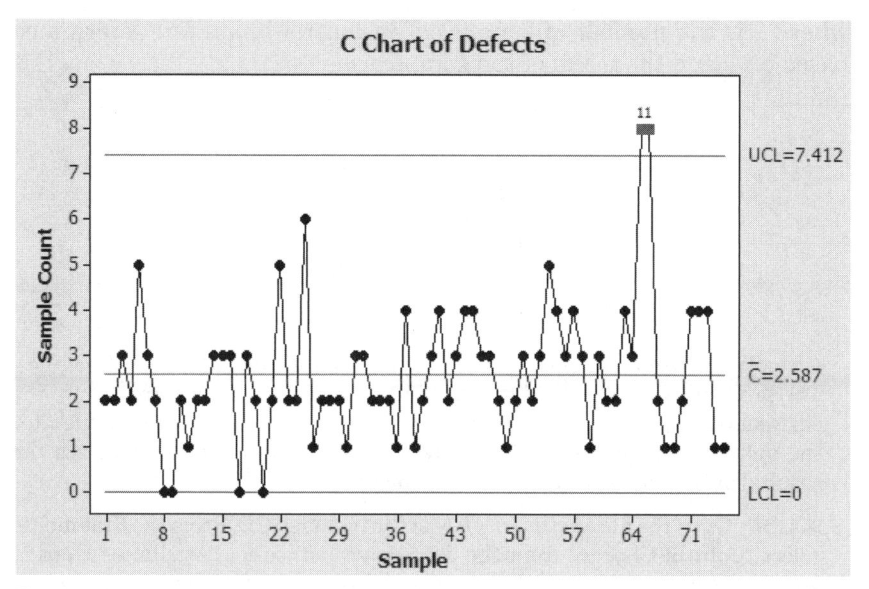

Figure 7.10

Figure 7.10 shows a stable and in-control process up to Sample 65. Sample 65 is beyond three standard deviations from the mean. Something special must have happened that caused it to be so far out of the control limits. The process must be investigated to determine the causes of that deviation and corrective actions taken to bring the process back under control.

The *u*-chart. One of the premises for a *c*-chart is that the sample sizes had to be the same. The sample sizes can vary when a *u*-chart is being used to monitor the quality of the production process, and the *u*-chart does not require any limit to the number of potential defects. Furthermore, for a *p*-chart or an *np*-chart the number of nonconformities cannot exceed the number of items on a sample, but for a *u*-chart it is conceivable because what is being addressed is not the number of defective items but the number of defects on the sample.

The first step in creating a *u*-chart is to calculate the number of defects per unit for each sample.

$$u = \frac{c}{n}$$

where u represents the average defect per sample, c is the total number of defects, and n is the sample size. Once all the averages are determined, a distribution of the means is created and the next step is to find the mean of the distribution—in other words, the *grand mean*,

$$\overline{u} = \frac{u_1 + u_2 + \cdots u_k}{k}$$

where k is the number of samples. The control limits are determined based on $\bar{u}$ and the mean of the samples, n:

$$UCL = \bar{u} + 3\sqrt{\frac{\bar{u}}{n}}$$
$$CL = \bar{u}$$
$$LCL = \bar{u} - 3\sqrt{\frac{\bar{u}}{n}}$$

Example Medina P&L manufactures pistons and liners for diesel engines. The products are assembled in kits of 70 per unit before they are sent to the customers. The quality manager wants to create a control chart to monitor the quality level of the products. He audits 35 units and summarizes the results on the file *Medina.mpj* on the included CD.

Solution Open the file *Medina.mpj* from the included CD. From the Stat menu, select "Control Charts," from the drop-down list, select "Attributes charts," and then select "U." Select "Defect" for the *Variables* field and for *Subgroup sizes,* select "Samples," and then select "OK." The graph should show similar to Figure 7.11.

Notice that the UCL is not a straight line. This is because the sample sizes are not equal and every time a sample statistic is plotted, adjustments are made to the control limits. The process has shown stability until Sample 27 is plotted. That sample is out of control.

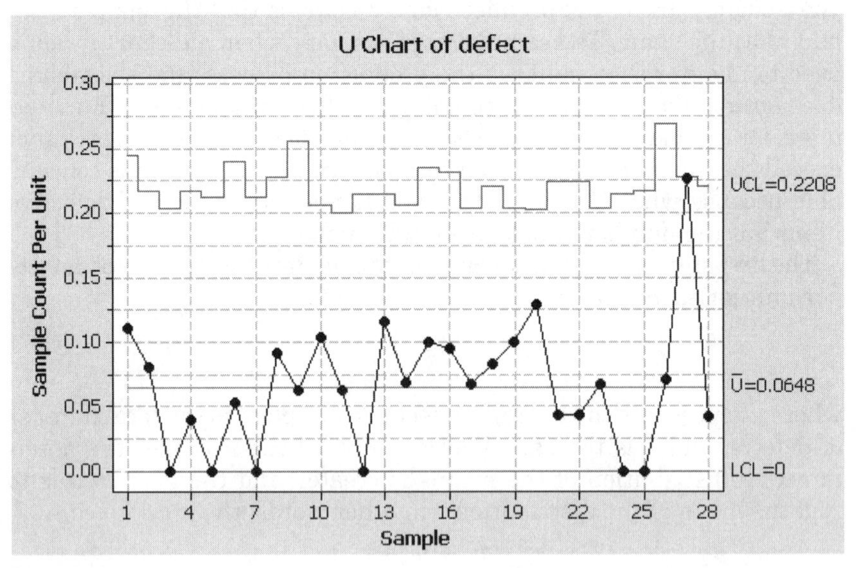

Figure 7.11

7.3.2 Variable control charts

Control charts monitor not only the means of the samples for CTQ characteristics but also the variability of those characteristics. When the characteristics are measured as variable data (length, weight, diameter, and so on), the $\overline{X}$-charts, S-charts, and R-charts are used.

These control charts are used more often and they are more efficient in providing feedback about the process performance. The principle underlying the building of the control charts for variables is the same as that of the attribute control charts. The whole idea is to determine the mean, the standard deviation, and the distance between the mean and the control limits based on the standard deviation.

$$UCL = \overline{\overline{X}} + 3\frac{\sigma}{\sqrt{n}}$$

$$CL = \overline{\overline{X}}$$

$$LCL = \overline{\overline{X}} - 3\frac{\sigma}{\sqrt{n}}$$

But because we do not know what the process population mean and standard deviation are, we cannot just plug numbers into these formulas to obtain a control chart. The standard deviation and the mean must be determined from sample statistics. The first chart that we will use will be the R-chart to determine whether the process is stable or not.

$\overline{X}$-charts and R-charts. The building of an $\overline{X}$-chart follows the same principle as for that of attribute control charts, with the difference that quantitative measurements are considered for the CTQ characteristics instead of qualitative attributes. $\overline{X}$-and R-charts are used together to monitor both the sample means and the variations within the samples through their spread. Samples are taken and measurements of the means $\overline{X}$ and the ranges R for each sample derived and plotted on two separate charts.

The CL is determined by averaging the $\overline{X}$s,

$$CL = \overline{\overline{X}} = \frac{\overline{X}_1 + \overline{X}_2 \cdots \overline{X}_n}{n}$$

where n represents the number of samples. The next step will be to determine the UCL and the LCL,

$$UCL = \overline{\overline{X}} + 3\sigma$$

$$CL = \overline{\overline{X}}$$

$$LCL = \overline{\overline{X}} - 3\sigma$$

We must determine the value of standard deviation σ for the population, which can be determined in several ways. One way to do this would be through the use of the standard error estimate $\sigma/\sqrt{n}$, and another would be the use of the *mean range*.

There is a special relationship between the mean range and the standard deviation for normally distributed data:

$$\sigma = \frac{R}{d_2}$$

where the constant d_2 is function of n. (see Table 7.3)

Standard error-based $\overline{X}$-chart The *standard error-based $\overline{X}$-chart* is straightforward. Based on the Central Limit Theorem, the standard deviation used for the control limits is nothing but the standard deviation of the process divided by the square root of the sample's size. Thus, we obtain

$$UCL = \overline{\overline{X}} + 3\left(\frac{\sigma}{\sqrt{n}}\right)$$

$$CL = \overline{\overline{X}}$$

$$LCL = \overline{\overline{X}} - 3\left(\frac{\sigma}{\sqrt{n}}\right)$$

Because the process standard deviation is not known, in theory these formulas make sense, but in actuality they are impractical. The alternative to this is the use of the mean range.

Mean range-based $\overline{X}$-chart. When the sample sizes are relatively small ($n \leq 10$), the variations within samples are likely to be small, so the range (the difference between the highest and the lowest observed values) can be used in lieu of the standard deviation when constructing a control chart.

$$\sigma = \frac{R}{d_2} \quad \text{or} \quad R = d_2\sigma$$

where R is called the *relative range*. The mean range is

$$\overline{R} = \frac{R_1 + R_2 \cdots R_k}{k}$$

where R_k is the range of the kth sample. Therefore, the estimator of σ is

$$\sigma = \frac{R}{d_2}$$

and the estimator of $\sigma/\sqrt{n}$ is

$$\frac{\sigma}{\sqrt{n}} = \frac{\overline{R}}{d_2\sqrt{n}}$$

Therefore,

$$UCL = \overline{X}\frac{3\overline{R}}{d_2\sqrt{n}}$$

$$CL = \overline{X}$$

$$LCL = \overline{X} - \frac{3\overline{R}}{d_2\sqrt{n}}$$

These equations can be simplified:

$$A_2 = \frac{3}{d_2\sqrt{n}}$$

The formulas for the control limits become

$$UCL = \overline{\overline{X}} + A_2\overline{R}$$

$$CL = \overline{\overline{X}}$$

$$LCL = \overline{\overline{X}} - A_2\overline{R}$$

***R*-chart.** For an R-chart, the center line will be $\overline{R}$ and the estimator of sigma is given as $\sigma_R = d_3\sigma$. Because

$$\sigma = \frac{R}{d_2}$$

we can replace σ with its value and therefore obtain

$$\sigma_R = \frac{d_3\overline{R}}{d_2}$$

Let

$$D_3 = \left(1 - 3\frac{d_3}{d_2}\right)$$

and

$$D_4 = \left(1 + 3\frac{d_3}{d_2}\right)$$

Therefore, the control limits become

$$UCL = D_4\overline{R}$$

$$CL = \overline{R}$$

$$LCL = D_3\overline{R}$$

Example Bamako Lightening is a company that manufactures chandeliers. The weight of each chandelier is critical to the quality of the product. The Quality Auditor monitors the production process using $\overline{X}$-and R-charts. Samples are taken of six chandeliers every hour and their means and ranges plotted on control charts. The data in Figure 7.12 represents samples taken over a period of 25 hours of production.

Microsoft Excel - Book1

	A	B	C	D	E	F	G	H	I
1								**Mean**	**Range**
2	**Hour 1**	9.9943	10.0196	9.9732	10.0088	9.9685	9.9544	**9.986467**	**0.0652**
3	**Hour 2**	9.9721	9.9643	9.9426	9.9712	10.0259	10.0177	**9.9823**	**0.0616**
4	**Hour 3**	9.9788	10.0213	9.9407	10.0696	10.0161	10.0709	**10.01623**	**0.1302**
5	**Hour 4**	10.0658	10.08	9.9514	9.9958	10.0338	10.0422	**10.02817**	**0.1286**
6	**Hour 5**	10.0136	9.9791	9.9449	9.9403	10.0124	10.0221	**9.9854**	**0.0818**
7	**Hour 6**	10.0295	10.0057	9.9715	10.0388	10.0019	9.9616	**10.0015**	**0.0772**
8	**Hour 7**	9.9144	10.0321	10.0361	10.0253	10.0109	9.9935	**10.00205**	**0.1217**
9	**Hour 8**	9.9748	9.9962	10.0369	9.9487	9.9996	9.9741	**9.988383**	**0.0882**
10	**Hour 9**	10.0063	10.0749	9.9971	9.978	9.9934	9.9555	**10.00087**	**0.1194**
11	**Hour 10**	10.084	9.9619	10.0063	10.0364	9.9907	9.9762	**10.00925**	**0.1221**
12	**Hour 11**	9.9436	10.0638	9.9617	10.0138	9.9672	9.9591	**9.984867**	**0.1202**
13	**Hour 12**	10.0074	9.9817	9.9753	10.0541	9.9845	9.9107	**9.985617**	**0.1434**
14	**Hour 13**	10.0069	10.0282	10.0385	9.9747	10.0323	9.986	**10.0111**	**0.0638**
15	**Hour 14**	9.9836	10.063	9.9824	9.9909	9.9684	10.0575	**10.00763**	**0.0946**
16	**Hour 15**	10.0317	10.0485	9.9659	10.0675	10.0047	10.0277	**10.02433**	**0.1016**
17	**Hour 16**	9.9756	9.9174	9.9076	9.9909	10.018	9.9979	**9.9679**	**0.1104**
18	**Hour 17**	9.9818	9.8962	9.9614	9.9612	9.9694	9.9349	**9.950817**	**0.0856**
19	**Hour 18**	10.0168	10.0043	10.0029	9.9549	10.0199	9.9691	**9.99465**	**0.065**
20	**Hour 19**	9.9857	10.0503	10.0381	9.9811	9.9603	9.9123	**9.987967**	**0.138**
21	**Hour 20**	9.9878	9.9841	9.9496	9.9754	10.025	10.0576	**9.996583**	**0.108**
22	**Hour 21**	10.0657	9.9838	9.9636	9.9779	10.0423	10.0547	**10.01467**	**0.1021**
23	**Hour 22**	9.9564	10.0184	9.989	9.9999	10.0146	10.0145	**9.9988**	**0.062**
24	**Hour 23**	10.0611	9.9443	9.9899	9.9833	9.9947	10.0033	**9.9961**	**0.1168**
25	**Hour 24**	9.9054	9.9149	9.987	10.058	9.9156	9.9782	**9.95985**	**0.1526**
26	**Hour 25**	9.9768	9.9954	9.9196	9.9341	9.9162	10.0721	**9.969033**	**0.1559**
27							**Means**	**9.994021**	**0.10464**

Figure 7.12

TABLE 7.2

n	A_2	A_3	D_3	D_4
2	1.88	2.657	—	3.269
3	1.023	1.954	—	2.575
4	0.729	1.628	—	2.282
5	0.577	1.427	—	2.115
6	0.483	1.287	—	2.004
7	0.419	1.182	0.076	1.924
8	0.373	1.099	0.136	1.864
9	0.337	1.032	0.184	1.816
10	0.308	0.975	0.223	1.777

Therefore,

$$\overline{\overline{R}} = 0.10464$$

$$\overline{X} = 9.994$$

From the control chart constant table (Table 7.2), we obtain the values of A_2, D_3, and D_4.

Because n is equal to 6,

$$A_2 = 0.483$$

$$D_3 = 0$$

$$D_2 = 2.004$$

Based on these results, we can find the values of the UCL and the LCL for both the $\overline{X}$-and R-charts. For the $\overline{X}$-chart,

$$UCL = \overline{\overline{X}} + A_2\overline{R} = 9.994021 + 0.10464 \times 0.483 = 10.04$$

$$CL = \overline{\overline{X}} = 9.994$$

$$LCL = \overline{\overline{X}} - A_2\overline{R} = 9.943$$

For the R-chart,

$$UCL = D_4\overline{R} = 2.004 \times 0.10464 = 2.09$$

$$CL = \overline{R} = 0.10464$$

$$LCL = D_3\overline{R} = 0 \times 0.10464 = 0$$

Using Minitab. Open the file *Bamako.mpj* on the included CD. The first thing that must be done is to stack the data. From the Data menu, select "Stack" and then select "Columns." A "Stack Columns" dialog box appears. Select columns C2 through C7 for the *Stack the following columns* text box before selecting "OK." A new worksheet appears with two columns. From the Stat menu, select "Quality Tools," from the drop-down list, select "Variable charts for subgroups," and then select "Xbar-R." Select "C2" for the text box under "All observations for a chart are in one column." In the *Subgroup sizes* field, enter "6." Select "OK" to see the graph in Figure 7.13.

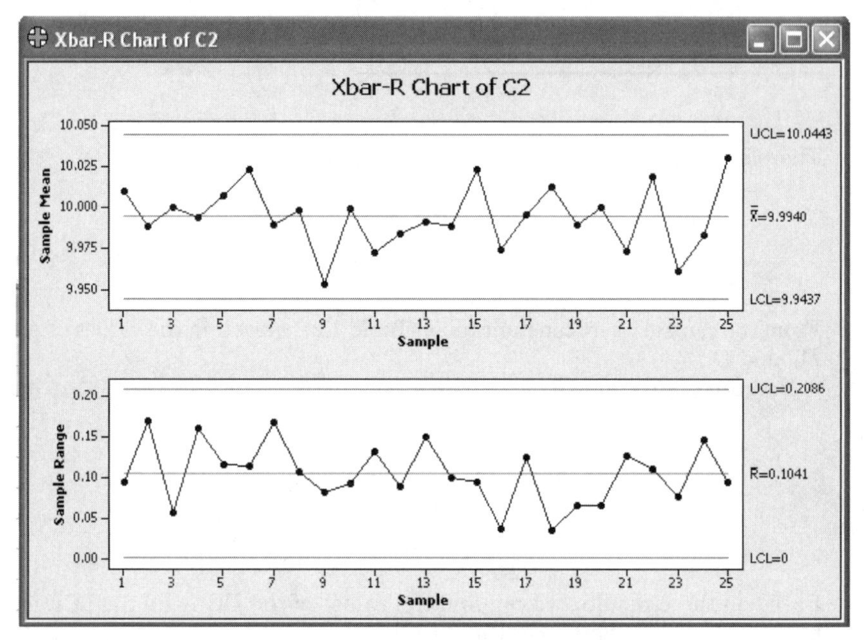

Figure 7.13

The results that we obtain prove our algebraic demonstration. On both charts, all the observations are within the control limits and the variations exhibit a random pattern, so we can conclude that the process is stable and under control.

$\overline{X}$-and *S*-control charts. The *S*-chart is used to determine if there is a significant level of variability in the process, so it plots the standard deviations of the samples taken at regular intervals. A strong variation in the data plots will indicate that the process is very unstable.

Because σ^2, the population's variance, is unknown it must be estimated using the samples' variance, S^2,

$$S^2 = \frac{\sum_{i=1}^{2}(x_i - \overline{x})^2}{n-1}$$

Therefore,

$$S = \sqrt{\frac{\sum_{i=1}^{2}(x_i - \overline{x})^2}{n-1}}$$

$$\overline{S} = \frac{1}{k}\sum_{i=1}^{k} S_i$$

for a number of k samples. But using S as an estimator for σ would lead to a biased result. Instead, $c_4\sigma$ is used, where c_4 is a constant that depends only on the sample size, n. If $\overline{S} = c_4\sigma$, then

$$\sigma = \frac{\overline{S}}{c_4}$$

The mean expected of the standard deviation (which is also the CL) will be $E(S) = c_4\sigma$, and the standard deviation of S is $\sigma\sqrt{1 - c_4^2}$. So the control limits will be as follows:

$$UCL = c_4\sigma + 3\sigma\sqrt{1 - c_4^2}$$

$$CL = c_4\sigma$$

$$LCL = c_4\sigma - 3\sigma\sqrt{1 - c_4^2}$$

These equations can be simplified using B_5 and B_6,

$$B_6 = c_4 + 3\sqrt{1 - c_4^2}$$

$$B_5 = c_4 - 3\sqrt{1 - c_4^2}$$

Therefore,

$$UCL = B_6\sigma$$

$$CL = c_4\sigma$$

$$LCL = B_5\sigma$$

Similarly,

$$UCL = \overline{S} + 3\frac{\overline{S}}{c_4}\sqrt{1 - c_4^2}$$

$$CL = \overline{S}$$

$$LCL = \overline{S} - 3\frac{\overline{S}}{c_4}\sqrt{1 - c_4^2}$$

These equations can be simplified:

$$B_3 = 1 - \frac{3}{c_4}\sqrt{1 - c_4^2}$$

$$B_4 = 1 + \frac{3}{c_4}\sqrt{1 - c_4^2}$$

Therefore,

$$UCL = B_4\overline{S}$$

$$CL = \overline{S}$$

$$LCL = B_3\overline{S}$$

The values of B_3 and B_4 are found in Table 7.3.

TABLE 7.3

Sample Size	A_2	A_3	B_3	B_4	d_2	d_3
2	1.88	2.659		3.267	1.128	0.853
3	1.023	1.954		2.568	1.693	0.888
4	0.729	1.628		2.266	2.059	0.88
5	0.577	1.427		2.089	2.326	0.864
6	0.483	1.287	0.030	1.970	2.534	0.848
7	0.419	1.182	0.118	1.882	2.704	0.833
8	0.373	1.099	0.185	1.815	2.847	0.820
9	0.337	1.032	0.239	1.761	2.970	0.808
10	0.308	0.975	0.284	1.716	3.078	0.797
11	0.285	0.927	0.321	1.679	3.173	0.787
12	0.266	0.886	0.354	1.646	3.258	0.778
13	0.249	0.850	0.382	1.618	3.336	0.770
14	0.235	0.817	0.406	1.594	3.407	0.763
15	0.223	0.789	0.428	1.572	3.472	0.756
16	0.212	0.763	0.448	1.552	3.532	0.750
17	0.203	0.739	0.466	1.534	3.588	0.744
18	0.194	0.718	0.482	1.518	3.640	0.739
19	0.187	0.698	0.497	1.503	3.689	0.734
20	0.18	0.68	0.51	1.49	3.735	0.729
21	0.173	0.663	0.523	1.477	3.778	0.724
22	0.167	0.647	0.534	1.466	3.819	0.72
23	0.162	0.633	0.545	1.455	3.858	0.716
24	0.157	0.619	0.555	1.455	3.895	0.712
25	0.153	0.606	0.565	1.435	3.031	0.708

Example Rufisque Housing manufactures plaster boards. The thickness of the boards is critical to quality, so samples of six boards are taken every hour to monitor the mean and standard deviation of the production process. The table in Figure 7.14 shows the measurements taken every hour.

B	C	D	E	F	G	H	I	J
							mean	Stdev
Hour 1	10.0211	9.9445	10.0692	10.0211	10.0384	9.9627	10.0095	0.047082
Hour 2	9.9937	10.0103	10.007	10.0651	9.965	10.0042	10.00755	0.03266
Hour 3	10.0217	10.0916	9.9439	9.925	9.994	9.9614	9.9896	0.060845
Hour 4	10.0096	9.9363	10.0478	10.1132	9.976	9.9619	10.00747	0.064718
Hour 5	10.0585	9.9751	9.9253	10.0006	10.0175	9.9739	9.991817	0.045185
Hour 6	9.9632	10.0051	9.9734	9.998	10.0696	9.9494	9.993117	0.042923
Hour 7	10.0454	9.9034	9.9971	10.1161	9.9666	9.944	9.995433	0.076167
Hour 8	9.9663	10.0692	10.0253	10.0754	9.9957	10.0604	10.03205	0.044166
Hour 9	9.9925	10.0179	10.0147	9.981	9.937	9.9472	9.981717	0.033778
Hour 10	9.9205	9.9555	9.8429	9.9982	9.9064	10.0179	9.940233	0.064239
Hour 11	10.0003	10.0036	9.977	9.9875	10.0303	10.0622	10.01015	0.0312
Hour 12	9.9415	9.9501	10.0649	10.0006	9.9585	10.039	9.992433	0.051038
Hour 13	10.0277	9.9405	10.0456	9.9131	10.0089	9.9824	9.986367	0.051405
Hour 14	9.9421	10.0064	9.9833	9.9915	9.9635	9.9134	9.9667	0.034418
Hour 15	10.0678	9.9854	9.9545	10.0493	9.9795	10.1045	10.0235	0.058886
						Totals	149.9276	0.738709
							9.995176	0.049247
B3 =0.03	B4 = 1.97		UCL = 0.097					
			LCL = 0.0015					

Figure 7.14

Using Minitab. The way the data are laid out on the worksheet does not lend itself to easy manipulation using Minitab. We will have to stack the data first before creating the control charts.

Worksheet 1 ***

↓	C1-T	C2	C3	C4	C5	C6	C7
1	Hour 1	10.0211	9.9445	10.0692	10.0211	10.0384	9.9627
2	Hour 2	9.9937	10.0103	10.0070	10.0651	9.9650	10.0042
3	Hour 3	10.0217	10.0916	9.9439	9.9250	9.9940	9.9614
4	Hour 4	10.0096	9.9363	10.0478	10.1132	9.9760	9.9619
5	Hour 5	10.0585	9.9751	9.9253	10.0006	10.0175	9.9739
6	Hour 6	9.9632	10.0051	9.9734	9.9980	10.0696	9.9494
7	Hour 7	10.0454	9.9034	9.9971	10.1161	9.9666	9.9440
8	Hour 8	9.9663	10.0692	10.0253	10.0754	9.9957	10.0604
9	Hour 9	9.9925	10.0179	10.0147	9.9810	9.9370	9.9472
10	Hour 10	9.9205	9.9555	9.8429	9.9982	9.9064	10.0179
11	Hour 11	10.0003	10.0036	9.9770	9.9875	10.0303	10.0622
12	Hour 12	9.9415	9.9501	10.0649	10.0006	9.9585	10.0390
13	Hour 13	10.0277	9.9405	10.0456	9.9131	10.0089	9.9824
14	Hour 14	9.9421	10.0064	9.9833	9.9915	9.9635	9.9134
15	Hour 15	10.0678	9.9854	9.9545	10.0493	9.9795	10.1045

After opening the file *Rufisque.mpj* on the included CD,from the Data menu, select "Stack" and then select "Columns." In the "Stack Columns" dialog box, select the columns C1, C2, C3, C4, C5, and C7 for the *Stack the following columns* text box before selecting "OK." A new worksheet will appear with the data stacked in two columns. Now from the Stat menu, select "Control Charts," from the drop-down list, select "Variable charts for subgroups," and then select "Xbar-S." Select "C2" for the text box under *All observations for a chart are in one column*. Enter "6" in the field *Subgroup size*. Select "OK" and the control charts appear.

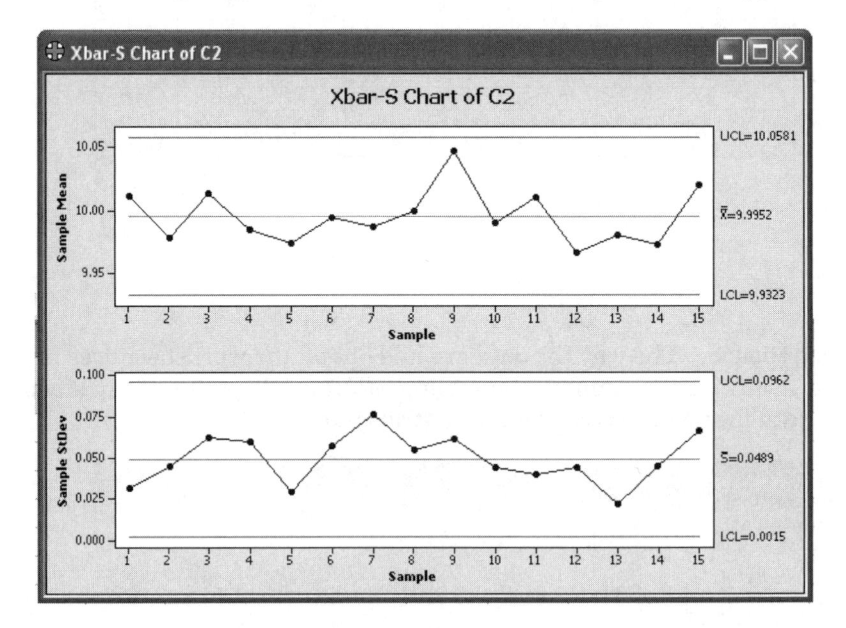

All the data plots for the two charts are well within the control limits and exhibit a random pattern, so we conclude that the process is table and under control.

Using Excel. We can find the mean and the standard deviations for the samples and construct the charts. We know that $n = 6$, so once we obtain the values of $\overline{\overline{X}}$ and $\overline{S}$ we look up the values of B_3 and B_4 on the control charts constant table (Table 7.3):

$$B_4 = 1.94$$

$$B_3 = 0.03$$

$$UCL = 1.97 \times 0.049 = 0.097$$

$$LCL = 0.03 \times 0.049 = 0.0015$$

Excel does not provide an easy way to generate control charts without adding macros, but because we know what the UCL and the LCL are, we can use the Chart Wizard to see the trends of the process.

Moving Range. When individual (samples composed of a single item) CTQ characteristics are collected, moving range control charts can be used to monitor production processes. The variability of the process is measured in terms of the distribution of the absolute values of the difference of every two successive observations.

Let x_i be the ith observation, and the moving average range MR will be

$$MR = \left|x_i - x_{i-1}\right|$$

and the mean MR will be

$$\overline{MR} = \frac{\sum_{i=1}^{n} \left|x_i - x_{i-1}\right|}{n}$$

The standard deviation S is obtained by dividing $\overline{MR}$ by the constant d_2. Because the moving range only involves two observations, n will be equal to 2 and therefore, for this case, d_2 will always be equal to 1.128.

$$UCL = \overline{x} + \frac{3}{d_2}\overline{MR}$$

$$CL = \overline{x}$$

$$LCL = \overline{x} - \frac{3}{d_2}\overline{MR}$$

Because d_2 is 1.128, these equations can be simplified:

$$UCL = \overline{x} + 2.66\overline{MR}$$

$$CL = \overline{x}$$

$$LCL = \overline{x} - 2.66\overline{MR}$$

Example The data in Table 7.4 represents diameter measurements of samples of bolts taken from a production line. Find the control limits.

TABLE 7.4

Sample Number	Measurements	Moving Range
1	9	—
2	7	2
3	11	4
4	8	3
5	8	0
6	7	1
7	10	3
8	9	1
9	12	3
10	11	1
Total	92	18
Mean	9.2	2

$$\bar{x} = 9.2$$
$$\overline{MR} = 2$$

Therefore, the control limits will be

$$UCL = 9.2 + (2.66 \times 2) = 14.52$$
$$CL = 9.2$$
$$LCL = 9.2 - (2.66 \times 2) = 3.88$$

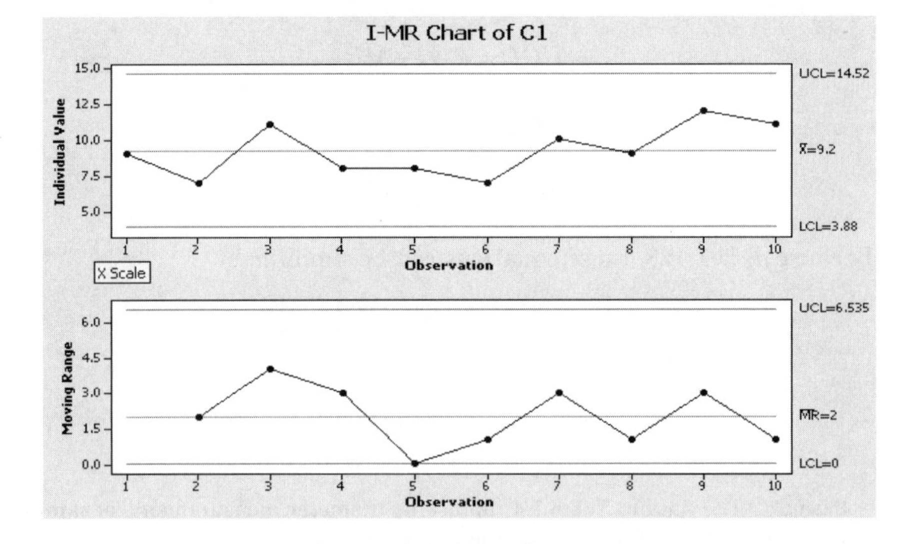

Chapter

8

Process Capability Analysis

Are Your Processes Yielding Products or Services That Meet Your Customers' Expectations?

Learning Objectives:

- Determine the difference between the purpose of a Statistical Process Control and the one of a process capability analysis
- Know how the process capability indices are generated
- Understand the difference between Taguchi's indices and the C_{pk} and C_p
- Analyze the capability of a process with normal and non-normal data

Two factors determine a company's ability to respond to market demand: the operating resources it has at its disposal, and the organizational structure it has elected to use. The operating resources establish the company's leverage and the maximum amount of products or services it is able to produce.

The organizational structure that is determined by the short- or long-term strategies of a company is composed of the multitude of processes that are used to generate the goods or services. How effective a company is in satisfying its customers' demand is measured in terms of the processes' capabilities, which are defined as the processes' abilities to generate products or services that meet or exceed customers' requirements.

Customer requirements are the significant features that the customers expect to find in a product or a service; the design engineers translate those requirements into CTQ characteristics of the products or services that they are about to produce. Those CTQs are fully integrated into the product design as measurable or attribute variables, and they are used as metrics to ensure that the production processes conform to customer requirements. Once the CTQs are assessed and quality targets are determined, the engineers specify the upper and lower limits within which those variables must fall.

While the production is in progress, the performance of the *production process* is monitored to detect and prevent possible variations. The tool frequently used to monitor a process performance while the production is in progress is the control chart. It helps detect assignable causes of variations and facilitate corrective actions.

But a control chart is not the correct tool to determine if the customers' requirements are met because it is only used to monitor the performance of production processes in progress, and an in-control process does not necessarily mean that all the products meet the customers' (or engineered) requirements. In other words, a process can be contained within the upper and lower control limits and still generate products that are outside the specified limits.

Suppose that we are monitoring a machine that produces rotor shafts, and that the length of the shaft is critical to quality. The $\overline{X}$ - R-charts of Figure 8.1 plot the measurements that were obtained while monitoring the production process. Remember that control charts are constructed

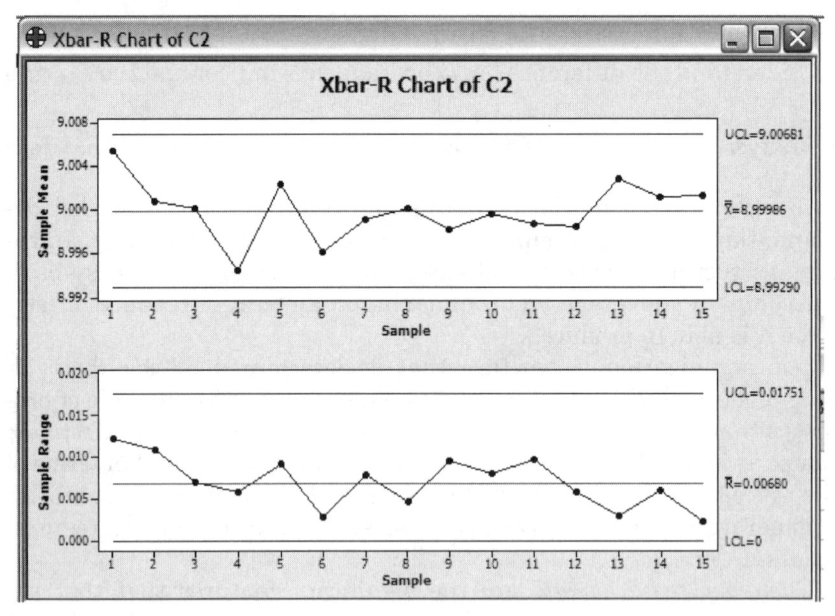

Figure 8.1

by taking samples from the production line at preset intervals and plotting the means of the samples on a chart. If the parts from the samples taken are considered defective, the measurements are still plotted and adjustments made on the machine to prevent further defects. But making adjustments on the machines does not mean that no more defects will be generated.

From the observation we make of the process that generated the graph of Figure 8.1, we can conclude that the process is acceptably stable and that the variations are within the control limits.

Yet we cannot conclude that all the output meets the customers' expectations. The output in this case is within the UCL (9.00681) and the LCL (8.99986). If the specified engineered limits were 9.01 for the *lower specified limit* (LSL) and 10.02 for the *upper specified limit* (USL), none of the parts produced by the machine would have met the customers' expectations. In other words, all the parts would have been considered as defective. So a stable and in-control production process does not necessarily mean that all the output meets customers' requirements. To dissipate any confusion, it is customary to relate the specified limits to the "voice of the customer," whereas the control limits are related to the "voice of the process."

The control charts do not relate the process performance to the customers' requirements because there is not any statistical or mathematical relationship between the engineered specified limits and the process control limits. The *process capability analysis* is the bridge between the two; it compares the variability of an in-control and stable production process with its engineered specifications and capability indices are generated to measure the level of the process performance as it relates to the customers' requirements.

A process is said to be *capable* if the process mean is centered to the specified target and the range of the specified limits is wider than the one of the actual production process variations (Control Limits), as in the graph in Figure 8.2. The Upper and Lower specified limits represent

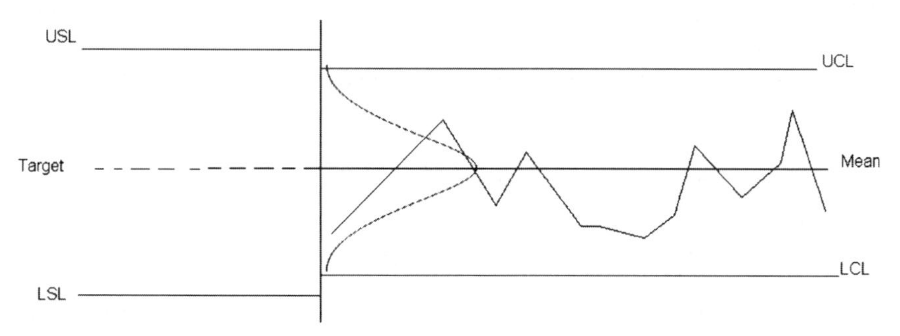

Figure 8.2

the engineered standards (or customer requirements) and the control chart on the right side depicts the production process performance. Because the control chart depicts the actual production process and all the output is within the control limits and the range of the control limits is smaller than that of the specified limits, we conclude that the output generated by the production process meets or exceeds the customers' expectations.

If the spread of the natural variations (control limits) is larger than the one of the specified limits, as in the example of Figure 8.3, the process is considered *incapable* because some of the parts produced are outside the engineered standards.

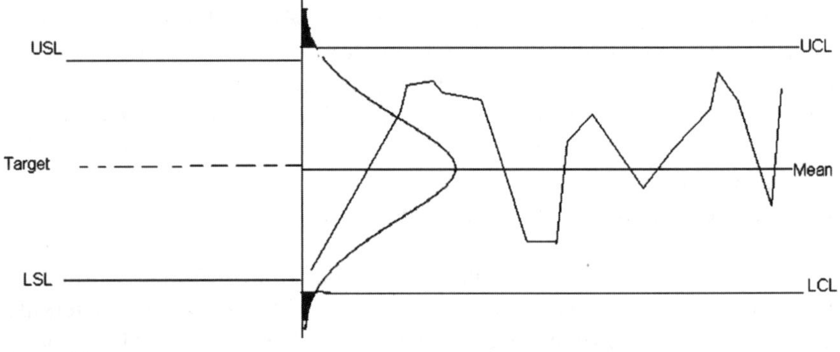

Figure 8.3

Process capability analysis assumptions. Process capability analysis assumes that the production process is in-control and stable. Because what are being compared are the specified limits and the control limits, if the process is out of control, some of the measurements might be outside the control limits and would not be taken into account.

The *stability* of the process refers to the ability of the process auditor to predict the process trends based on past experience. A process is said to be stable if all the variables used to measure the process' performance have a constant mean and a constant variance over a sufficiently long period of time.

Process capabilities in a Six Sigma project are usually assessed at the "Measure" phase of the project to determine the level of conformance to customer requirements before changes are made, and at the "Improve" phase to measure the quality level after changes are made.

8.1 Process Capability with Normal Data

The outputs of most production processes are normally distributed. When a probability distribution is normal, most of the data being analyzed are concentrated around the mean. For a sigma-scaled normal graph, 99.73 percent of the observations would be concentrated between $\pm 3\sigma$ from the mean.

Testing your data for normality. Minitab offers several ways to test the normality of data. That test can be done through the "Individual Distribution Identification" option under the "Quality tools" or through the "Normality test" under "Basic Statistics." Because the normality test has already been used in the previous chapters, we will only concern ourselves with the Individual Distribution Identification. The purpose of this option is to help the experimenter determine the type of distribution the data at hand follow. The experimenter can a *priori* select several types of distributions and test the data for all of them at the same time. Based on the p-values that the test generates, one can assess the nature of the distribution.

Open the file *normalitytest.mpj* on the included CD. From the Stat menu, select "Quality tools," and from the drop-down list, select "Individual Distribution Identification." Select "C1" for *Single Column.* Select *Use all distributions* and then select "OK." A probability plot for each distribution should appear. An observation of the plots and the p-values shows that the distribution is fairly normal. The Anderson-Darling normality test shows a p-value of 0.415, which indicates that the data are normally distributed for an α-level of 0.05.

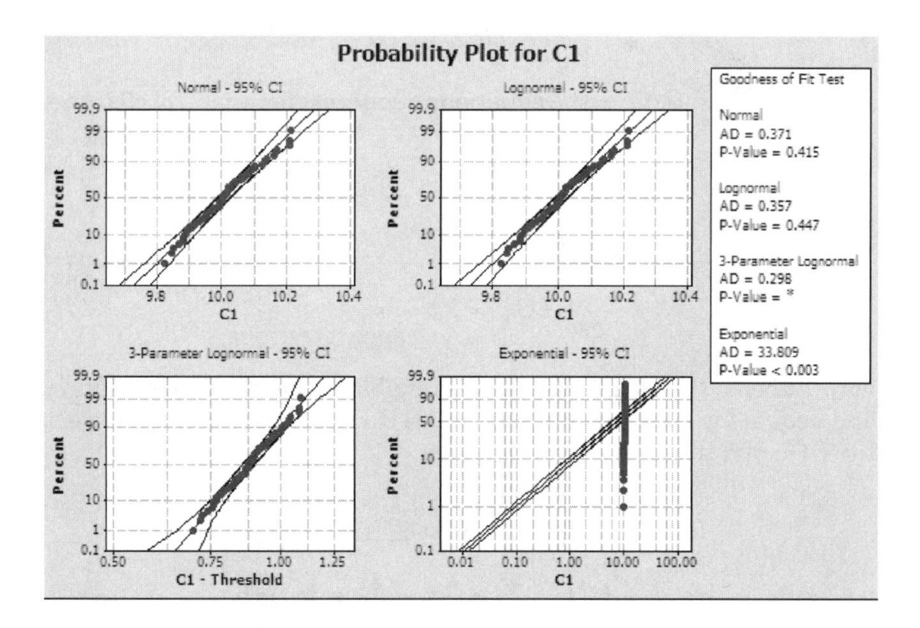

The process capability analysis compares the spread of the specified limits to the spread of the natural variations of the process (control chart). The most commonly used process capability indices are: C_p, C_{pk}, C_r, P_p, and P_{pk}. The process capability indices are unitless; that is, they are not expressed in terms of a predetermined unit of measurement.

8.1.1 Potential capabilities vs. actual capabilities

Process capability indices can be divided into two groups: the indices that measure the processes potential capabilities, and the ones that measure their actual capabilities. The potential *capability* indices determine how capable a process is if certain conditions are met—essentially, if the mean of the process' natural variability is centered to the target of the engineered specifications. The *actual capability* indices do not require the process to be centered to be accurate.

Short-term potential capabilities, C_p and C_r. A process is said to be capable if the spread of the natural variations fits in the spread of the specified limits. This is so when the ratio of the specified range to the control limits is greater than one. In other words, the following ratio should be greater than 1:

$$C_p = \frac{USL - LSL}{UCL - LCL}$$

For a sample statistic y, the equations of interest for a control chart are

$$UCL = \overline{\overline{X}}_y + k\sigma_y$$

$$CL = \overline{\overline{X}}_y$$

$$LCL = \overline{\overline{X}}_y - k\sigma_y$$

with $\overline{\overline{X}}_y$ being the mean of the process, σ_y being the standard deviation, and k equal to 3. The range of the control chart is the difference between the UCL and the LCL and is given as

$$UCL - LCL = (\overline{\overline{X}}_y + 3\sigma_y) - (\overline{\overline{X}}_y - 3\sigma_y)$$

$$UCL - LCL = \overline{\overline{X}}_y + 3\sigma_y - \overline{\overline{X}}_y + 3\sigma_y = 6\sigma_y$$

Therefore,

$$C_p = \frac{USL - LSL}{UCL - LCL} = \frac{USL - LSL}{6\sigma}$$

The value of $C_p = 1$ if the specified range equals the range of the natural variations of the process, in which case the process is said to be *barely capable*; it has the potential to only produce nondefective products if the process mean is centered to the specified target. Approximately 0.27 percent, or 2700 parts per million, are defective.

The value of $C_p > 1$ if the specified range is greater than the range of the control limits, in which case the process is *potentially capable*—if the process mean is centered to the engineered specified target—and is (probably) producing products that meet or exceed the customers' requirements.

The value of $C_p < 1$ if the specified range is smaller than the range of the control limits. The process is said to be *incapable*; in other words, the company is producing junk.

Example The specified limits for a product are 75 for the upper limits and 69 for the lower limit with a standard deviation of 1.79. Find C_p. What can we say about the process' capabilities?

Solution

$$C_p = \frac{USL - LSL}{6\sigma} = \frac{75 - 69}{6 \times 1.79} = \frac{6}{10.74} = 0.56$$

Because C_p is less than 1, we have to conclude that the process will generate nonconforming products.

Capability ratios. Another way of expressing the short-term potential capability would be the use of the *capability ratio*, C_r. It determines the proportion or percentage of the specified spread that is needed to contain the process range. Note that it is not the proportion that is necessarily occupied. If the process mean is not centered to the target, the range of the control chart may not be contained within the specified limits.

$$C_r = \frac{1}{C_p} = \frac{UCL - LCL}{USL - LSL} = \frac{6\sigma}{USL - LSL}$$

Example What is the capability ratio for the previous example? How do we interpret it?

Solution Because

$$C_p = 0.56$$

$$C_r = \frac{1}{C_p} = \frac{1}{0.56} = 1.79$$

The proportion of the specified spread needed to accommodate the spread of the process range is 1.79. It is greater than 1, and therefore the production process is not capable.

Process performance, long-term potential process capabilities. Control charts are built using samples taken while the production is in progress. The process mean is the mean of the samples' means. Because of the common (and also special) causes of variation, both the process mean and the process variance will tend to drift from their original positions in the long term. A *long-term process performance* takes into account the possibility of a shift of the process mean and variance.

P_p and P_r are the indices used to measure the long-term process capabilities. They are computed the same way the C_p and the C_r are computed,

$$P_p = \frac{USL - LSL}{6\sigma_{LT}}$$

$$P_r = \frac{6\sigma_{LT}}{USL - LSL}$$

where σ_{LT} is the long-term standard deviation. The interpretation of these equations is also the same as in the case of the short-term capabilities.

8.1.2 Actual process capability indices

The reason why $C_p > 1$ does not necessarily mean that the process is not producing defects is that the range of the control limits might be smaller than the one of the specified limits, but if the process mean is not centered to the specified target, one side of the control chart might exceed the specified limits, as in the case of the graph of Figure 8.4, and defects are being produced.

If the process mean is not centered to the specified target, C_p would not be very informative because it would only tell which of the two ranges (process control limits and engineered specified limits) is wider, but it would not be able to inform on whether the process is generating defects or not. In that case, another capability index is used to determine a process' ability to respond to customers' requirements.

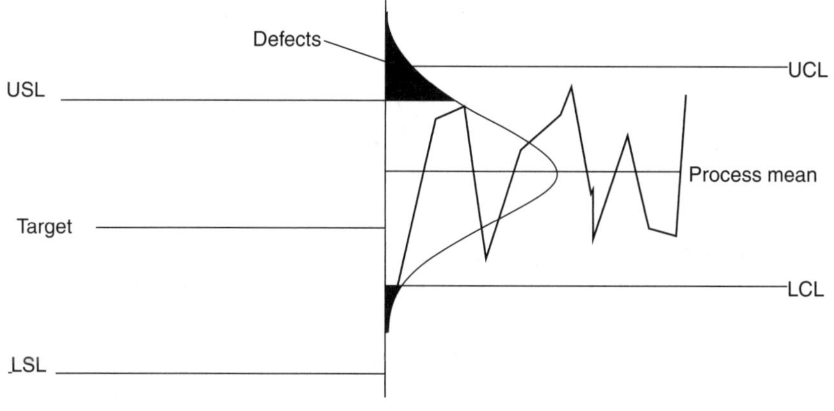

Figure 8.4

The C_{pk} measures how much of the production process really conforms to the engineered specifications. The k in C_{pk} is called the k-factor; it measures the level of deviation of the process mean from the specified target.

$$C_{pk} = (1 - k)C_p$$

To find out how to derive C_p from this formula, see Appendix C. With

$$k = \frac{\left|(USL + LSL)/2 - \overline{\overline{X}}\right|}{(USL - LSL)/2}$$

(USL + LSL)/2 is nothing but the target T, so k becomes

$$k = \frac{\left|T - \overline{\overline{X}}\right|}{(USL - LSL)/2}$$

$$k = \frac{|((USL + LSL)/2) - \overline{\overline{X}}|}{(USL - LSL)/2}$$

Therefore, if $\dfrac{USL + LSL}{2} > \overline{\overline{X}}$ in other words, if $T > \overline{\overline{X}}$

then $k = \dfrac{((USL + LSL)/2) - \overline{\overline{X}}}{(USL - LSL)/2}$

if $\dfrac{USL + LSL}{2} < \overline{\overline{X}}$ then $k = \dfrac{\overline{\overline{X}} - (USL + LSL)/2}{(USL - LSL)/2}$

A little algebraic manipulation can help demonstrate that $C_{pk} = (1 - k)C_p$

$$\text{if } k = \frac{((USL + LSL)/2) - \overline{\overline{X}}}{(USL - LSL)/2} = \frac{USL + LSL}{USL - LSL} - \frac{2\overline{\overline{X}}}{USL - LSL}$$

Since

$$C_p = \frac{USL - LSL}{6\sigma}, \quad (1-k)C_p = \left(1 - \left(\frac{USL + LSL}{USL - LSL} - \frac{2\overline{\overline{X}}}{USL - LSL}\right)\right) \times \left(\frac{USL - LSL}{6\sigma}\right)$$

We can develop this equation a little further

$$C_{pk} = (1-k)C_p = \left(\frac{USL - LSL}{USL - LSL} - \frac{USL + LSL}{USL - LSL} + \frac{2\overline{\overline{X}}}{USL - LSL}\right) \times \left(\frac{USL - LSL}{6\sigma}\right)$$

$$C_{pk} = (1-k)C_p = \left(\frac{USL - LSL}{6\sigma} - \frac{USL - LSL}{6\sigma} + \frac{2\overline{\overline{X}}}{6\sigma}\right)$$

$$= \frac{2\overline{\overline{X}} - 2LSL}{6\sigma} = \frac{\overline{\overline{X}} - LSL}{3\sigma}$$

$$k = \frac{((USL + LSL)/2) - \overline{\overline{X}}}{(USL - LSL)/2}$$

if $\frac{USL+LSL}{2} < \overline{\overline{X}}$, then $k = \frac{\overline{\overline{X}} - (USL+LSL)/2)}{(USL-LSL)/2}$

$$C_{pk} = (1-k)C_p, \text{ therefore } C_{pk} = \left(1 - \left(\frac{2\overline{\overline{X}}}{USL - LSL} - \frac{USL + LSL}{USL - LSL}\right)\right) \times \left(\frac{USL - LSL}{6\sigma}\right)$$

$$C_{pk} = (1-k)C_p = \left(\frac{USL - LSL}{6\sigma} - \frac{2\overline{\overline{X}}}{6\sigma} + \frac{USL - LSL}{6\sigma}\right)$$

$$= \frac{2(USL - \overline{\overline{X}})}{6\sigma} = \frac{USL - \overline{\overline{X}}}{3\sigma}$$

A result of $k = 0$ means that the process is perfectly centered, and therefore $C_{pk} = C_p$.

$$C_{pk} = (1 - k)C_p$$

$$1 - k = \frac{C_{pk}}{C_p}$$

If $C_{pk} = C_p$,

$$1 - k = 1$$

$$k = 0$$

If $k \neq 0$, then

$$3C_{pk} = \min\{Z_{ul}, Z_{ll}\} \quad \text{or} \quad C_{pk} = \min\left\{\frac{1}{3}Z_{ul}, \frac{1}{3}Z_{ll}\right\}$$

with

$$Z_{ul} = \frac{USL - \overline{\overline{X}}}{\sigma}$$

and

$$Z_{ll} = \frac{\overline{\overline{X}} - LSL}{\sigma}$$

Call T the engineering specified target:

$$T = \frac{USL + LSL}{2}$$

If $T < \overline{\overline{X}}$, then

$$C_{pk} = \frac{1}{3}Z_{ul} = \frac{USL - \overline{\overline{X}}}{3\sigma}$$

If $T > \overline{\overline{X}}$, then

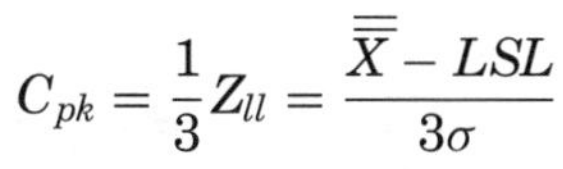

$$C_{pk} = \frac{1}{3}Z_{ll} = \frac{\overline{\overline{X}} - LSL}{3\sigma}$$

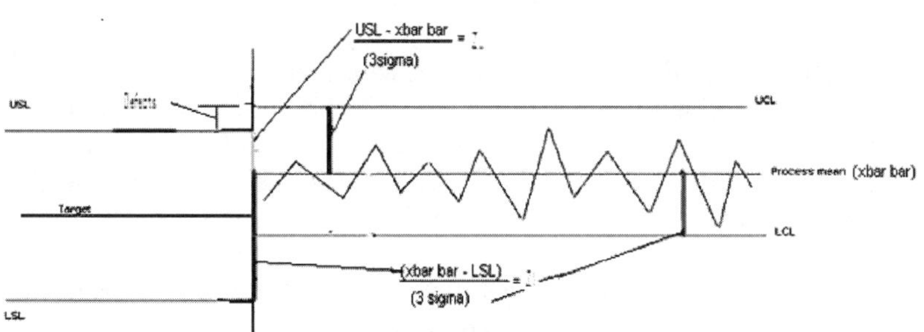

Just as in the case of the short-term capability indices, the actual long-term process capability takes into account the possibility of a drift in both the mean and the variance of the production process.

P_{pk} is the index used to measure the long-term process capability:

$$P_{pk} = \min\left\{\frac{1}{3}Z_{ULT}, \frac{1}{3}Z_{LLT}\right\}$$

$$Z_{LLT} = \frac{\overline{\overline{X}}_{LT} - LSL}{\sigma_{LT}}$$

$$Z_{ULT} = \frac{USL - \overline{\overline{X}}_{LT}}{\sigma_{LT}}$$

Consider the Minitab process capability analysis output of Figure 8.5:

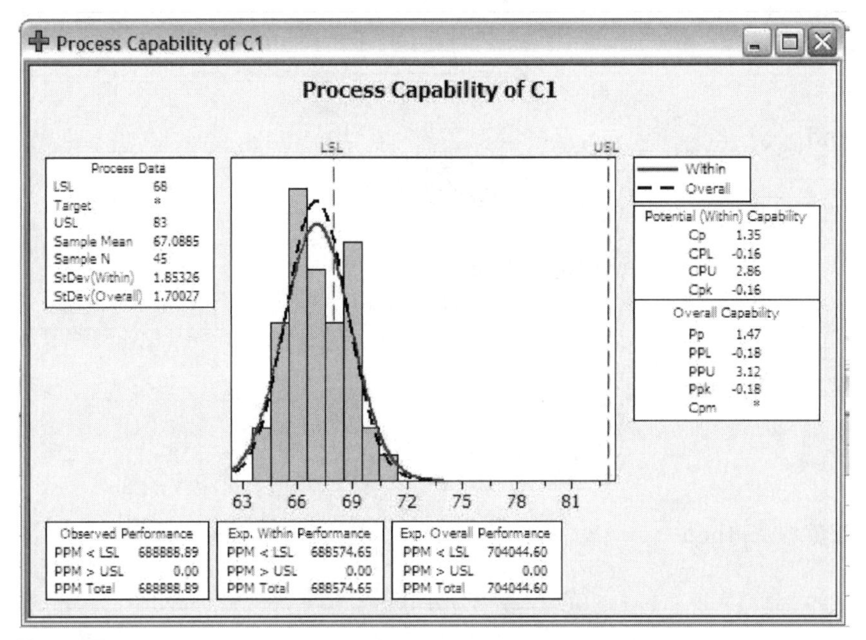

Figure 8.5

The histogram and the normal curves represent in this case the process output, and the specified limits specify the engineered standards. The spread of the engineered specified limits is a lot wider than that of the process control limits; that's why C_p is greater than one. But because the process is not centered to the engineered specified target, more than half the output is outside (to the left) the specified limits. The C_{pk} is extremely small: -0.16 indicates that there is a great deal of opportunities for process improvement.

Because C_{pk} is a better measure of capability than C_p, why not just use C_{pk} instead of C_p? The variable C_{pk} only shows the spread between the process mean and the closest specified limit; therefore, it will not reveal the spread of the process controls.

Example The average call time at a call center is 7.27 minutes. No lower specification is set, and the upper specification is set at 9.9 minutes. What is the maximum standard deviation if a C_{pk} greater than 1.67 is required?

Solution We already know the formula for calculating the C_{pk}:

$$C_{pk} = \frac{USL - \overline{\overline{X}}}{3\sigma} = \frac{9.9 - 7.27}{3\sigma}$$

Therefore,

$$\sigma \leq \frac{9.9 - 7.4}{3 \times 1.67}$$

$$\sigma \leq 0.524$$

So the maximum standard deviation must be 0.499.

8.2 Taguchi's Capability Indices C_{PM} and P_{PM}

So far all the indices that were used (C_p, C_{pk}, P_p, P_{pk}, and C_r) only considered the specified limits, the standard deviation, and—in the case of C_{pk} and P_{pk}—the production process mean. None of these indices take into account the variations within tolerance, the variations when the process mean fails to meet the specified target but is still within the engineered specified limits. Because of Taguchi's approach to tolerance around the engineered target (see Chapter 12), the definition and approach to capability measures differ from that of the traditional process capability analysis.

Taguchi's approach suggests that any variation from the engineered target, be it within or outside the specified limits, is a source of defects and a loss to society. That loss is proportional to the distance between the production process mean and the specified target. Taguchi's loss function measures the loss that society incurs as a result of not producing products that match the engineered targets. It quantifies the deviation from the target and assigns a financial cost to the deviations:

$$l(y) = k(y - T)^2$$

with

$$k = \frac{\Delta}{m^2}$$

where Δ is the cost of a defective product, m is the difference between the specified limit and the target T, and y is the process mean.

Because Taguchi considers both the process standard deviation and the position of the process mean, both C_{pm} and P_{pm} will take into account these variables. The formulas for the capability indices therefore become

$$C_{pm} = \frac{USL - LSL}{6\tau_{ST}}$$

and

$$P_{pm} = \frac{USL - LSL}{6\tau_{LT}}$$

where τ depends on the variance and the deviation of the process mean from the engineered standards.

$$\tau = \sqrt{\sigma^2 + (T - M)^2}$$

where T is the target and M is the process mean.

Example A machine produces parts with the following specified limits: USL = 16, LSL = 12, and specified target = 14. The standard deviation is determined to be 0.55 and the process mean 15. Find the value of C_{pm}. Compare the C_{pm} with C_{pk}.

Solution

$$C_{pm} = \frac{USL - LSL}{6\sqrt{\sigma^2 + (T - M)^2}} = \frac{16 - 12}{6\sqrt{0.55^2 + (15 - 14)^2}}$$

$$= \frac{4}{6 \times 1.141271} = \frac{4}{6.85} = 0.584$$

$$C_u = \frac{16 - 15}{3 \times 0.55} = \frac{1}{1.65} = 0.61$$

$$C_l = \frac{15 - 12}{3 \times 0.55} = \frac{3}{1.65} = 1.82$$

So $C_{pk} = 0.61$. Even though both the C_{pm} and the C_{pk} show that the production process is incapable, $C_{pk} > C_{pm}$.

Example The amount of inventory kept at Touba's warehouse is critical to the performance of that plant. The objective is to have an average of 15.5 DSI with a tolerance of an USL of 16.5 and a LSL of 14.5. The data on the file *Touba warehouse.mpj* on the included CD represent a sample of the DSI.

a. Run a capability analysis to determine if the production process used so far has been capable.
b. Is there a difference between C_{pm} and C_{pk}? Why?
c. The tolerance limits have been changed to USL = 16 and LSL = 14 and the target set at 15 DSI. What effect did that change have on the C_{pm} and the C_{pk}?

Solution Open the file *Touba warehousei.mpj* on the included CD. From the Stat menu, select "Quality Tools," then select "Capability Analysis," and then select "Normal." Fill out the dialog box as indicated in Figure 8.6.

Select "Options..." and in the "Capability Analysis Options" dialog box, enter "15.5" into the *Target (add C_{pm} totable)* field. Leave the value "6" in the *K* field and select the option *Include confidence intervals*. Select "OK" and then select "OK" again to get the output shown in Figure 8.7.

Interpretation. The data plot shows that all the observations are well within the specified limits and not a single one comes anywhere close to any one of the limits, yet all of them are concentrated between the LSL and the target. The fact that not a single observation is outside the specified limits generated a PPM (defective Parts Per Million) equal to zero for the observed performance. A result of $C_{pk} = 1.07$ suggests that the process is barely capable.

But from Taguchi's approach, the process with a $C_{pm} = 0.64$ is absolutely not capable because even though all the observations are within the specified limits, very few (outliers) of them meet the target.

After the specified limits and the target have been changed, we obtain the output shown in Figure 8.8. Note that the production process is now centered and the observations are closely clustered around the target. The PPM is still equal to zero but $C_{pm} = 2.13$ and $C_{pk} = 2.08$ are closer than before the changes were made.

8.3 Process Capability and PPM

The process capability indices enable an accurate assessment of the production process' ability to meet customer expectations, but because they are unitless, it is not always easy to interpret changes that result from improvement on the production processes based solely on the

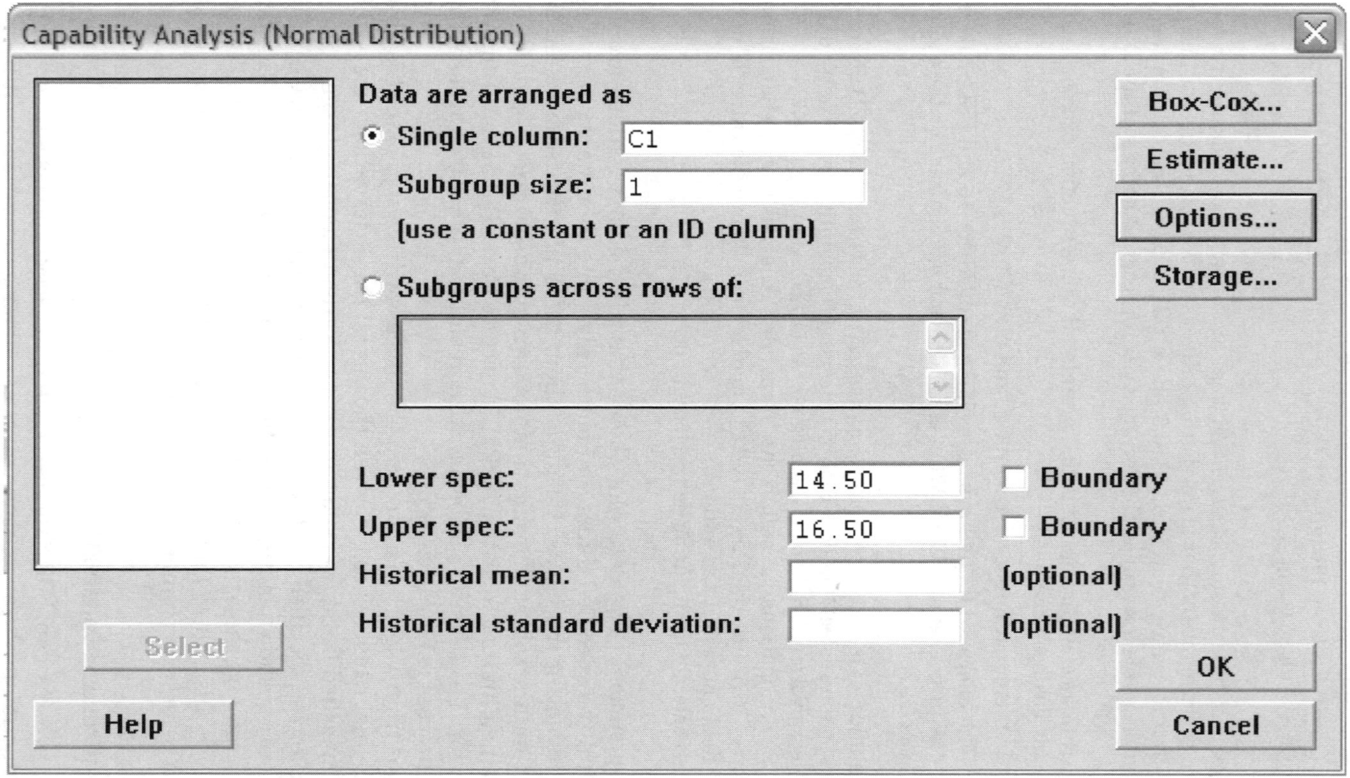

Capability Analysis (Normal Distribution)
Data are arranged as
Single column: C1
Subgroup size: 1
(use a constant or an ID column)
Subgroups across rows of:
Lower spec: 14.50 Boundary
Upper spec: 16.50 Boundary
Historical mean: (optional)
Historical standard deviation: (optional)
Select
Help
Box-Cox...
Estimate...
Options...
Storage...
OK
Cancel

Figure 8.6

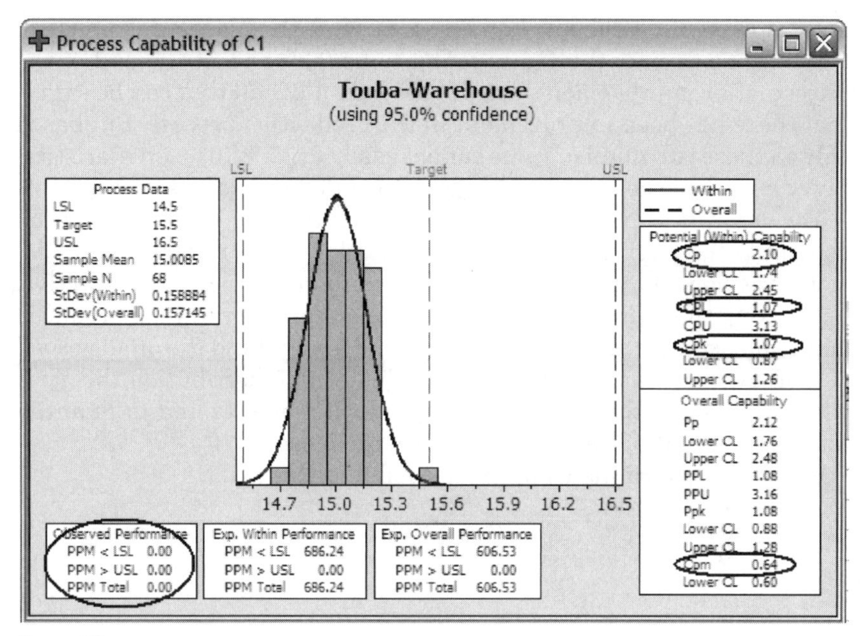

Figure 8.7

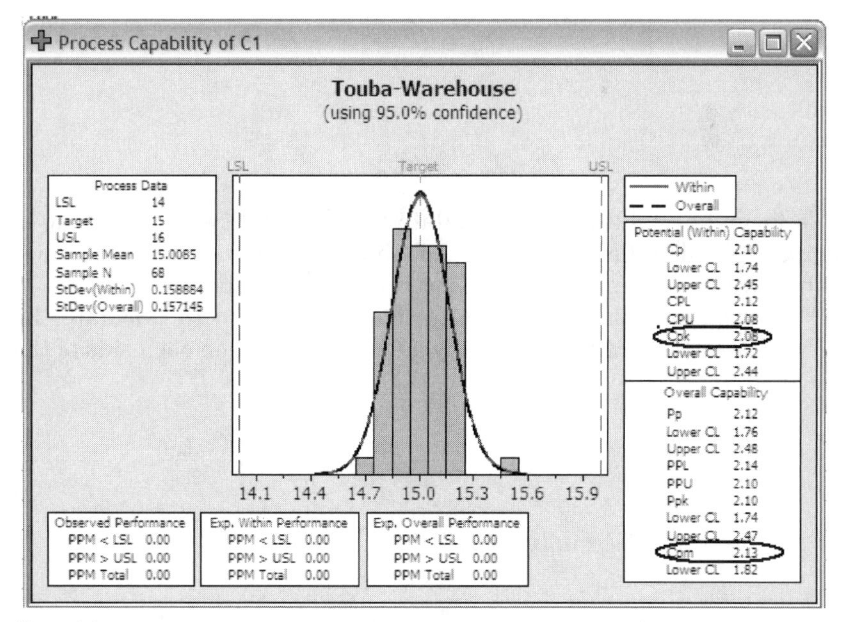

Figure 8.8

process capability indices. For instance, if at the "Measure" phase of a Six Sigma project the C_{pk} is found to be 0.89 and at the end of the project, after improvement, the C_{pk} becomes 1.32, all that can be said is that there has been improvement in the production process. But based only on these two numbers, one cannot easily explain to a non-statistics-savvy audience the amount of reduction of defects from the process.

The quantification of the *parts per million* that fall outside the specified limits can help alleviate that shortcoming. *Parts per million* (PPM) measures how many parts out of every million produced are defective. Estimating the number of defective parts out of every million produced makes it easier for anyone to visualize and understand the quality level of a production process. Here again, the normal distribution theory is used to estimate the probability of producing defects and to quantify those defects out of every million parts produced.

Recall the Z formula from the normal distribution:

$$Z = \frac{\bar{X} - \mu}{\sigma}$$

If

$$C_{pk} = \frac{\bar{\bar{X}} - LSL}{3\sigma}$$

then

$$C_{pk} = \frac{Z_{\min}}{3}$$

and

$$Z_{\min} = 3C_{pk}$$

This formula enables us to calculate the probability for an event to happen and also the cumulative probability for the event to take place if the data being considered are normally distributed.

The same formula (with minute changes) is used to calculate the PPM. The total PPM is obtained by adding the PPM on each side of the specified limits,

$$PPM_{LL} = \Omega(Z_{LL}) \times 10^6 = \Omega\left(\frac{\bar{\bar{X}} - LSL}{\sigma}\right) \times 10^6$$

for the lower specified limit and

$$PPM_{UL} = \Omega(Z_{UL}) \times 10^6 = \Omega\left(\frac{USL - \bar{\bar{X}}}{\sigma}\right) \times 10^6$$

for the upper specified limit. The quantities $\Omega(Z_{LL})$ and $\Omega(Z_{UL})$ represent the values of Z_{LL} and Z_{UL} obtained from the normal probability table.

$$PPM = PPM_{LL} + PPM_{UL}$$

There is a constant relationship between C_{pk}, Z_{min}, and PPM when the process is centered.

C_{pk}	Z_{min}	PPM
0.50	1.50	133,600
0.52	1.56	118,760
0.55	1.64	100,000
0.78	2.33	20,000
0.83	2.50	12,400
1.00	3.00	2,700
1.10	3.29	1,000
1.20	3.60	318
1.24	3.72	200
1.27	3.80	145
1.30	3.89	100
1.33	4.00	63
1.40	4.20	27
1.47	4.42	10
1.50	4.40	7.00
1.58	4.75	2.00
1.63	4.89	1.00
1.67	5.00	0.600
1.73	5.20	0.200
2.00	6.00	0.002

Example The data on the worksheet in the file *Tirethread.mpj* on the included CD measures the depth of the threads of manufactured tires. The USL and LSL are given as 5.011 mm and 4.92 mm, respectively.

a. Find the value of C_p.

b. Find the values of C_u and C_l.

c. Find the value of C_{pk}.

d. Find the number of defective parts generated for every million parts produced.

e. What is the value of the k-factor?

Solution

$$\overline{\overline{X}} = 5.00008$$

$$LSL = 4.92 \qquad USL = 5.011$$

$$\sigma = 0.0046196$$

a. The value of C_p:

$$C_p = \frac{USL - LSL}{6\sigma} = \frac{5.011 - 4.92}{6 \times 0.0046196} = \frac{0.091}{0.0277} = 3.2852$$

b. The values of C_u and C_l:

$$C_u = \frac{USL - \overline{\overline{X}}}{3\sigma} = \frac{5.011 - 5.00008}{3 \times 0.0046196} = \frac{0.01092}{0.0.138588} = 0.79$$

$$C_l = \frac{\overline{\overline{X}} - LSL}{3\sigma} = \frac{5.00008 - 4.92}{3 \times 0.0046196} = \frac{0.08008}{0.0.138588} = 5.7782$$

c. The value of C_{pk}:
The values of Z_{UL} and Z_{LL} are obtained using the same method as the z-transformation for the normal distribution,

$$Z_{UL} = \frac{USL - \overline{\overline{X}}}{\sigma} = \frac{5.011 - 5.00008}{0.0046196} = 2.364$$

and

$$Z_{LL} = \frac{\overline{\overline{X}} - LSL}{\sigma} = \frac{5.00008 - 4.92}{0.0046196} = 17.335$$

$Z_{UL} > Z_{LL}$, therefore

$$C_{pk} = \frac{1}{3} Z_{UL} = \frac{2.364}{3} = 0.79$$

d. Defective parts produced per million (PPM): From the normal distribution table, we obtain

$$\Omega(Z_{LL}) = \Omega(17.335) \approx 0.5$$

The value 0.5 represents half the area under the normal curve. The equation $\Omega(Z_{LL}) = \Omega(17.335) \approx 0.5$ represents the area under the normal curve that is within the specified limits on the left side. So the area outside the specified limits on the left side should be $0.5 - 0.5 = 0$. On the right side of the specified limits, $\Omega(Z_{UL}) = \Omega(2.364) \approx 0.491$ represents the area under the curve; the area outside the curve is approximately $0.5 - 0.491 = 0.009$. So the number of PPM on the right side should be approximately $0.009 \times 10^6 \approx 9000$.

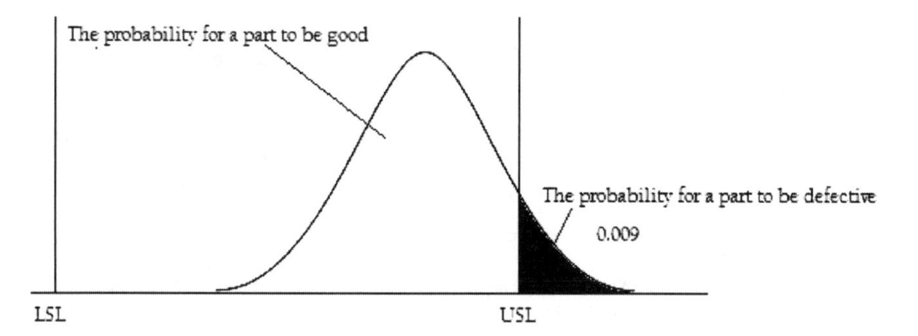

The total PPM $\approx PPM_{LL} + PPM_{UL} \approx 0 + 9000 \approx 9000$. The Minitab output for the analysis of this example is given in Figure 8.9. Note that there is a slight difference of 34.7. This difference is due to the rounding of the results.

e. The k-factor:

$$C_{pk} = (1 - k)\,C_p$$

$$k = 1 - \frac{C_{pk}}{C_p} = 1 - \frac{0.79}{3.28} = 1 - 0.241 = 0.759$$

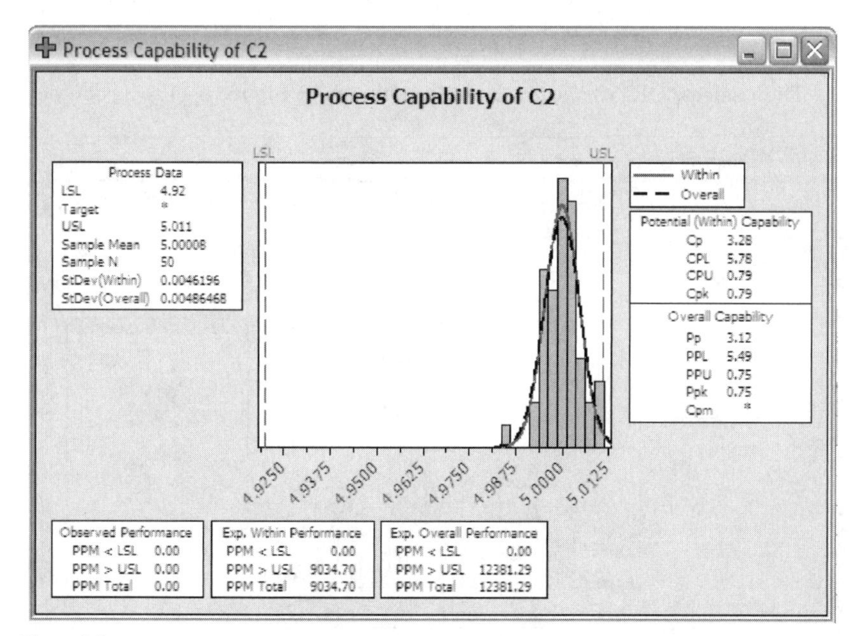

Figure 8.9

Example Using the data in the file *Fuel pump.mpj* on the included CD, find the C_{pk} given that the USL is 81.3 and the LSL is 73.9. The data are already subgrouped. Is the process capable? Is the process centered? What can we say about the long-term process capability? How many parts out of every million are defective?

Solution Using Minitab, open the file *Fuel pump.mpj* on the included CD.From the Stat menu, select "Quality Tools," select "Capability Analysis," and then select "Normal." Fill out the dialog box as shown in Figure 8.10.

Capability Analysis (Normal Distribution)

C2 Diameter

Data are arranged as

Single column: Value

Subgroup size:

[use a constant or an ID column]

Subgroups across rows of:

Diameter

Lower spec: 73.9 Boundary

Upper spec: 83.1 Boundary

Historical mean: (optional)

Historical standard deviation: (optional)

Box-Cox... Estimate... Options... Storage...

Select Help OK Cancel

Figure 8.10

Then select "OK" to obtain the output shown in Figure 8.11.

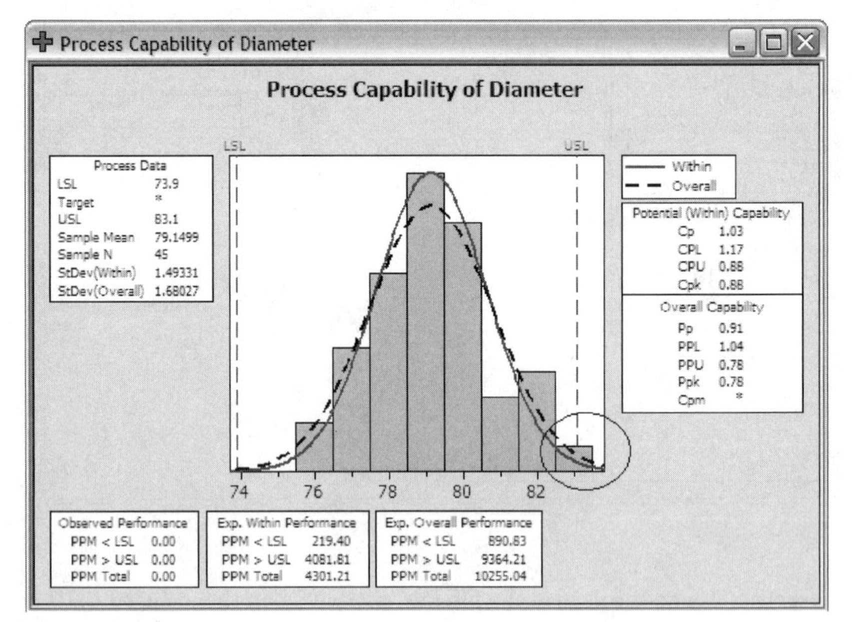

Figure 8.11

Is the process capable? The histogram represents the data being analyzed; its shape indicates that the data are normally distributed. A process is said to be capable if all the parts it produces are within the specified limits. In this case, even though $C_p = 1.03$ is greater than 1, $C_{pk} = 0.88$ is lower than 1. This result suggests that not only is the process not centered but it is also incapable. The graph shows that part of the histogram is outside the USL.

A result of $P_{pk} = 0.78$ indicates that the process must be adjusted and centered. PPM indicates how many parts are outside the specified limits for every million produced. For every one million parts produced for the *Overall Performance*, 890.83 parts will be outside the specified limits on the LSL side and 9364.21 will be outside the specified limits on the USL. The total overall PPM is the sum of the two values, which in this case is equal to 10,255.04.

8.4 Capability Sixpack for Normally Distributed Data

Two essential conditions among others were set for a process capability analysis to be valid: the production process must be in-control and stable. In addition, when selecting the type of analysis to conduct we must determine the probability distribution that the data follows. If we think that the data are normally distributed and we run a capability analysis and, unfortunately, it happens not to follow the normal distribution, the results obtained would be wrong.

Minitab's Capability Sixpack offers a way to run the test and verify if the data are normally distributed, and if the production process is stable and in-control. As its name indicates, it generates six graphs that help assess if the predetermined conditions are met. It shows the *Xbar Chart*, the *Normal Probability Plot*, the *R-Chart*, the last 25 subgroups, and a summary of the analysis.

Example The weight of a rotor is critical to the quality of an electric generator. Samples of rotors have been taken and their weight tracked on the worksheet in the file *Rotor weight.mpj*. Open the file *Rotor weight.mpj* on the included CD. From the Stat menu, select "Quality Tools," from the drop-down list, select "Capability Sixpack," and then select "Normal." A dialog box identical to the "Capability Analysis" dialog box pops up. For *Data are arranged as*, select "Single column" and select "Rotor weight." In the *Subgroup* field, enter "3." For *Lower spec*, enter "4.90" and enter "5.10" for *Upper spec*. Then select "OK" and the output box of Figure 8.12 pops up.

The *Xbar* and the *Rbar* control charts show that the production process is stable and under control. The *Normal Probability Plot* shows that the data are normally distributed, and the *Capability Histogram* shows that the process is capable, almost centered and well within specification.

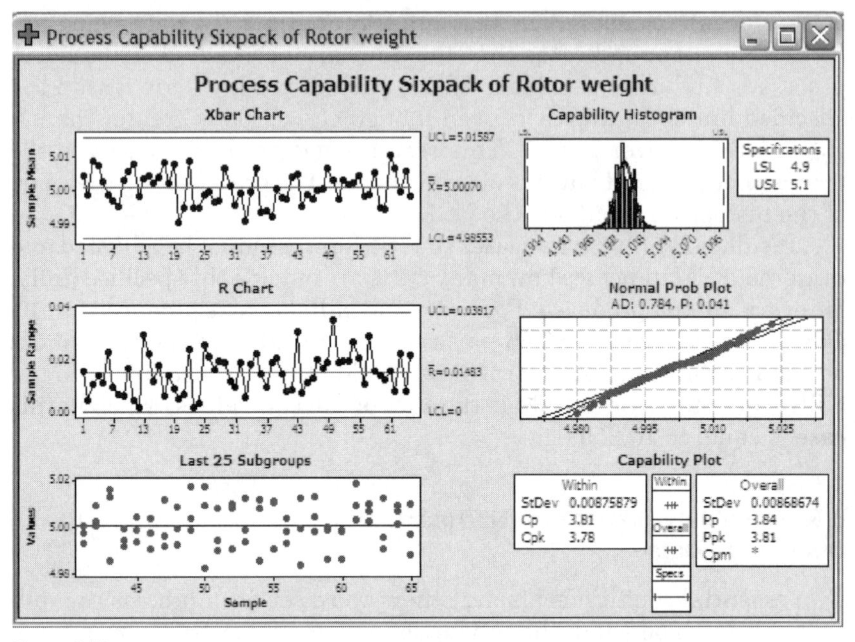

Figure 8.12

The *Capability Plot* summary shows that $C_{pk} = 3.78$ and $C_p = 3.81$. These two numbers and fairly close, which suggests that the process is almost centered.

8.5 Process Capability Analysis with Non-Normal Data

So far, one of the assumptions for a process capability analysis has been the normality of the data. The values of C_{pk}, P_{pk}, and PPM were calculated using the z-transformation, therefore assuming that the data being analyzed were normally distributed. If we elect to use normal option for process capability analysis and the normality assumption is violated because the data are skewed in one way or another, the resulting values of C_{pk}, C_p, P_p, P_{pk}, and PPM would not reflect the actual process capability.

Not all process outputs are normally distributed. For instance, the daily number of calls or the call times at a call center are generally not normally distributed unless a special event makes it so. In a distribution center where dozens of employees pick, pack, and ship products, the overall error rate at inspection is not normally distributed because it depends on a lot of factors, such as training, the mood of the pickers, the SOPs, and so on. It is advised to test the normality of the data being assessed before conducting a capability analysis.

There are several ways process capabilities can be assessed when the data are not normal:

- If the subsets that compose the data can are normal, the capabilities of the subsets can be assessed and their PPM aggregated.
- If the subsets are not normal and the data can be transformed using the Box-Cox, natural log for parametric data, or Logit transformations for binary data, transform the data before conducting the analysis.
- Use other distributions to calculate the PPM.

8.5.1 Normality assumption and Box-Cox transformation

One way to overcome the non-normal nature of the data is to through the use of the Box-Cox transformation. The Box-Cox transformation converts the observations into an approximately normal set of data.

The formula for the transformation is given as

$$T(y) = \frac{y^{\lambda} - 1}{\lambda}$$

If $\lambda = 0$, the denominator would equal zero, and to avoid that hurdle, the natural log will be used instead.

Example Transform the data included in the file *Boxcoxtrans.mpj.*

Solution Open the file *Boxcoxtrans.mpj* on the included CD. From the Stat menu, select "Control Charts" and then select "Box-Cox Transformation." In the "Box-Cox Transformation" dialog box, leave *All observations for the chart are in one column* in the first field. Select "C1" for the second textbox. Enter "1" into the *Subgroup size* field. Select "Options" and enter "C2" in *Store transformed data in:*. Select "OK" and select "OK" again.

The system should generate a second column that contains the data yielded by the transformation process. The normality of the data in column C2 can be tested using the probability plot. The graph shown in Figure 8.13 plots the data before and after transformation.

The Anderson-Darling hypothesis testing for normality shows an infinitesimal p-value of less than 0.005 for C1 (before transformation), which indicates that the data are not normally distributed. The same hypothesis testing for C2 (after transformation) shows a p-value of 0.819 and the graph clearly shows normality.

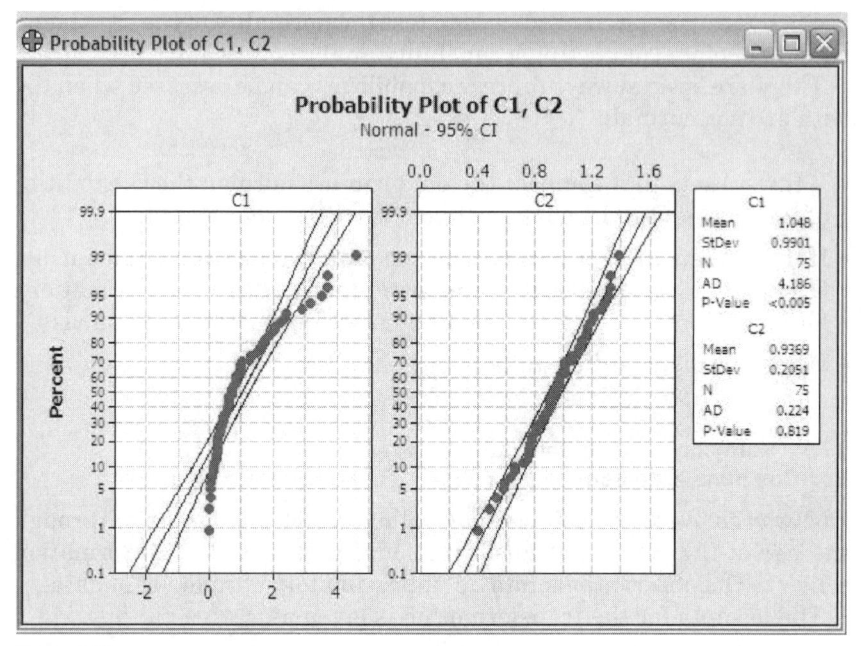

Figure 8.13

8.5.2 Process capability using Box-Cox transformation

The data in the file *Downtime.mpj* measure the time between machine breakdowns. A normality test has revealed that the data are far from being normally distributed; in fact, they follow an exponential distribution. Yet if the engineered specified limits for the downtimes are set at zero for the LSL and 25 for the USL and if we run a capability test assuming normality, we would end up with the results shown in Figure 8.14.

It is clear that no matter what type of unit of measurement is being used, the time between machine breakdowns cannot be negative. Set the LSL at zero but the PPM for the lower specification is 34,399.27 for the *Within Performance* and 74,581.98 for the *Overall Performance*. This result suggests that some machines might break down at negative units of time measurements. This is because the normal probability z-transformation was used to calculate the probability for the machine breakdowns to occur even though the distribution is exponential.

One way to correct this problem is through the transformation, the normalization of the data. For this example, we will use the Box-Cox transformation and instead of setting the lower limit at zero, we increase it to one unit of time measurement. The process of estimating

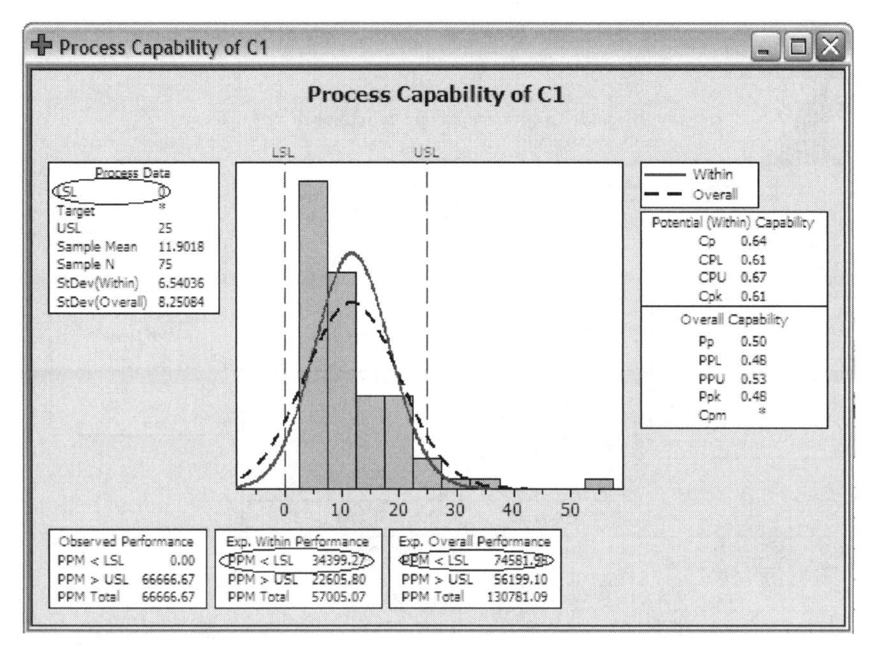

Figure 8.14

the process capabilities using Minitab is the same as the one we performed previously with the exception that we:

- Select the "Box-Cox . . . " option
- Select the "Box-Cox Power Transformation (W = Y**Lambda)" option
- Leave the option checked at *Use Optimal Lambda* and select "OK" button to obtain the output shown in Figure 8.15.

The process is still incapable, but in this case the transformation has yielded a PPM equal to zero for the lower specification. In other words, the probability for the process to generate machine breakdowns at less than zero units of measurements is zero.

Example WuroSogui Stream is a call center that processes customer complaints over the phone. The longer the customer services associates stay on the phone with the customers, the more associates will be needed to cater to the customers' needs, which would result in extra operating cost for the center. The quality control department set the specifications for the time that the associates are required to stay on the phone with the customers. They are expected to expedite the customers concerns in 10 minutes or less. So in this case, there is no lower specification and the USL is 10 minutes with a target of 5 minutes.

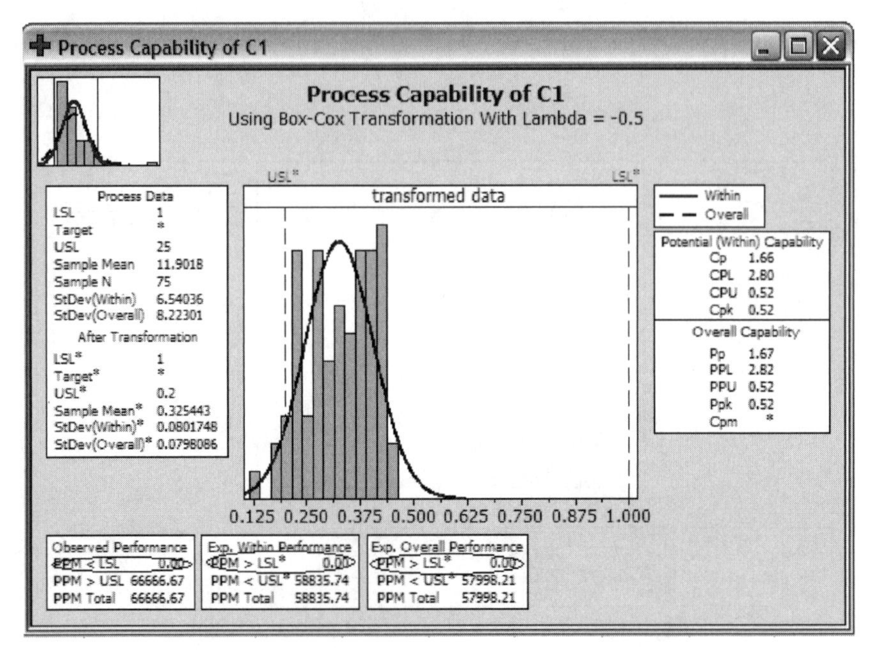

Figure 8.15

The file *Wurossogui.mpj* on the included CD contains data used to create a control chart to monitor production process at the call center.

a. What can be said about the normality of the distribution?
b. What happens if a normal process capability analysis is conducted?
c. If the data are not normally distributed, run a process capability analysis with a Box-Cox transformation.
d. Is the process capable?
e. If the organization operates under Taguchi's principles, what could we say about the process capabilities?
f. Compare C_{pk} with C_{pm}.
g. What percentage (not PPM) of the parts produced is likely to be defective for the overall performance?

Solution

a. The normality of the data can be tested in several ways. The easiest way would be through the probability plot.

   From the Graph menu, select "Probability plot." The *Single* option should be selected, so just select "OK." The "Probability Plot — Single" dialog box pops up, select "C1" for the *Graph Variable* textbox before selecting "OK." The graph in Figure 8.16 pops up.

   The graph itself shows that the data are not normally distributed for a confidence interval of 95 percent. A lot of the dots are scattered outside

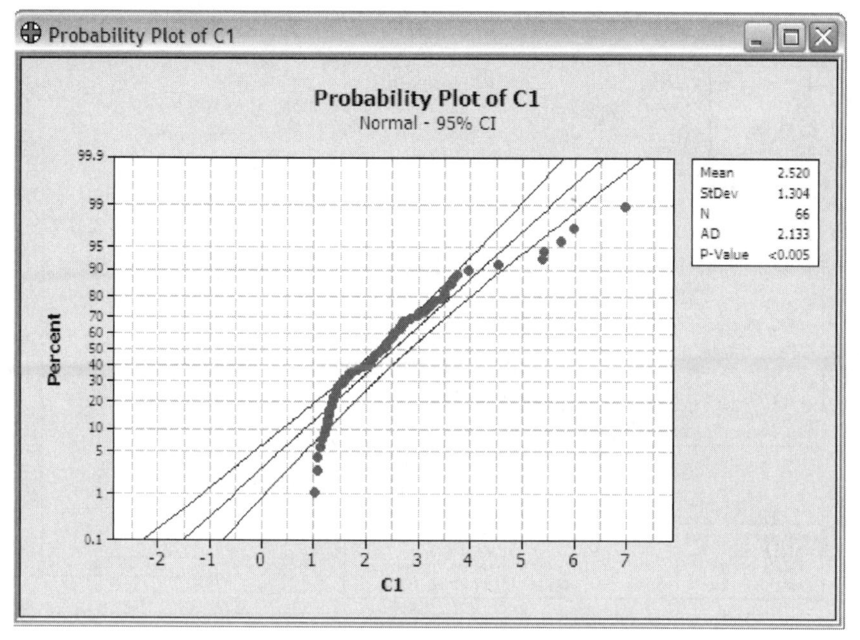

Figure 8.16

confidence limits and the Anderson-Darling null hypothesis for normality yielded an infinitesimal p-value of less than 0.005; therefore, we must conclude that the data are not normally distributed.

b. If we conduct a process normal capability analysis, we will obtain a C_{pk} and PPM that were calculated based on the normal z-transformation. Because the z-transformation cannot be used to calculate a process capability for non-normal data unless the data have been normalized, the results obtained would be misleading.

c. Open the file *Wurossogui.mpj* on the included CD. From the Stat menu, select "Quality Tools," then select "Capability Analysis" from the drop-down list, and select "Normal." Select the *Single Column* option and select "C1" for that field. For *Subgroup Size*, enter "1." Leave the *Lower Spec* field empty and enter "10" in the *Upper Spec* field. Select the *Box-Cox* option, and select the *Box-Cox Power Transformation (W = Y**Lambda)* and then select "OK." Select "Options" and enter "5" in the *Target (adds CPM to table)* field. Select the *Include Confidence Interval* option and then select "OK." Select "OK" again to obtain the graph of Figure 8.17.

d. Based on the value of $C_{pk} = 1.07$, we can conclude that the process is barely capable even though the results show opportunities for improvement.

e. If the organization operates under Taguchi's principles, we would have to conclude that the process is absolutely incapable because $C_{pm} = 0.25$, and this is because while all the observations are within the specified limits, most of them do not match the target value of 5.

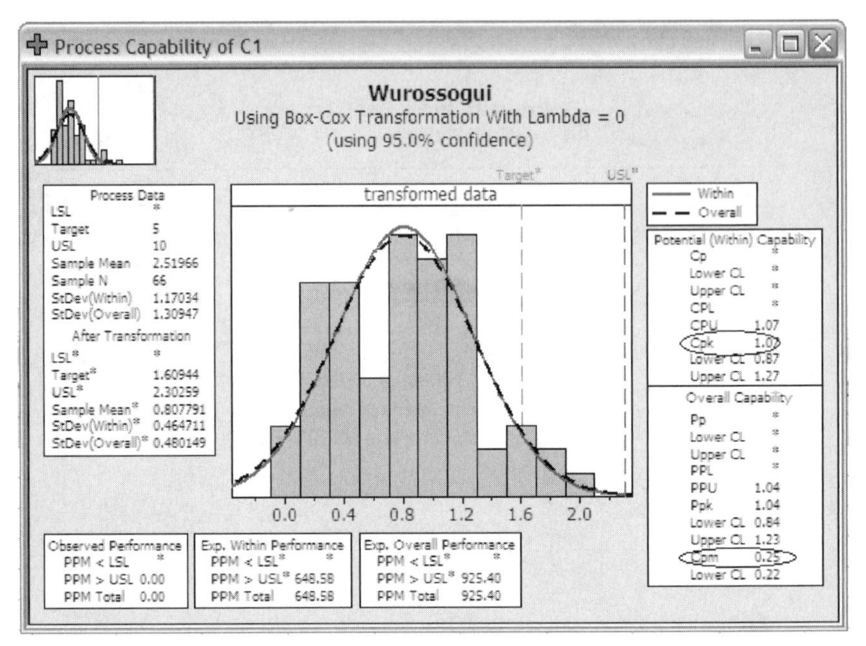

Figure 8.17

f. $C_{pk} = 1.07$ and $C_{pm} = 0.25$. The difference is explained by the fact that Taguchi's approach is very restrictive, and because most of the observations do not meet the target, the process is not considered capable.

g. For the overall performance PPM = 925.4, and the percentage of the part that are expected to be defective will be

$$925.4 \times \frac{100}{10^6} = 925.4 \times 10^{-4} = 0.09254$$

A total of 0.09254 percent of the parts are expected to be defective.

8.5.3 Process capability using a non-normal distribution

If the data being analyzed are not normally distributed, an alternative to using a transformation process to run a capability analysis as if the data were normal would be to use the probability distribution that the data actually follow. For instance, if the data being used to assess capability follow a Weibull or log-normal distribution, it is possible to run a test with Minitab. In these cases, the analysis will not be done using the z-transformation and therefore C_{pk} will not be provided because it is based on the Z formula. The values of P_p and P_{pk} are not obtained based on the mean and the standard deviation but rather on the parameters of the particular distributions that the observations follow. For instance, in the case of the Weibull distribution the shape and the scale of the observations are used to estimate the probability of the event being considered to happen.

Example Futa-Toro Electronics manufactures circuit boards. The engineered specification of the failure time of the embedded processors is no less than 45 months. Samples of circuit boards have been taken for testing and they have generated the data. The file *Futa Toro.mpj* on the included CD gives the lifetime of the processors. The observations have proved to follow a Weibull distribution. Without transforming the data, what is the expected overall capability of the process that generated the processors? What is the expected PPM?

Solution The process has only one specified limit because the lifetime of the processors is expected to last more than 45 months, so there is no upper specification. The capability analysis will be conducted using the Weibull option.

Open the file *Futa Toro.MPJ* on the included CD. From the Stat menu, select "Quality tools," from the drop-down list, select "Capability Analysis," and then select "Nonnormal ... " In the "Capability Analysis (Nonnormal Distribution)" dialog box, select "C1 Lifetime" for the *Single Column* field, select "Weibull" from the *Distribution:* drop-down list, and enter "45" in the *Lower Spec* field, leaving the *Upper Spec* field empty. Then select "OK." The graph in Figure 8.18 should pop up.

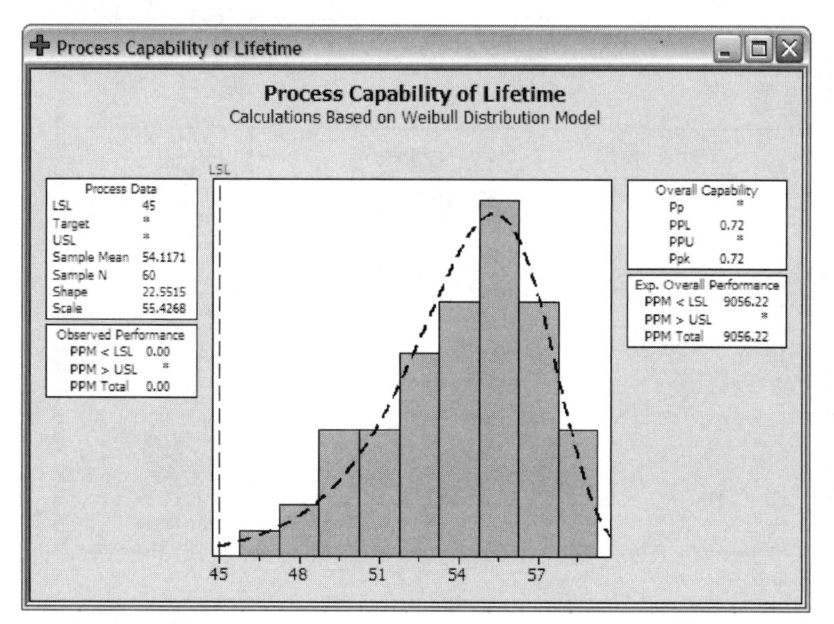

Figure 8.18

Because there is only one specified limit and it is the LSL, the P_{pk} will therefore be based solely on the PPL, which is equal to 0.72. The value of P_{pk} is much lower than the threshold, 1.33. We must conclude that the process is not capable. The expected overall PPM is 9056.2.

Example The purity level of a metal alloy produced at the Sabadola Gold Mines is critical to the quality of the metal. The engineered specifications have been set to 99.0715 percent or more. The data contained in the file *Sabadola.mpj* on the included CD represent samples taken to monitor the production process at Sabadola Gold Mines. The data have proved to have a log-normal distribution. How capable is the production process and what is the overall expected PPM?

Solution Open the file *Sabadola.mpj* on the included CD. From the Stat menu, select "Quality tools," from the drop-down list, select "Capability Analysis," and then select "Nonnormal..." In the "Capability Analysis (Nonnormal Distribution)" dialog box, select "C1" for the *Single Column* field, select "Lognormal" from the *Distribution:* drop-down list, and enter "99.0715" into the *Lower Spec* field, leaving the *Upper Spec* field empty. Then select "OK" and the graph of Figure 8.19 should pop up.

The overall capability is $P_{pk} = PPL = 1.02$, therefore the production process is barely capable and shows opportunity for improvement. The PPM yielded by such a process is 1128.17.

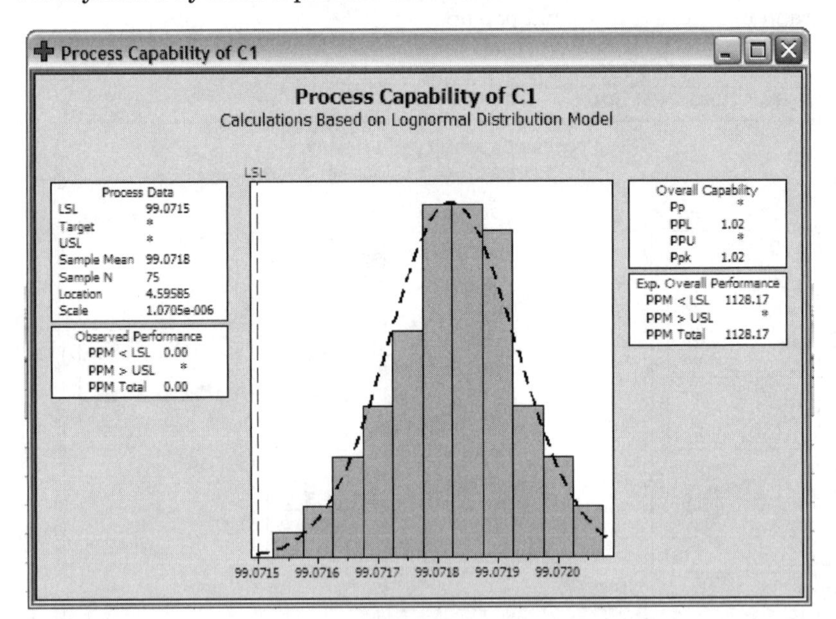

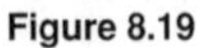
Figure 8.19

Chapter

9

Analysis of Variance

9.1 ANOVA and Hypothesis Testing

The standard error-based $\sigma/\sqrt{n}$ t-test can be used to determine if there is a difference between two population means. But what happens if we want to make an inference about more than two population means, say three or five means?

Suppose we have three methods of soldering chips on a circuit board and we want to know which one will perform better with the CPUs that we are using. We determine that the difference between the three methods depends on the amount of solder they leave on the board. A total of 21 circuit boards are used for the study, at the end of which we will determine if the methods of soldering have an impact on the heat generated by the CPU.

In this experiment, we are concerned with only one *treatment* (or *factor*), which is the amount of solder left on the circuit boards with three *levels* (the small quantity, medium quantity, or heavy quantity of solder) and the response variable, which is the heat generated by the CPU. The intensity of the factor (which values are under control and are varied by the experimenter) determines the levels.

One way to determine the best method would be to use t-tests, comparing two methods at a time: Method I will be compared to Method II, then to Method III, then Method II is compared to Method III. Not only is this procedure too long but it is also prone to multiply the Type I errors. We have seen that if $\alpha = 0.05$ for a hypothesis testing, there is a five percent chance that the null hypothesis is rejected when it is true. If multiple tests are conducted, chances are that the Type I error will be made several times.

Another way of doing it would be the *Analysis Of Variance (ANOVA)*. This method is used to pinpoint the sources of variation from one or more possible factors. It helps determine whether the variations are due to variability between or within methods. The within-method variations are variations due to individual variation within treatment groups, whereas the between-method variations are due to differences between the methods. In other words, it helps assess the sources of variation that can be linked to the independent variables and determine how those variables interact and affect the predicted variable.

The ANOVA is based on the following assumptions:

- The treatment data must be normally distributed.
- The variance must be the same for all treatments.
- All samples are randomly selected.
- All the samples are independent.

But a violation of these prerequisites does not necessarily lead to false conclusions. The probability of a Type I error will still be lower than if the different methods were compared to one another using the standard error-based $\sigma/\sqrt{n}$ *t*-test.

Analysis of variance tests the null hypothesis that all the population means are equal at a significance level α: The null hypothesis will be

$$H_0 : \mu_1 = \mu_2 = \mu_3$$

where μ_1 is the mean for Method I.

9.2 Completely Randomized Experimental Design (One-Way ANOVA)

When performing the *one-way ANOVA*, a single input factor is varied at different levels with the objective of comparing the means of replications of the experiments. This will enable us to determine the proportion of the variations of the data that are due to the factor level and the variability due to random error (within-group variation). The within-group variations are variations due to individual variation within treatment groups. The null hypothesis is rejected when the variation in the response variable is not due to random errors but to variation between treatment levels.

The variability of a set of data depends on the sum of square of the deviations,

$$\sum_{i=1}^{n}(x_1 - \overline{x})^2$$

In ANOVA, the total variance is subdivided into two independent variances: the variance due to the treatment and the variance due to random error,

$$SSk = \sum_{j=1}^{k} n_j(\overline{X}_j - \overline{X})^2$$

$$SSE = \sum_{i=1}^{n_{ij}}\sum_{j=1}^{k}(X_{ij} - \overline{X}_j)^2$$

$$TSS = \sum_{i=1}^{n_{ij}}\sum_{j=1}^{k}(X_{ij} - \overline{X})^2$$

$$TSS = SSk + SSE$$

where i is a given part of a treatment level, j is a treatment level, k is the number of treatment levels, n_j is the number of observations in a treatment level, $\overline{X}$ is the grand mean, $\overline{X}_j$ is the mean of a treatment group level, and X_{ij} is a particular observation. So the computation of the ANOVA is done through the sums of squares of the treatments, the errors, and their total.

SSk measures the variations between factors; the SSE is the sum of squares for errors and measures the within-treatment variations. These two variations (the variation between mean and the variation within samples) determine the difference between μ_1 and μ_2. A greater SSk compared to SSE indicates evidence of a difference between μ_1 and μ_2.

The rejection or nonrejection of the null hypothesis depends on the F statistic, which is based on the F probability distribution. If the calculated F value is greater than the critical F value, then the null hypothesis is rejected. So the test statistic for the null hypothesis ($H_0 : \mu_1 = \mu_2$) will be based on $F = \frac{MSk}{MSE}$, where MSk represents the mean square for the treatment and MSE represents the mean square for the error. The variable F is equal to one when MSk and MSE have the same value because both of them are estimates of the same quantity. This would

TABLE 9.1

Source of variation	Sum of squares	Degrees of freedom	Mean square	F-statistic
Between treatments	SSk	$k-1$	$MSk = SSk/(k-1)$	$F = MSk/MSE$
Error	SSE	$N-k$	$MSE = SSE/(N-k)$	
Total	TSS	$N-1$		

SSk = sum of squares between treatments
SSE = sum of squares due to error
TSS = total sum of squares
MSk = mean square for treatments
MSE = mean square for error
t = number of treatment levels
n = number of runs at a particular level
N = total number of runs
F = the calculated F statistic with $t-1$ and $N-t$ are the degrees of freedom

imply that both the means and the variances are equal, therefore the null hypothesis cannot be rejected.

These two mean squares are ratios of the sum of squares of the treatment and the sum of squares of the error to their respective degrees of freedom. The one-way ANOVA table is shown in Table 9.1.

If the calculated F value is significantly greater than the critical F value, then the null hypothesis is rejected. The critical value of F for $\alpha = 0.05$ can be obtained from the F Table (Appendix 6), which is based on the degrees of freedom between treatments and the error.

9.2.1 Degrees of freedom

The concept of *degrees of freedom* is better explained through an example. Suppose that a person has \$10 to spend on 10 different items that cost \$1 each. At first, his degree of freedom is 10 because he has the freedom to spend the \$10 however he wants, but after he has spent \$9 his degree of freedom becomes 1 because he does not have more than one choice.

The concept of degrees of freedom is widely used in statistics to derive an unbiased estimator. The degrees of freedom between treatment is $k-1$; it is the number of treatments minus one. The degrees of freedom for the error is $N-k$. The total degrees of freedom is $N-1$.

Example Suppose that we have a soap manufacturing machine that is used by employees grouped in three shifts composed of an equal number of employees. We want to know if there is a difference in productivity between the three shifts.

Had it been two shifts, we would have used the t-based hypothesis testing and determine if a difference exists, but because we have three shifts using the t-based hypothesis testing would be prone to increase the probability for making mistakes. In either case, we will formulate a hypothesis about the productivity of the three shifts before proceeding with the testing. The hypothesis for this particular case will stipulate that there is no difference between the productivity of the three groups.

The null hypothesis will be

$$H_0 : \textit{Productivity of the first shift} = \textit{productivity of second shift}$$
$$= \textit{productivity of third shift}$$

and the alternate hypothesis will be

$$H_1 : \textit{There is a difference between the productivity of at least two shifts.}$$

Some conditions must be met for the results derived from the test to be valid:

- The treatment data must be normally distributed.
- The variance must be the same for all treatments.
- All samples are randomly selected.
- All the samples are independent.

Seven samples of data have been taken for every shift and summarized in Table 9.2. What we are comparing is not the productivity by day but the productivity by shift; the days are just levels. In this case, the shifts are called *treatments*, the days are called *levels*, and the daily productivities are the *factors*.

The objective is to determine if the differences are due to random errors (individual variations within the groups) or to variations between the groups.

If the differences are due to variations between the three shifts, we reject the hypothesis. If it is due to variations within treatments, we cannot reject the hypothesis. Note that statisticians do not accept the null hypothesis—a hypothesis is either rejected or the experimenter fails to reject it

There are several ways to build the table; we will use two of them. First, we will use the previous formulas step by step.

TABLE 9.2

	First shift	Second shift	Third shift
Monday	78	77	88
Tuesday	88	75	86
Wednesday	90	80	79
Thursday	77	83	93
Friday	85	87	79
Saturday	88	90	83
Sunday	79	85	79

TABLE 9.3

	First shift	Second shift	Third shift
Monday	78	77	88
Tuesday	88	75	86
Wednesday	90	80	79
Thursday	77	83	93
Friday	85	87	79
Saturday	88	90	83
Sunday	79	85	79

First method First, calculate SSk, the sum of squares between treatments:

$$SSk = \sum_{j=1}^{k} n_j \left(\overline{X}_j - \overline{X}\right)^2$$

The table is presented under the form of

	First shift	Second shift	Third shift
Monday	a_{ii}	......	a_{ij}
Tuesday	—	—	
Wednesday	—	—	
Thursday	—		
Friday	—		
Saturday	—		
Sunday	a_{ij}		a_{ij}

with $i = 3$ and $j = 7$. $\overline{X}$ is the mean of all the observed data. It is equal to the sum of all the observations divided by 21. $\overline{X}_j$ is the mean of each treatment. For the first shift, it is equal to 83.571; for the second shift, it is 82.429; and for the third shift, it is 83.857.

$\overline{X}_j - \overline{X}$ represents the difference between the mean for each treatment and the mean of all the observations.

First shift	Second shift	Third shift
78	77	88
88	75	86
90	80	79
77	83	93
85	87	79
88	90	83
79	85	79

$\overline{X} = 83.28571$

mean$\overline{X}_j$		83.571	82.429	83.857
$\overline{X}_j - \overline{X}$		0.286	−0.857	0.571
$(\overline{X}_j - \overline{X})^2$	=	0.081632653	0.73469388	0.326530612
$\sum_{j=1}^{k} (\overline{X}_j - \overline{X})^2 =$		1.142857		$n_j = 7$

$$SSk = \sum_{j=1}^{k} n_J (\overline{X}_j - \overline{X})^2 = 7 \times 1.142857 = 8$$

The sum of squares between treatments is therefore equal to 8.

Now we will find the SSE, the sum of squares for the error:

$$SSE = \sum_{i=1}^{nj} \sum_{j=1}^{k} (X_{ij} - \overline{X}_j)^2$$

	First shift	Second shift	Third shift
	78	77	88
	88	75	86
	90	80	79
	77	83	93
	85	87	79
	88	90	83
	79	85	79
$\cdot\overline{X}_j =$	83.5714	82.4286	83.8571

Now we find the difference between each observation and its treatment mean:

$$(X_{ij} - \overline{X}_j)$$

First shift	Second shift	Third shift
−5.5714	−5,4286	4,1429
4.4286	−7,4286	2,1429
6.4286	−2,4286	4,8571
−6.5714	0,5714	9,1429
1.4286	4,5714	4,8571
4.4286	7,5714	−0,8571
−4.5714	2,5714	−4,8571

The next step will consist of finding the square of the data:

$$(X_{ij} - \overline{X}_j)^2$$

First shift	Second shift	Third shift
31.0408	29.4694	17.1633
19.6122	55.1837	4.5918
41.3265	5.8980	23.5918
43.1837	0.3265	83.5918
2.0408	20.8980	23.5918
19.6122	57.3265	0.7347
20.8980	6.6122	23.5918

$$SSE = \sum_{i=1}^{nj} \sum_{j=1}^{k} (\overline{X}_{ij} - \overline{X}_j)^2 = 177.7143 + 175.7143 + 176.8571 = \underline{530.2857}$$

Now we can find the total sum of squares, TSS.

$$\sum_{i=1}^{nj} \sum_{j=1}^{k} (X_{ij} - \overline{X}_j)^2$$

	First shift	Second shift	Third shift
	78	77	88
	88	75	86
	90	80	79
	77	83	93
	85	87	79
	88	90	83
	79	85	79
$\overline{X}_j$	83.5714	82.4286	83.8571

Recall the value of $\overline{X}$:

$$\boxed{\overline{X} = 83.28571}$$

We then subtract the value of $\overline{X}$ from every observation:

$$(X_{ij} - \overline{X})^2$$

First	Second	Third
−5.2857	−6.2857	4.7143
4.7143	−8.2857	2.7143
6.7143	−3.2857	−4.2857
−6.2857	−0.2857	9.7143
1.7143	3.7143	−4.2857
−4.7143	6.7143	−0.2857
−4.2857	1.7143	−4.2857

The next step will consist of squaring all the data. The TSS will be the sum of all the following data:

$(X_{ij} - \overline{X})^2$

First	Second	Third
27.9388	39.5102	22.2245
22.2245	68.6531	7.3673
45.0816	10.7959	18.3673
39.5102	0.0816	94.3673
2.9388	13.7959	18.3673
22.2245	45.0816	0.0816
18.3673	2.9388	18.3673

$$\text{TSS} = \sum_{i=1}^{nj} \sum_{j=1}^{k} (X_{ij} - \overline{X})^2 = 178.2857 + 180.8571 + 179.1429 = \underline{538.2857}$$

Now that we have solved the most difficult problems, we can find the degrees of freedom.

Because we have three treatments, the degrees of freedom between treatments will be two (three minus one). We have 21 factors, so the degrees of freedom for the error will be 18 (the number of factors minus the number of treatments, 21 minus 3).

The mean square for the treatment will be the ratio of the sum of squares to the degrees of freedom (8/2). The mean square for the error will be the ratio of the sum of squares for the error to its degrees of freedom (530.2857/18). The F-statistic is the ratio of the "Between Treatment" value of Table 9.4 to the error ($4/29.4603 = 0.13578$).

The F-statistic by itself does not provide grounds for rejection or nonrejection of the null hypothesis. It must be compared with the *critical* F-*value,* which is found on a separate F-table (Appendix 6). If the calculated F value is greater than the critical F value on the F-table, then the null hypothesis is rejected; if not, we cannot reject the hypothesis. In our case, from the F-table the critical value of F for $\alpha = 0.05$ with the degrees of freedom $\nu_1 = 2$ and $\nu_2 = 18$ is 3.55. Because 3.55 is greater

TABLE 9.4

Source of variation	Sum of squares	Degrees of freedom	Mean square	F-statistic
Between treatments	8	2	4	0.13578
Error	530.2857	18	29.4603	
Total	538.2857	20		

than 0.13578, we cannot reject the null hypothesis. We conclude that there is not a statistically significant difference between the means of the three shifts.

Using Minitab. Open the file *Productivity.mpj* on the included CD. From the Stat menu, select "ANOVA" and then select "One-Way-Unstacked." Select "C2," "C3," and "C4" (in separate columns) for the *Responses* text box. Select "OK" to obtain the Minitab output of Figure 9.1.

One-way ANOVA: First Shift, Second Shift, Third shift

```
Source  DF     SS    MS     F      P
Factor   2    8.0   4.0  0.14  0.874
Error   18  530.3  29.5
Total   20  538.3

S = 5.428   R-Sq = 1.49%   R-Sq(adj) = 0.00%


                                  Individual 95% CIs For Mean Based on Pooled
                                  StDev
Level          N    Mean  StDev   ---------+---------+---------+---------
First Shift    7  83.571  5.442       (--------------*-------------)
Second Shift   7  82.429  5.412   (--------------*-------------)
Third shift    7  83.857  5.429        (--------------*-------------)
                                  ---------+---------+---------+---------
                                  78.0      81.0      84.0      87.0


Pooled StDev = 5.428
```

Figure 9.1

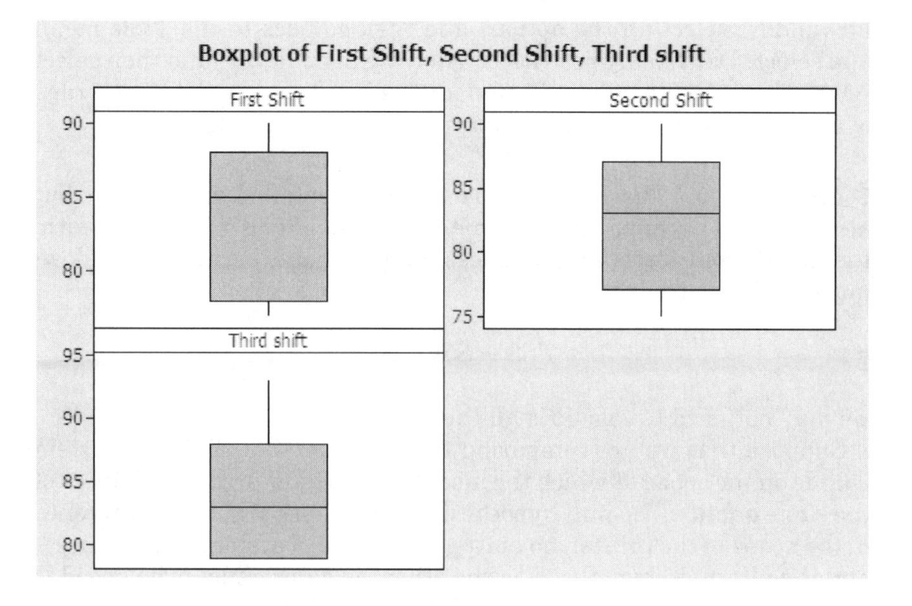

Using Excel. Had we chosen to use Excel, we would have had the table shown in Figure 9.2. To use Excel, we must have Data Analysis installed. If it is not, follow these steps: Open the file *Productivity.xls* on the included CD. From the Tools menu, select "Add- ins." On the pop

Microsoft Excel - ANOVA

File Edit View Insert Format Tools Data Window Help Adobe PDF

F22

	A	B	C	D	E	F	G
1	Anova: Single Factor						
2							
3	SUMMARY						
4	*Groups*	*Count*	*Sum*	*Average*	*Variance*		
5	Column 1	7	585	83.57142857	29.61904762		
6	Column 2	7	577	82.42857143	29.28571429		
7	Column 3	7	587	83.85714286	29.47619048		
8							
9							
10	ANOVA						
11	*Source of Variation*	*SS*	*df*	*MS*	*F*	*P-value*	*F crit*
12	Between Groups	8	2	4	0.135775862	0.873924	3.554561
13	Within Groups	530.2857143	18	29.46031746			
14							
15	Total	538.2857143	20				
16							

Figure 9.2

up window, select all the options and then go back to the Tools menu and select "Data Analysis." Select "Anova: Single factor" and then select "OK." The "ANOVA Single Factor" dialog box pops up. Select the rage of data to be inserted in the *Input range* field. Then select "OK."

Second method. The first method was very detailed and explicit but also long and perhaps cumbersome. There should be a way to perform the calculations faster and with less complications. Because the following equation is true,

$$TSS = SSk + SSE$$

we may not need to calculate all the variables.

Suppose that we are comparing the effects of four different system boards on the speed at which the model XYT printer prints out papers of the same quality. The null hypothesis is that there is not any difference in the speed of the printer, no matter what type of system board is used, and the alternate hypothesis is that there is a difference. The speed is measured in seconds and the samples shown in Table 9.5 were taken at random for each system board:

TABLE 9.5

Sys I	Sys II	Sys III	Sys IV
7	4	8	7
4	4	6	3
5	5	4	6
7	5	3	3
6	3	5	6
4	8	5	6
3	5	5	5

The sum of all the observations is 142 and the sum of the squares of the 28 observations is 780, so the *TSS* will be

$$TSS = 780 - \frac{142^2}{28} = 59.8571$$

The totals of the seven observations of the four different system boards are respectively 36, 34, 36, and 36 The sum of squares between treatments will be

$$\begin{aligned} SSk &= \frac{36^2}{7} + \frac{34^2}{7} + \frac{36^2}{7} + \frac{36^2}{7} - \frac{142^2}{28} = 720.5714281 - 720.14286 \\ &= 0.428568143 \end{aligned}$$

Now that we have the *TSS* and the *SSK*, we can find the *SSE* by just subtracting the *SSK* from the *TSS*. Therefore,

$$SSE = TSS - SSk = 59.8571 - 0.428568143 = 59.4285$$

Now that we have the *TSS*, the sum of squares between treatments (SSK), and the *SSE*, we must determine the degrees of freedom. Because we have four treatments, the degrees of freedom between treatments will be 3. We have 28 observations, so the degrees of freedom for the error will be 24–28 minus the number of treatments, which is 4. The total degrees of freedom will be 27 (24 plus 3).

The next step will be the determination of the mean squares. The mean square for treatment (*MSK*) will be the ratio of the *SSK* to its degrees of freedom.

$$\text{MSK} = \frac{SSk}{\delta f} = \frac{0.428}{3} = 0.14286$$

The *MSE* will be the ratio of the *SSE* to its degree of freedom:

$$\text{MSE} = \frac{SSE}{\delta f} = \frac{59.428}{24} = 2.476$$

Now that we have the *MSK* and the *MSE*, we can easily determine the calculated *F*-statistic:

$$\text{F-Stat} = \frac{0.14286}{2.476} = 0.05769$$

We can now put all the results into the ANOVA table, Table 9.6:

TABLE 9.6

Source of variation	Sum of squares	Degrees of freedom	Mean square	*F*-statistic	*F*-critical
Between treatments	0.4285	3	0.14286	0.05769	3.01
Error	59.4285	24	2.476		
Total	59.8571	27			

Based on the information, we cannot reject or not reject the null hypothesis until we compare the calculated *F*-statistic with the critical *F*-value found on the F table (Appendix 6). The critical *F*-value from the table for *df2* equal to 3 and *df1* equal to 24 is 3.01, which is greater than 0.05769, the calculated *F*-statistic. Therefore, we cannot reject the null hypothesis. There is not a significant difference between the system boards.

Minitab would have given the output shown in Figure 9.3.

One-way ANOVA: Sys I, Sys II, Sys III, Sys IV

```
Source  DF     SS    MS     F      P
Factor   3   0.43  0.14  0.06  0.981
Error   24  59.43  2.48
Total   27  59.86

S = 1.574   R-Sq = 0.72%   R-Sq(adj) = 0.00%


                                 Individual 95% CIs For Mean Based on
                                 Pooled StDev
Level    N   Mean  StDev  --------+---------+---------+---------+-
Sys I    7  5.143  1.574      (----------------*-----------------)
Sys II   7  4.857  1.574  (----------------*-----------------)
Sys III  7  5.143  1.574      (----------------*-----------------)
Sys IV   7  5.143  1.574      (----------------*-----------------)
                          --------+---------+---------+---------+-
                                4.20      4.90      5.60      6.30

Pooled StDev = 1.574
```

Figure 9.3

Had we used Excel, we would have had the table shown in Figure 9.4.

Example Consider the example of the solder on the circuit boards. The Table 9.7 summarizes the temperatures in degrees Celsius generated by the CPUs after a half-hour of usage.

The sum of all the observations is 1607, and the sum of the squares of the 21 observations is 123,031, so the TSS will be

$$TSS = 123031 - \frac{1607^2}{21} = 57.24$$

The sums of the seven observations of the three different methods are respectively 526, 536, and 545. The sum of squares between treatments will be

$$SSk = \frac{526^2}{7} + \frac{536^2}{7} + \frac{545^2}{7} - \frac{1607^2}{21} = 122999.6 - 122973.8 = 25.8$$

Because SSE is nothing but the difference between TSS and SSk,

$$SSE = 57.24 - 25.8 = 31.44$$

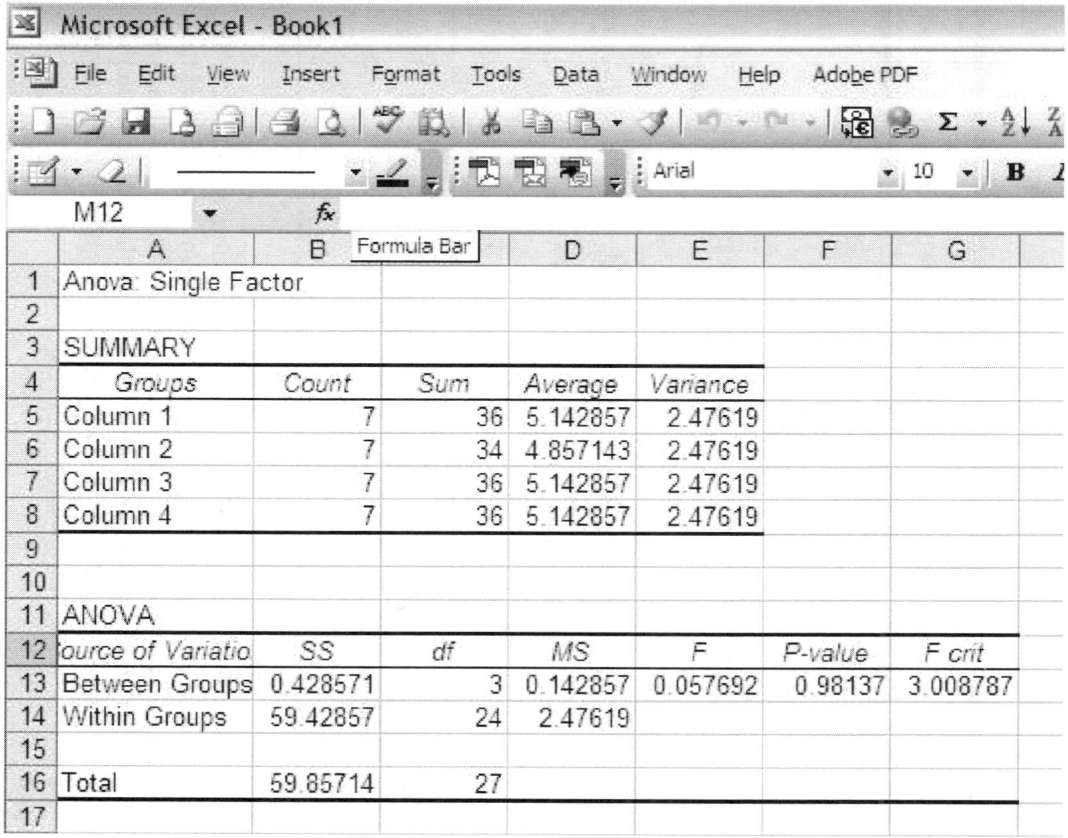

Microsoft Excel - Book1

File Edit View Insert Format Tools Data Window Help Adobe PDF

M12 fx Formula Bar

	A	B	C	D	E	F	G
1	Anova: Single Factor						
2							
3	SUMMARY						
4	*Groups*	*Count*	*Sum*	*Average*	*Variance*		
5	Column 1	7	36	5.142857	2.47619		
6	Column 2	7	34	4.857143	2.47619		
7	Column 3	7	36	5.142857	2.47619		
8	Column 4	7	36	5.142857	2.47619		
9							
10							
11	ANOVA						
12	*ource of Variatio*	*SS*	*df*	*MS*	*F*	*P-value*	*F crit*
13	Between Groups	0.428571	3	0.142857	0.057692	0.98137	3.008787
14	Within Groups	59.42857	24	2.47619			
15							
16	Total	59.85714	27				
17							

Figure 9.4

TABLE 9.7

Method I	Method II	Method III
75	76	78
74	76	79
76	75	78
75	79	76
77	75	77
76	78	78
73	77	79

We have three treatments, so the degree of freedom between treatments will be 2(3 – 1) and the total number of observations is 21; therefore, the degree of freedom for the errors will be 18(21 – 3).

$$MSk = \frac{25.8}{2} = 12.9$$

$$MSE = \frac{31.44}{18} = 1.746$$

$$F = \frac{12.9}{1.746} = 7.39$$

From the F table, the critical value of F for $\alpha = 0.05$ with the degrees of freedom $\nu_1 = 2$ and $\nu_2 = 18$ is 3.55.

We can now plot the statistics obtained in an ANOVA chart, as shown in Table 9.8.

TABLE 9.8

Source of variation	Sum of squares	Degrees of freedom	Mean square	F-statistic	F-critical
Between treatments	25.80	2	12.9	7.39	3.55
Error	31.44	18	1.746		
Total	57.24	20			

Using Excel. Open the file *CPU.xls* from the included CD. From the Tools menu, select "Data Analysis." Select "ANOVA: Single factor" from the listbox on the "Data Analysis" dialog screen and then select "OK." In the "ANOVA: Single factor" dialog box, select all the data including the labels in the *Input Range* field. Select the *Labels in First Row* option and leave *Alpha* at "0.05." Select "OK" to get the Excel output shown in Figure 9.5.

9.2.2 Multiple comparison tests

The reason we used ANOVA instead of conducting multiple pair testing was to avoid wasting too much time and, above all, to avoid multiplying the Type I errors. But after conducting the ANOVA and determining

Microsoft Excel - CPU

	A	B	C	D	E	F	G
1	Anova: Single Factor						
2							
3	SUMMARY						
4	*Groups*	*Count*	*Sum*	*Average*	*Variance*		
5	Method I	7	526	75.14286	1.809524		
6	Method II	7	536	76.57143	2.285714		
7	Method III	7	545	77.85714	1.142857		
8							
9							
10	ANOVA						
11	*rce of Varia*	*SS*	*df*	*MS*	*F*	*P-value*	*F crit*
12	Between G	25.80952	2	12.90476	7.390909	0.004537	3.554557
13	Within Gro	31.42857	18	1.746032			
14							
15	Total	57.2381	20				

Figure 9.5

that there is a difference in the means, it becomes necessary to figure out where the difference lies. To make that determination without reverting to the multiple pair analyses, we can use a technique known as *multiple comparison testing*. The multiple comparisons are made after the ANOVA has determined that there is a difference in the samples' means.

Tukey's Honestly significant difference (HSD) test. The T (Tukey) method is a pair-wise *a posteriori* test that requires an equality of the sample sizes. The purpose of the test is to determine the critical difference necessary between any two treatment levels' means to be significantly different. The T method considers the number of treatment levels, the mean square error, and the sample must be independent and be of the same size.

The HSD (for the T method) is determined by the formula:

$$\omega = q_{\alpha,t,v}\sqrt{\frac{MSE}{n}}$$

where α is the protection level covering all possible comparisons, n is the number of observation in each treatment, ν is the degree of freedom of the MSE, and t is the number of treatments. The values are computed from the q-table and two means are said to be significantly different if they differ by ω or more.

In the previous example, the degrees of freedom was 18, the number of treatments was 3 and α was equal to 0.05, which yields 3.61 from the q - table.

For $\alpha = 0.05$

		Number of populations		
Degree of freedom	2	3	4	5
10	3.15	3.88	4.33	4.65
11	3.11	3.82	4.26	4.57
12	3.08	3.77	4.2	4.51
13	3.06	3.73	4.15	4.45
14	3.03	3.7	4.11	4.41
15	3.01	3.67	4.08	4.37
16	3.00	3.65	4.05	4.33
17	2.98	3.65	4.02	4.3
18	2.97	3.61	4.00	4.28
19	2.96	3.59	3.98	4.25
20	2.95	3.58	3.96	4.23

Using the formula,

$$\omega = 3.61\sqrt{\frac{1.746}{7}} = 1.803$$

the treatment means are

For Method I	75.14286
For Method II	76.57143
For Method III	77.85714

The absolute values of the differences will be as follows:

$$|Method\ I - Method\ II| = 1.42857$$
$$|Method\ I - Method\ III| = 2.71428$$
$$|Method\ II - Method\ III| = 1.28571$$

Only the absolute value of the difference between the means of Method I and Method III is greater than 1.803, so only the means between these two methods are significantly different.

Using Minitab. Open the file *CPU.mpj* from the included CD. From the Data menu, select "Stack" and then select "Columns." Select "C1," "C2," and "C3" for the field *Stack the following columns* and then select "OK." The data should be stacked on the new worksheet. From the Stat menu, select "ANOVA" and then select "One-way." In the "One-Way Analysis of Variance" dialog box, select "C2" for *Response*. Select "Comparisons." Select the *Tukey's Family of Error Rate* option. Select "OK" and then select "OK" again. The output of Figure 9.6 should appear.

Because the computed value of $F = 7.39$ exceeds the critical value $F_{0.05,2,18} = 3.55$, we reject the null hypothesis and conclude that there

```
One-way ANOVA: C2 versus Subscripts

Source      DF     SS     MS     F      P
Subscripts   2  25.81  12.90  7.39  0.005
Error       18  31.43   1.75
Total       20  57.24

S = 1.321   R-Sq = 45.09%   R-Sq(adj) = 38.99%


                                Individual 95% CIs For Mean Based on
                                Pooled StDev
Level       N    Mean  StDev  ------+---------+---------+---------+---
Method I    7  75.143  1.345  (------*------)
Method II   7  76.571  1.512          (------*------)
Method III  7  77.857  1.069                  (------*------)
                              ------+---------+---------+---------+---
                                  75.0      76.5      78.0      79.5

Pooled StDev = 1.321


Tukey 95% Simultaneous Confidence Intervals
All Pairwise Comparisons among Levels of Subscripts

Individual confidence level = 98.00%


Subscripts = Method I subtracted from:

Subscripts   Lower  Center  Upper  -----+---------+---------+---------+----
Method II   -0.374   1.429  3.232            (--------*--------)
Method III   0.911   2.714  4.517                   (--------*--------)
                                   -----+---------+---------+---------+----
                                      -2.0       0.0       2.0       4.0


Subscripts = Method II subtracted from:

Subscripts   Lower  Center  Upper  -----+---------+---------+---------+----
Method III  -0.517   1.286  3.089            (--------*--------)
                                   -----+---------+---------+---------+----
                                      -2.0       0.0       2.0       4.0
```

Figure 9.6

is a difference in the means and that one of the methods is likely to cause the CPU to generate less heat than the other two.

9.3 Randomized Block Design

In the previous example, we only considered the three methods of soldering and concluded their difference had an impact on the heat generated by the CPU. But other factors that were not included in the analysis (such as the power supply, the heat sink, the fan, and so on) could well have influenced the results.

In *randomized block design*, these variables, referred to as *blocking variables*, are included in the experiment. Because the experimental units are not all homogeneous, homogeneous materials can be found and grouped into blocks so that the means in each block related to the treatment being considered may be compared. Because the comparisons are made within blocks, the error variation does not contain the effects of the blocks or the block-to-block variations.

If the randomized block design is to be used for the three methods of soldering, we can subdivide the 21 units into three blocks (A, B, and C), and each block will use all three methods. Each cell in Table 9.9 displays the average temperature generated by the associated method for every block.

The variables that must be considered in this experiment are two: the blocks and the treatments (the methods, in this case). So the TSS of the deviations of the predicted variable is divided into three parts:

TABLE 9.9

	Block A	Block B	Block C
Method I	75	76	73
Method II	76	77	75
Method III	77	79	79

- The sum of squares of the treatment (SST)
- The sum of squares of the blocks (SSB)
- The sum of squares of the errors (SSE)

$$TSS = SST + SSB + SSE$$

with

$$SST = n\sum_{j=1}^{t}(\overline{X}_j - \overline{X})^2$$

$$SSB = T\sum_{t=1}^{n}(\overline{X}_t - \overline{X})^2$$

$$SSE = \sum_{t=1}^{n}\sum_{j=1}^{t}(X_{ij} - \overline{X}_j - \overline{X}_t + \overline{\overline{X}})^2$$

$$TSS = \sum_{t=1}^{n}\sum_{j=1}^{t}(X_{ij} - \overline{\overline{X}})^2$$

where i is the block group, j is the treatment level, T is the number of treatment levels, n is the number of observations in each treatment level, X_{ij} is the individual observation, $\overline{X}_j$ is the treatment mean, $\overline{\overline{X}}$ is the grand mean, and N is the total number of observations.

As in the case of the completely randomized experimental design, the mean squares for the blocks, the treatments, and the errors are obtained by dividing their sums of squares by their respective degrees of freedom. The degrees of freedom for the treatments and the blocks are fairly straightforward: they will be the total number of treatments minus one and the total number of blocks minus one, respectively; the error will be the product of these two degrees of freedom:

$$df_e = (T-1)(n-1) = N - n - T + 1$$

$$MST = \frac{SST}{T-1}$$

$$MSB = \frac{SSB}{n-1}$$

$$MSE = \frac{SSE}{N-n-T+1}$$

$$F_T = \frac{MST}{MSE}$$

$$F_B = \frac{MSB}{MSE}$$

where F_T is the F-value for the treatments and F_B is the F-value for the blocks. We can summarize this information in an ANOVA table, Table 9.10:

TABLE 9.10

Source of variation	*SS*	*df*	*MS (Mean square)*	*F*
Treatment	SST	$T-1$	$SST/(T-1)$	MST/MSE
Block	SSB	$n-1$	$SSB/(n-1)$	MSB/MSE
Error	SSE	$N-n-T+1$	$SSE/(N-n-T+1)$	
Total	TSS			

The null hypothesis for the randomized block design is

$$H_0 : \mu_A = \mu_B = \mu_C$$

The F-value for the treatment is compared to the critical F-value from the table. If it is greater than the value on the table, the null hypothesis is rejected for the set α value. Use Table 9.10 as an example.

	Block A	Block B	Block C	Treatment Means
Method I	75	76	73	74.667
Method II	76	77	75	76
Method III	77	79	79	78.3333
Block means	76	77.3333	75.6667	76.33333

$$\overline{X} = 76.33333$$

$$T = 3; n = 3; N = 9$$

$$SST = n\sum_{j=1}^{t}(\overline{X}_j - \overline{X})^2 = 3[(74.667 - 76.333)^2 + (76 - 76.3333)^2$$
$$+ (78.3333 - 76.3333)^2] = 20.6667$$

$$SSB = T\sum_{i=1}^{n}(\overline{X}_i - \overline{X})^2 = 3[(76 - 76.3333)^2 + (77.3333 - 76.3333)^2$$
$$+ (75.6667 - 76.3333)^2] = 4.66667$$

$$SSE = \sum_{i=1}^{n}\sum_{j=1}^{T}(X_{ij} - \overline{X}_j - \overline{X}_i + \overline{X})^2 = 4.6667$$

$$TSS = \sum_{i=1}^{n}\sum_{j=1}^{t}(X_{ij} - \overline{X})^2 = 30$$

We can verify that

$$TSS = SSB + SSE + SST = 4.6667 + 4.6667 + 20.6667 = 30$$

Because the number of treatments and the number of blocks are equal, the degrees of freedom for the blocks and the treatments will be the same: $3 - 1 = 2$. The degree of freedom for the SSE will be

$9 - 3 - 3 + 1 = 4.$

$$MST = \frac{SST}{T-1} = \frac{20.66667}{2} = 10.333333$$

$$MSB = \frac{SSB}{n-1} = \frac{4.66667}{2} = 2.3333333$$

$$MSE = \frac{SSE}{N-n-T+1} = \frac{4.66667}{4} = 1.166667$$

$$F_T = \frac{MST}{MSE} = \frac{10.3333333}{1.1666667} = 8.86$$

$$F_B = \frac{MSB}{MSE} = \frac{2.3333333}{1.1666667} = 2$$

TABLE 9.11

Source of variation	*SS*	*df*	*MS*	*F*
Treatment	20.66667	2	10.33333	8.86
Block	4.66667	2	2.3333	2
Error	4.66667	4	1.166667	
Total	30	8		

The critical value of F obtained from the F-table is 9.28, and that value is greater than the observed value of F for treatment; therefore, the null hypothesis should not be rejected. In other words, there is not a significant difference between the means that would justify rejecting the null hypothesis.

Using Minitab. Open the file *CPUblocked.mpj* from the included CD. From the Stat menu, select ANOVA and then select "Two-Way." In the "Two-Way Analysis of Variance" dialog box, select "Response" for the *Response* field, "Method" for the *Row Factor*, and "Block" for the *Column Factor*. Then select "OK."

Two-way ANOVA: Response versus Method, Block

```
Source  DF       SS       MS     F      P
Method   2  20.6667  10.3333  8.86  0.034
Block    2   4.6667   2.3333  2.00  0.250
Error    4   4.6667   1.1667
Total    8  30.0000

S = 1.080   R-Sq = 84.44%   R-Sq(adj) = 68.89%
```

Using Excel. Open the file *CPUblocked.xls* from the included CD. From the Tools menu, select "Data Analysis." From the "Data Analysis" dialog box, select "Anova: Two Factor without Replication" and then select "OK." Select all the cells for *Input Range*. Select the *Labels* option and then select "OK."

	A	B	C	D	E	F	G
1	Anova: Two-Factor Without Replication						
2							
3	*SUMMARY*	*Count*	*Sum*	*Average*	*Variance*		
4	Method I	3	224	74.66667	2.333333		
5	Method II	3	228	76	1		
6	Method III	3	235	78.33333	1.333333		
7							
8	Block A	3	228	76	1		
9	Block B	3	232	77.33333	2.333333		
10	Block C	3	227	75.66667	9.333333		
11							
12							
13	ANOVA						
14	*Source of Variatior*	*SS*	*df*	*MS*	*F*	*P-value*	*F crit*
15	Rows	20.6666667	2	10.33333	8.857143	0.033934	6.944272
16	Columns	4.66666667	2	2.333333	2	0.25	6.944272
17	Error	4.66666667	4	1.166667			
18							
19	Total	30	8				

9.4 Analysis of Means (ANOM)

ANOVA is a good tool to determine if there is a difference between several sample means, but it does not determine from where the difference comes, if there is any difference. It does not show what samples are so disparate that the null hypothesis must be rejected. To know where the difference originates from, it is necessary to conduct further analyses after rejecting the null hypothesis. Tukey, Fisher, and Dunnett are examples of comparisons that can help situate the sources of variations between means.

A simpler way to determine if the sample means are equal and, at the same time, visually determine where the difference is coming from (if there is any) would be the *analysis of means* (ANOM). ANOM is a lot simpler and easier to conduct than ANOVA, and it provides an easy-to-interpret visual representation of the results.

When conducting ANOM, what we want to achieve is to determine the upper and lower decision limits. If all the sample means fall within these boundaries, we can say with confidence that there are no grounds to reject the null hypothesis, i.e., there are no significant differences between the samples' means. If at least one mean falls outside these limits, we reject the null hypothesis.

The upper and lower decision limits depend on several factors:

- The samples' means
- The mean of all the observed data (the mean of the samples' means)
- The standard deviation
- The alpha level
- The number of samples
- The sample sizes (to determine the degrees of freedom)

ANOM compares the natural variability of every sample mean with the mean of all the sample means.

If we have j samples and n treatment levels, then the sample means are given as

$$\bar{x}_j = \frac{\sum_{i=1}^{n} x_i}{n}$$

and the mean of all the sample means is

$$\bar{\bar{x}} = \frac{\sum \bar{x}_j}{j}$$

Call N the number of all observed data and s the standard deviation. Then the variance for the treatments would be

$$s_i^2 = \frac{\sum_{i=1}^{n} (x_i - \bar{x})^2}{n - 1}$$

the overall standard deviation would be

$$s = \sqrt{\frac{\sum_{i=1}^{n} s_i^2}{j}}$$

and the upper and lower decision limits would be

$$UDL = \bar{\bar{x}} + h_o 3\sqrt{\frac{j-1}{N}}$$

$$LDL = \bar{\bar{x}} - h_o 3\sqrt{\frac{j-1}{N}}$$

where α represents the significance level

TABLE 9.12

	First shift	Second shift	Third shift
Monday	78	77	88
Tuesday	88	75	86
Wednesday	90	80	79
Thursday	77	83	93
Friday	85	87	79
Saturday	88	90	83
Sunday	79	85	79

In our previous example for ANOVA, we wanted to know if there was a difference between the productivity of the three shifts. After conducting the test, we concluded that there was not a significant difference between them. Take the same example again and this time, use the ANOM. Unfortunately, Excel does not have the capabilities to conduct ANOM, so we will use only Minitab.

Using Minitab, we need to first stack the data. Open the file *Productivity.mpj* from the included CD. From the Data menu, select "Stack" and then select "Columns." Select "First shift," "Second shift," and "Third shift" for the field *Stack the Following Columns*. Then select "OK" and a new worksheet appears with the stacked data.

Now that we have the stacked data, we can conduct the ANOM. From the Stat menu, select "ANOVA" and then "Analysis of Means." For

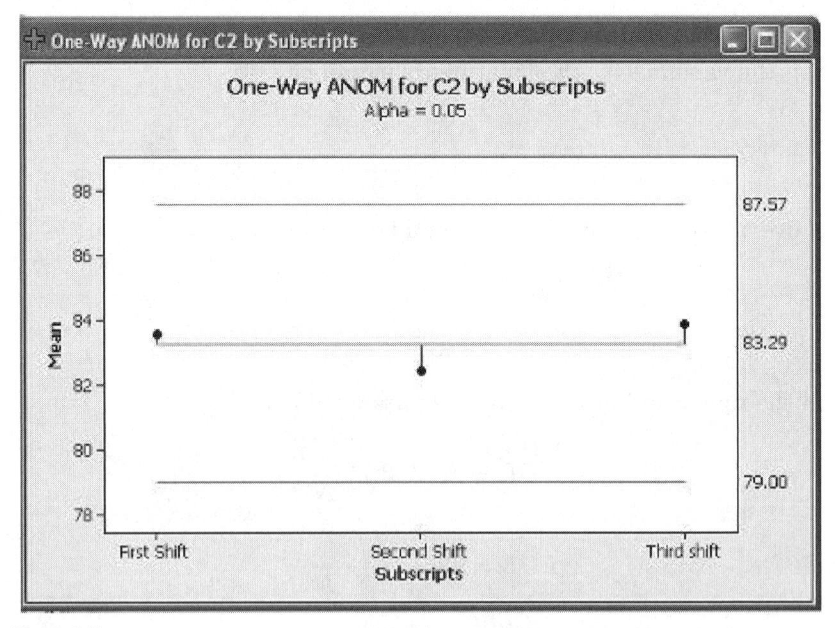

Figure 9.7

Response, select "C2" because that is where the data we are looking for resides, and then double-click on "Subscripts" for *Factor 1* if the *Normal* option is selected. Remember that "Subscripts" is the default column title for the treatments titles. The default for the *Alpha* level is "0.05." We can change this value, but for the sake of this example, leave it as it is. When we select "OK," the graph of Figure 9.7 pops up.

Because all the points are within the decision boundaries, we conclude that there is not enough evidence to reject the null hypothesis. The difference between the three means is insignificant at an alpha level of 0.05.

This is the same conclusion we reached when we conducted an Analysis Of Variance with the same data.

Exercise. Complete Table 9.13.

TABLE 9.13

ANOVA						
Source of variation	*SS*	*df*	*MS*	*F*	*P-value*	*F-critical*
Between groups		4			0.465027	2.412682
Within groups	190.83532		0.867433			
Total	193.95656	224				

Exercise. Using the data in Table 9.14, show that the null hypothesis should not be rejected at an alpha level equal to 0.05. The same data are contained in the files *Rooftile.xls* and *Rooftile.mpj* on the included CD. Compare the ANOM results to the one-way ANOVA.

TABLE 9.14

71.7923	70.5991	70.4748	68.2488
71.0687	70.7309	70.3751	68.5724
68.9859	69.0848	68.4265	68.2465
67.6239	68.6249	70.4857	68.9361
67.9830	70.1810	69.2576	72.0380
69.5726	71.6446	67.6044	72.2734
72.4664	68.9734	69.9449	67.1732
69.5111	68.2361	71.5813	71.5326
69.3777	71.5434	68.8229	72.1982
72.5865	71.9704	69.8774	68.4985

Exercise. Open the files *Machineheat.xls* and *Machineheat.mpj* from the included CD and run a one-way ANOVA. Run an ANOM. Should the null hypothesis be rejected? What can be said of the normality of the data?

Chapter

10

Regression Analysis

Learning Objectives:

- Build a mathematical model that shows the relationship between several quantitative variables
- Identify and select significant variables for model building
- Determine the significance of the variables in the model
- Use the model to make predictions
- Measure the strength of the relationship between quantitative variables
- Determine what proportion in the change of one variable is explained by changes in another variable

A good and reliable business decision-making process is always founded on a clear understanding on how a change in one variable can affect all the other variables that are in one way or another associated with it.

- How would the volume of sales react if the budget of the marketing department is cut in half?
- How does the quality level of the products affect the volume of returned goods?
- Does an increase in the R&D budget necessarily lead to an increase the price the customers must pay for our products?
- How do changes in the attributes of a given product affect its sales?

Regression analysis is the part of statistics that analyzes the relationship between quantitative variables. It helps predict the changes in a response variable when the value of a related input variable changes.

The objective here is to determine how the predicted or *dependent variable* (the response variable, the variable to be estimated) reacts to the variations of the predicator or *independent variable* (the variable that explains the change). The first step should be to determine whether there is any relationship between the independent and dependent variables, and if there is any, how important it is.

The covariance, the correlation coefficient, and the coefficient of determination can determine that relationship and its level of importance. But these alone cannot help make accurate predictions on how variations in the independent variables impact the response variables. The objective of regression analysis is to build a mathematical model that will help make predictions about the impact of variable variations.

It is obvious that in most cases, there is more than one independent variable that can cause the variations of a dependent variable. For instance, there is more than one factor that can explain the changes in the volume of cars sold by a given car maker. Among other factors, we can name the price of the cars, the gas mileage, the warranty, the comfort, the reliability, the population growth, the competing companies, and so on. But the importance of all these factors in the variation of the dependent variable (the number of cars sold) is disproportional. So in some cases, it is more beneficial to concentrate on one important factor instead of analyzing all the competing factors.

When building a regression model, if more than one independent variable is being considered, we call it a *multiple regression analysis*, if only one independent variable is being considered, the analysis is a *simple linear regression.* In our quest for that model, we will start with the model that enables us to find the relatedness between two variables.

10.1 Building a Model with Only Two Variables: Simple Linear Regression

Simple regression analysis is a bivariate regression because it involves only two variables: the independent and the dependent variables. The model that we will attempt to build will be a *simple linear equation* that will show the relationship between the two variables. We will attempt to build a model that will enable us to predict the volume of defective batteries returned by customers when the in-house quality failure rate varies.

Six Sigma case study

Project background. For several years, the *in-house failure rate* (IHFR) has been used by the Quality Control department of Dakar Automotive to estimate the projected defective batteries sent to customers. For instance, if after auditing the batteries before they are shipped, two percent of the samples taken (the samples taken are three percent of the total shipment) fail to pass audit and 7000 batteries are shipped, the Quality Control department would estimate the projected defective batteries sent to customers to be 140 (two percent of 7000).

The *projected volume of defective* (PVD) products sent to customers is a metric used by the Customer Services division to estimate the volume of calls expected from the customers and for the planning of the financial and human resources to satisfactorily answer the customers' calls. The same metric is used by Returned Goods department to estimate the volume of returned products from the customers to estimate the volume of the necessary new products for replacement and the financial resources for refunds.

Yet there has historically always been a big discrepancy between the volume of customer complaints and the PVD products that Quality Control sends to Customer Services. This situation has caused the Customer Services department to have difficulties planning their resources to face the expected call volume from unhappy customers.

Upper management of Dakar Automotive initiated a Six Sigma project to investigate the relevance of IHQF as a metric to estimate the PVD products sent to customers. If it is a relevant metric, the Black Belt is expected to find a way to better align it to the Customer Services' needs; if not, he is expected to determine a better metric for the scorecard.

Project Execution. In the "Analyze" phase of the project, the Black Belt decides to build a model that will help determine if the expected number of batteries returned by the customers is indeed related to failure rate changes. To build his regression model, he considers a sample of 14 days of operations. He tabulates the proportions of the returned batteries and the failure rate before the batteries were shipped and obtains the data shown in Table 10.1. The table can be found in *ReturnAccuracy.xls* and *ReturnAccuracy.mpj* on the included CD.

In this case, "Return" is the y variable; it is supposed to be explained by "Accuracy," which is the x variable. The equation that expresses the relationship between x and y will be under the form of

$$\hat{y} = f(x)$$

10.1.1 Plotting the combination of x and y to visualize the relationship: scatter plot

We can preempt the results of the regression analysis by using a graph that plots the relationship between the xs and the ys. A scatter plot can help visualize the relationship between the two variables.

TABLE 10.1

Return	Accuracy
0.050	0.9980
0.050	0.9970
0.004	0.9950
0.003	0.9960
0.006	0.9900
0.060	0.9970
0.009	0.9905
0.010	0.9980
0.050	0.9907
0.004	0.9990
0.050	0.9951
0.040	0.9980
0.005	0.9980
0.005	0.9970

Using Minitab. After pasting Table 10.1 into a Minitab Worksheet, from the Graph menu select "Scatterplot . . ."

The dialog box of Figure 10.1 pops up and select the *With Regression* option.

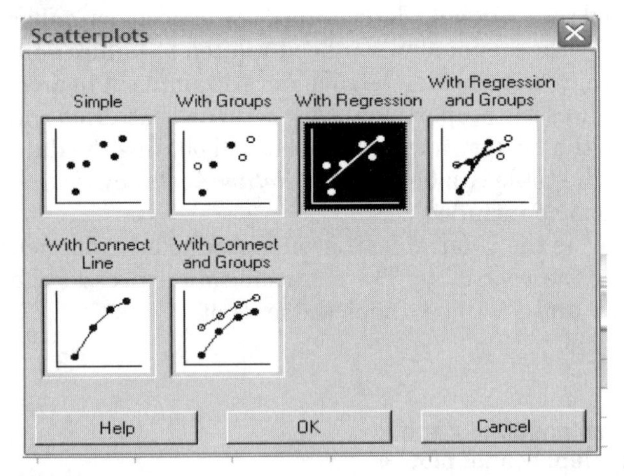

Figure 10.1

In the "Scatterplot" dialog box, enter "Return" and "Accuracy" in the appropriate fields and select "OK."

Scatterplot - With Regression

C1 Return
C2 Accuracy

	Y variables	X variables
1	Return	Accuracy
2		
3		
4		
5		
6		
7		

Scale... Labels... Data View...
Multiple Graphs... Data Options...
Select
Help OK Cancel

Every point on the graph represents a vector of x and y.

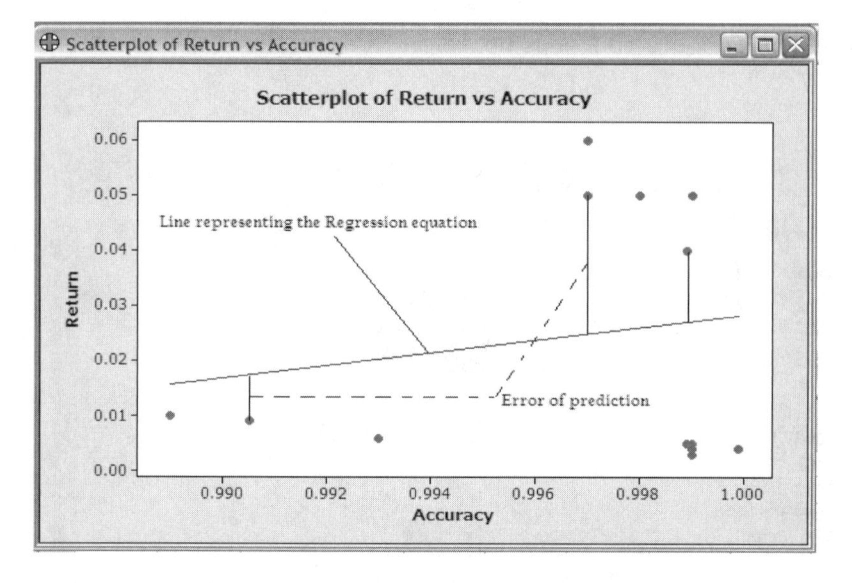

By looking at the spread of the points on the graph, we can conclude that an increase of the accuracy rate does not necessarily lead to an increase or a decrease in the return rate.

The equation we are about to derive from the data will determine the line that passes through the points that represent the vector "Accuracy-Return." The vertical distance between the line and each point is called the *error of prediction*. An unlimited number of lines could be plotted between the points, but there is only one regression line and it would be the one that minimizes the distance between the points and the line.

Using Excel. We can use Excel to not only plot the vectors but also add the equation of the regression line and the coefficient of determination. Select the fields you want to plot and from the **Insert** menu, select "**Chart...**"

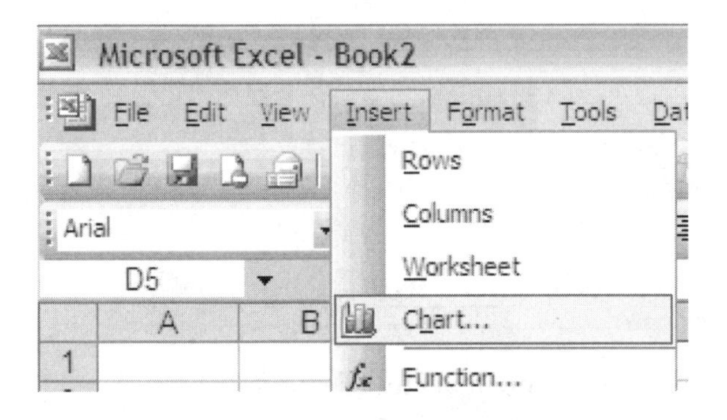

The box of Figure 10.2 pops up, select "XY (Scatter)," and then select "Next >."

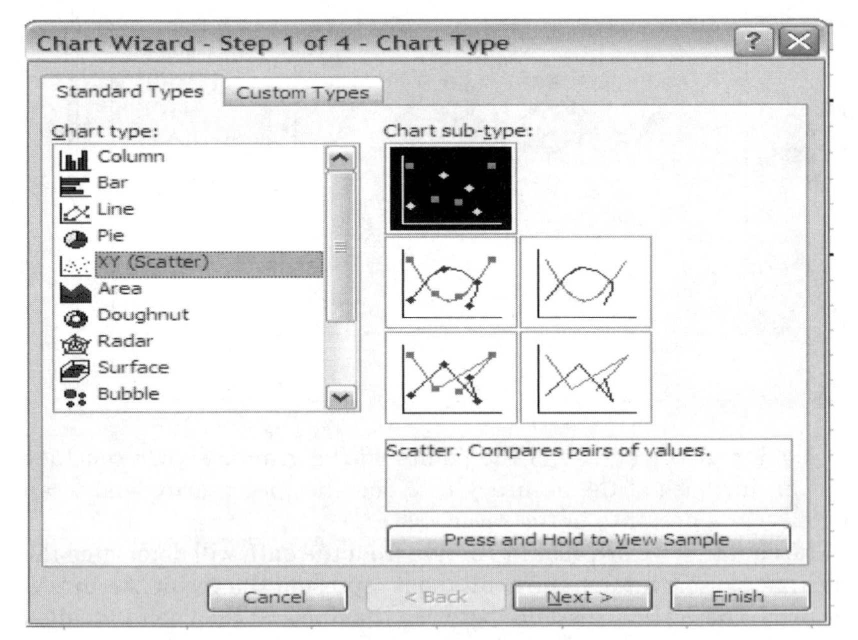

Figure 10.2

The scatter plot appears with a grid:

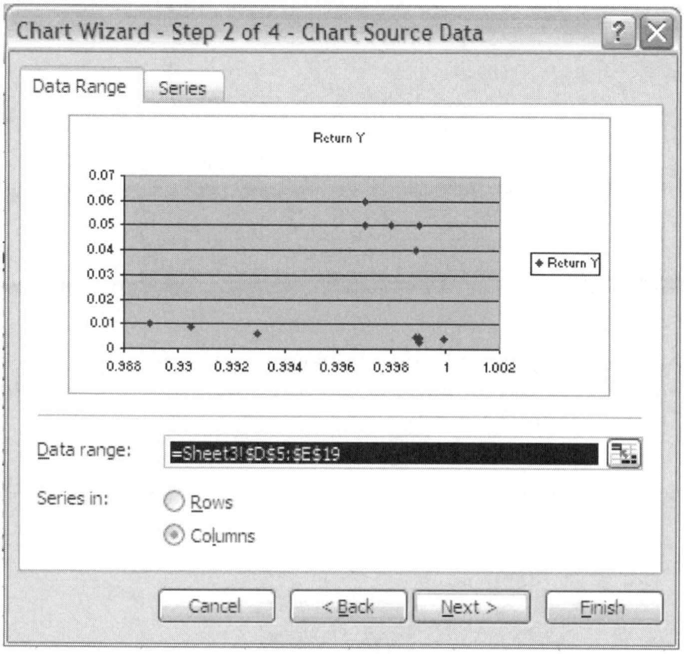

To remove the grid, select "Gridlines" and uncheck all the options.

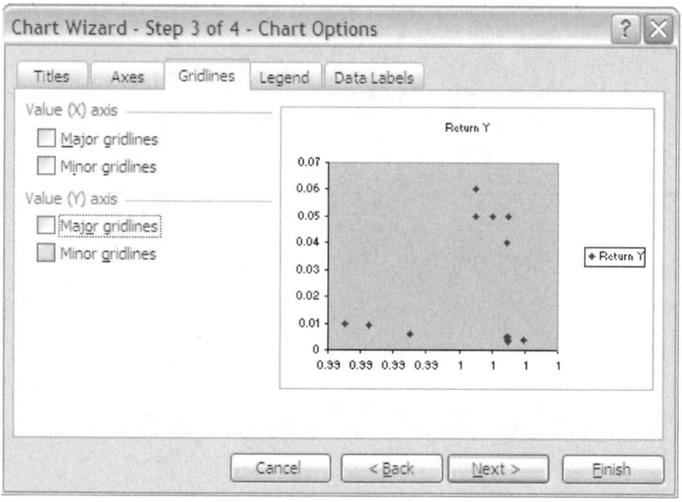

Then, select "Next >."

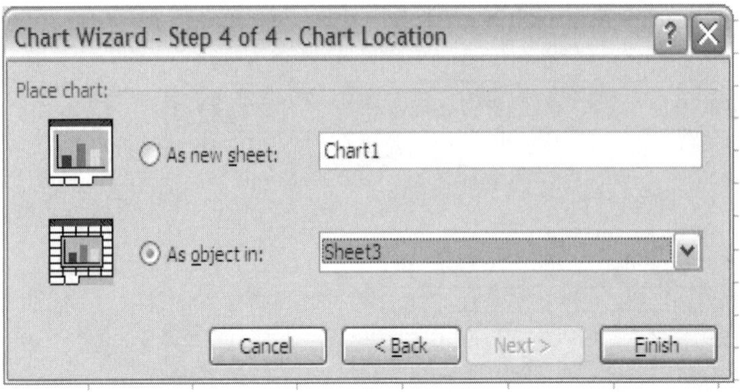

The plotted surface appears without the regression line. To add the line along with its equation, right-click on any point and select "Add Trendline. . . " from the drop-down list.

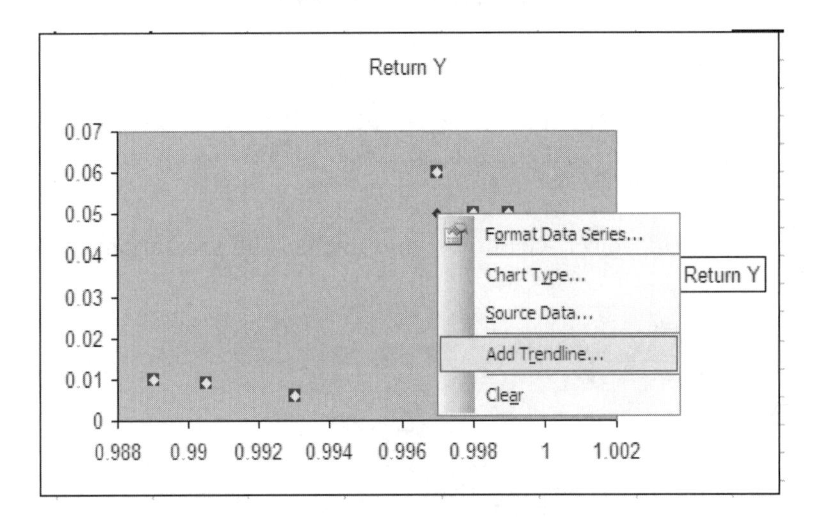

Then, select the "Linear" option.

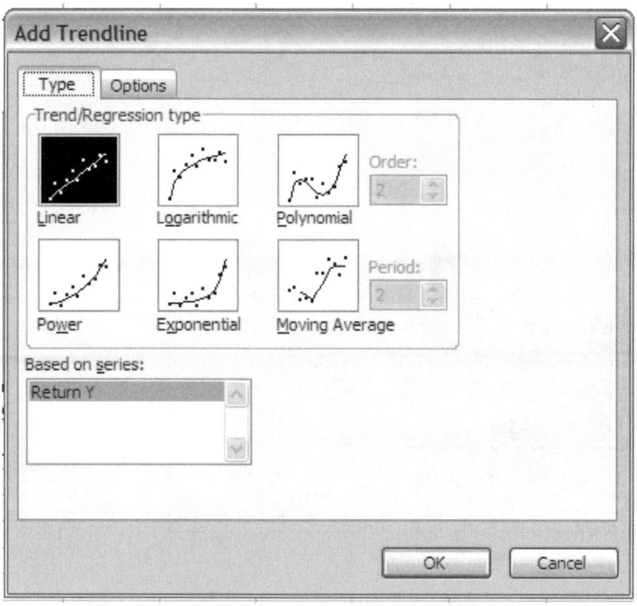

Then select the "Options" tab and select the options "Display equation on chart" and "Display R-squared on chart."

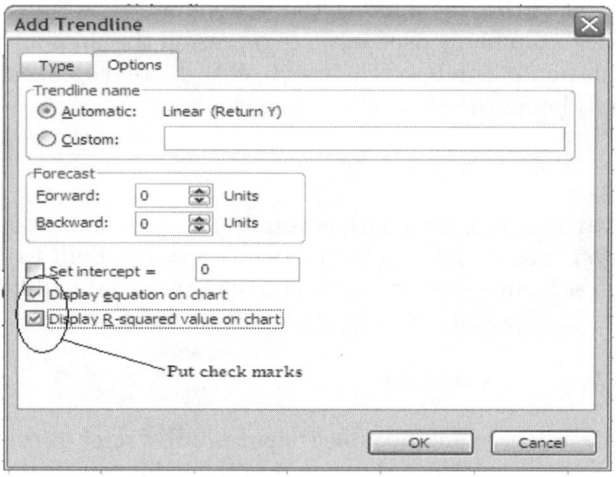

Select "OK" and the chart appears with the regression equation.

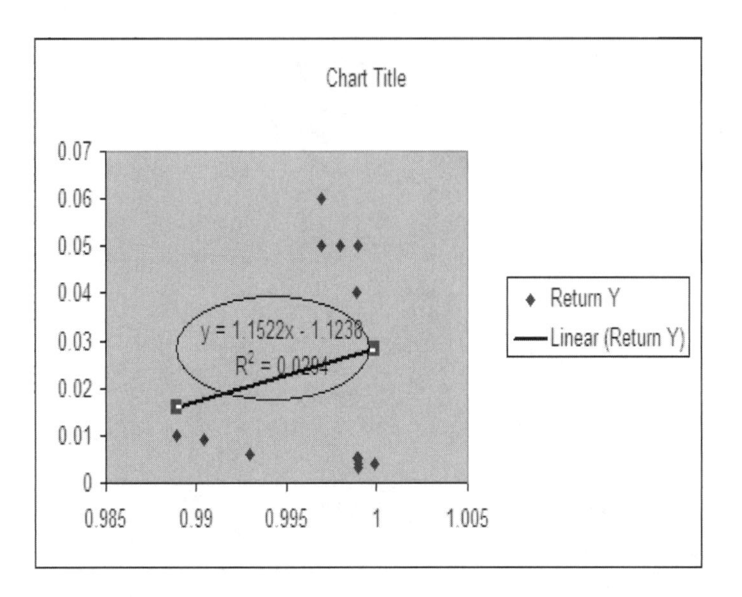

10.1.2 The regression equation

The regression equation $\hat{y} = f(x)$ that we are looking for will be a first degree polynomial function under the form of $\hat{y} = ax + b$, and it will yield two points of interest: the slope of the line and the y-intercept. The value a is the slope of the line and b is the y-intercept.

In statistics, the most commonly used letter to represent the slope and intercept for a population is the Greek letter β. With β_0 representing the y-intercept and β_1 being the slope of the line, we have

$$\hat{y} = \beta_1 x + \beta_0$$

If the independent variable is known with certainty and only that variable can affect the response variable $\hat{y}$, the model that will be built will generate an exact predictable output. In that case, the model will be called a *deterministic model* and it will be under the form of:

$$\hat{y} = \beta_1 x + \beta_0$$

But in most cases, the independent variable is not the only factor affecting y, so the value of $\hat{y}$ will not always be equal to the value generated by the equation for a given x. This is why an error term is added to the deterministic model to take into account the uncertainty. The equation for the probabilistic model is:

$$\hat{y} = \beta_1 x + \beta_0 + \varepsilon$$

for a population, or

$$\hat{y} = b_1 x + b_0 + \varepsilon$$

for a sample, where ε represents the error term.

10.1.3 Least squares method

To determine the equation of the model, what we are looking for are the values of b_1 and b_0. The method that will be used for that purpose is called the *least squares method*. As mentioned earlier, the vertical distance between each point and the line is called the error of prediction. The line that generates the smallest error of predictions will be the least squares regression line.

The values of b_1 and b_0 are obtained from the following formula:

$$b_1 = \frac{\sum_{t=1}^{n}(x_t - \overline{x})(y_t - \overline{y})}{\sum_{t=1}^{n}(x_t - \overline{x})^2}$$

In other words,

$$b_1 = \frac{\sum xy - \frac{\left(\sum x\right)\left(\sum y\right)}{n}}{\sum x^2 - \frac{\left(\sum x\right)^2}{n}}$$

$$b_1 = \frac{SS_{xy}}{SS_{xx}}$$

The value of b_1 can be rewritten as:

$$b_1 = \frac{\mathrm{cov}(X, Y)}{S_x^2}$$

The y-intercept b_0 is obtained from the following equation:

$$b_0 = \overline{Y} - b_1\overline{X}$$

Now that we have the formula for the parameters of the equation, we can build the Return-Accuracy model.

We will need to add a few columns to the two that we had. Remember that y is the response variable, in this case "Return," and x is the independent variable, "Accuracy."

	Return	Accuracy	$(x-\overline{x})$	$(y-\overline{y})$	$(x-\overline{x})(y-\overline{y})$	$(x-\overline{x})^2$
	0.05	0.998	0.0012	0.025285714	0.000030343	0.0000014
	0.05	0.997	0.0002	0.025285714	0.000005057	0.0000000
	0.004	0.999	0.0022	−0.02071429	−0.000045571	0.0000048
	0.003	0.999	0.0022	−0.02171429	−0.000047771	0.0000048
	0.006	0.993	−0.0038	−0.01871429	0.000071114	0.0000144
	0.06	0.997	0.0002	0.035285714	0.000007057	0.0000000
	0.009	0.9905	−0.0063	−0.01571429	0.000099000	0.0000397
	0.01	0.989	−0.0078	−0.01471429	0.000114771	0.0000608
	0.05	0.997	0.0002	0.025285714	0.000005057	0.0000000
	0.004	0.9999	0.0031	−0.02071429	−0.000064214	0.0000096
	0.05	0.999	0.0022	0.025285714	0.000055629	0.0000048
	0.04	0.9989	0.0021	0.015285714	0.000032100	0.0000044
	0.005	0.999	0.0022	−0.01971429	−0.000043371	0.0000048
	0.005	0.9989	0.0021	−0.01971429	−0.000041400	0.0000044
Mean	0.024714	0.9968		Totals	0.000177800	0.0001543

$$\overline{x} = 0.9968$$

$$\overline{y} = 0.024714$$

$$b_1 = \frac{0.0001778}{0.0001543} = 1.152151$$

$$b_0 = 0.024714 - (1.152151 \times 0.9968) = -1.12375$$

For a deterministic model,

$$\hat{y} = 1.152151x - 1.12375$$

or

$$Return = (1.1521 \times Accuracy) - 1.12375$$

To determine the expected return, all we need to do is replace "Accuracy" by a given value.

Assumptions for least squares regression. For the least squares regression analysis to be reliable for prediction, it must fit the following assumptions:

- The error term ε has a constant variance.
- At each value of x, the error terms ε follow the normal distribution.
- The model is linear.
- At each possible value of x, the error terms are independent.

Using Minitab to find the regression equation. After pasting the data into the worksheet, from the Stat menu, select "Regression" and then "Regression. . . " again.

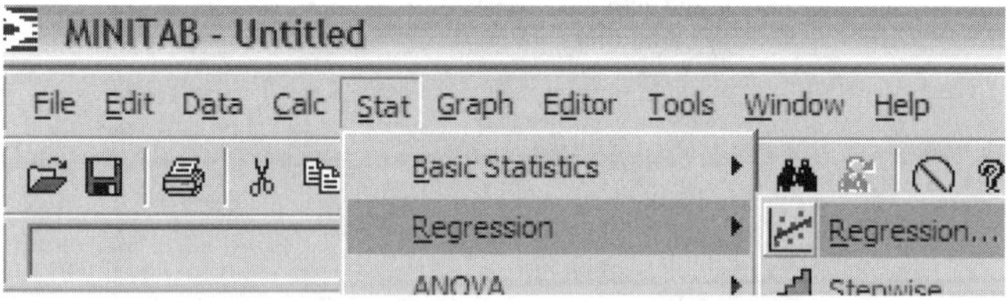

The dialog box of Figure 10.3 should appear.

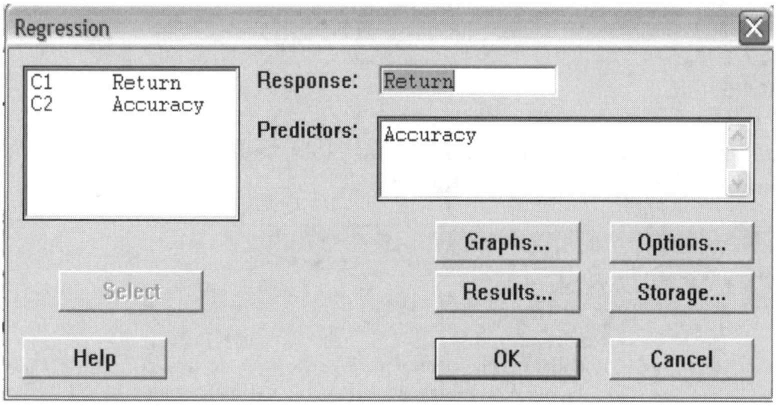

Figure 10.3

Select "Return" for the *Response* box and "Accuracy" for the *Predictors*. Then select "Graph. . . " and the dialog box of Figure 10.4 appears.

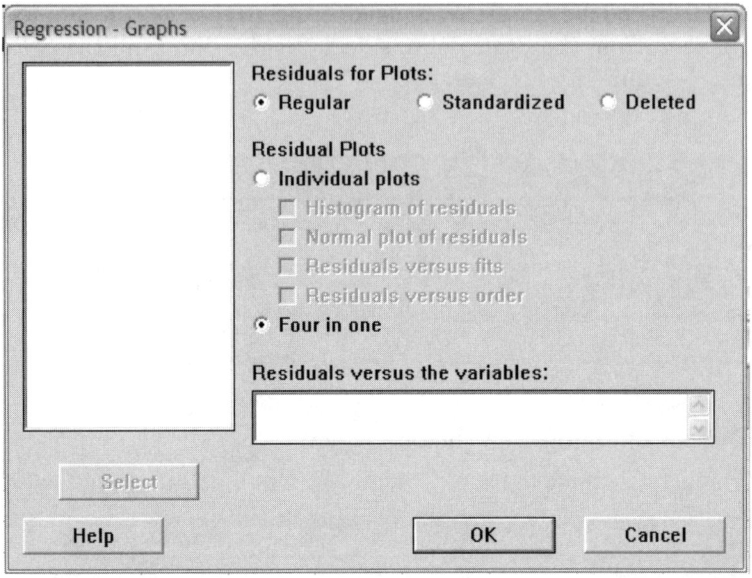

Figure 10.4

Because we want to have all the residual plots, select "Four in one" and then select "OK." Then select "Options. . ."

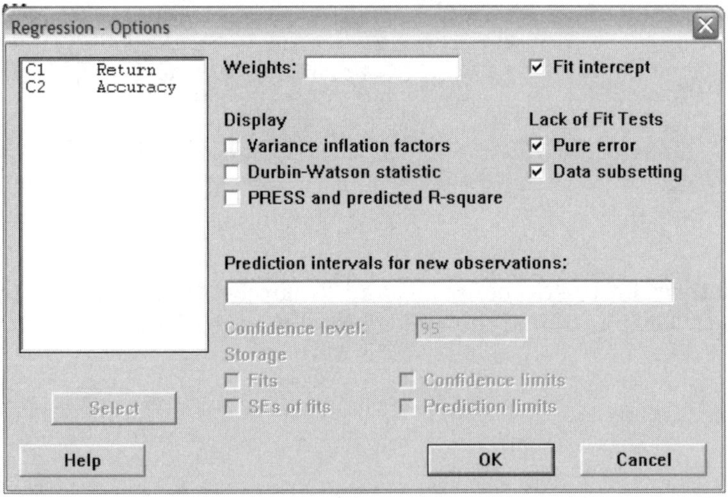

In the "Options" dialog box, select the *Fit intercept*, *Pure error* and *Data subsetting* options and then select "OK" to get back to the "Regression" dialog box.

Now select **"Results . . ."** By default, the third option should be selected; leave it as is.

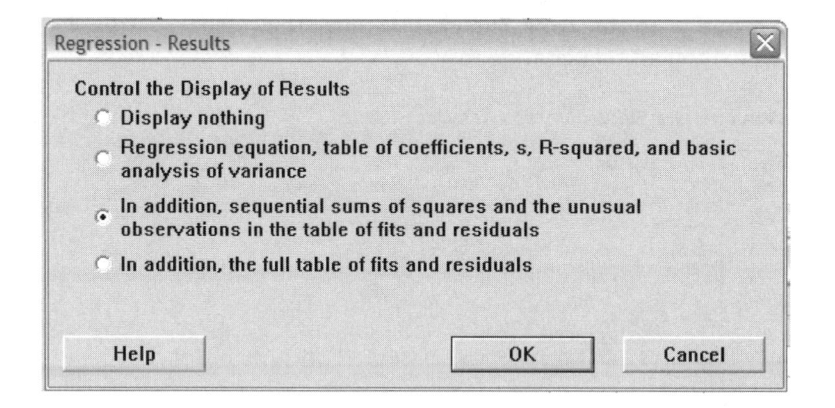

Select "OK" to get back to the "Regression" dialog box.

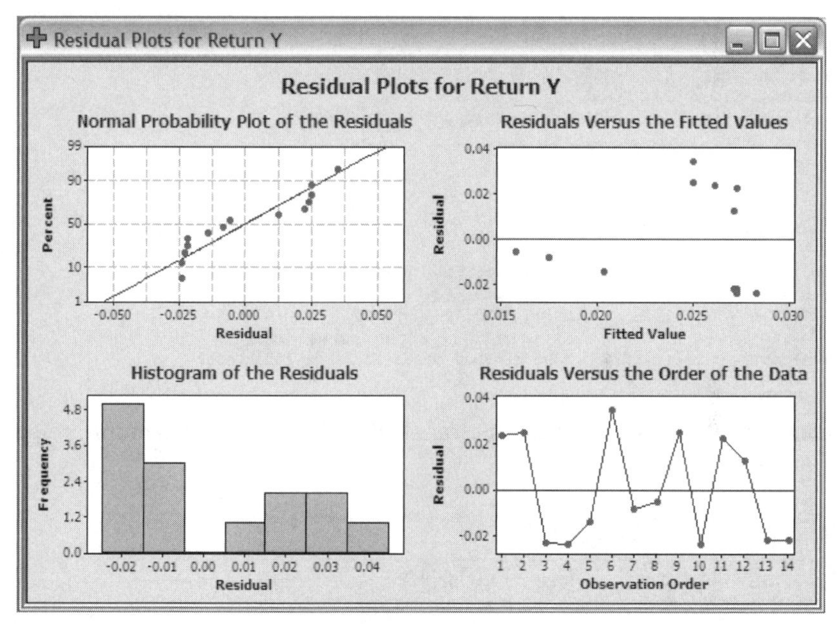

Figure 10.5

The histogram at the bottom left corner and the scatter plot at the top left corner of Figure 10.5 show if the residuals are normally distributed. If the "Residuals" were normally distributed, the histogram would have been symmetrically spread in a way that a bell-shaped curve could have been drawn through the center tops of the bars. On the probability plot of the residuals, the points would have been very close to the line in a steady pattern, which is not the case. So we can conclude that the normality assumption is violated. The top right plot shows the relationship between x and y. In this case, the graph shows that a change in x does not necessarily lead to a change in y.

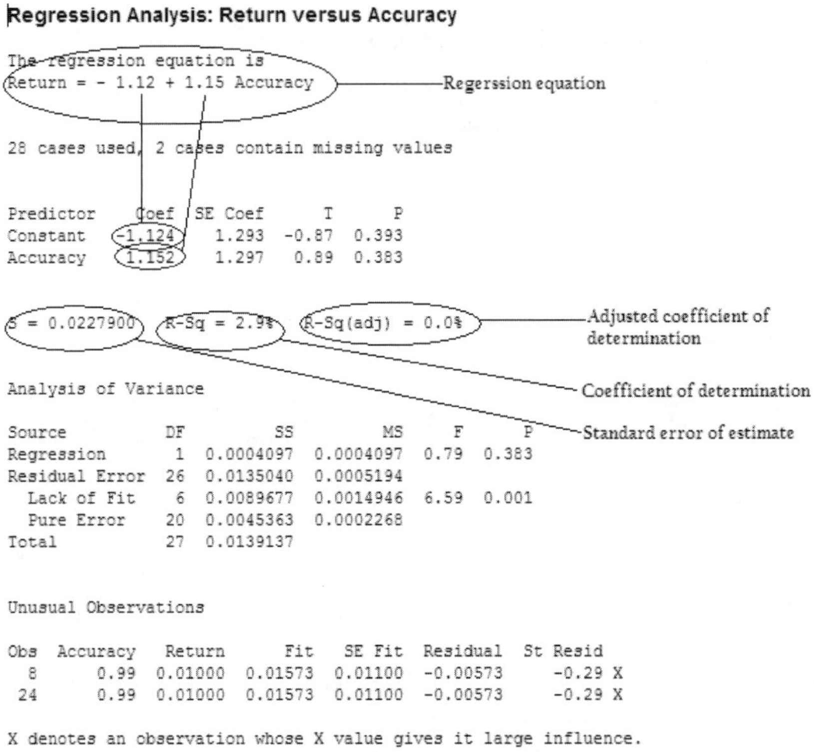

Regression Analysis: Return versus Accuracy

```
The regression equation is
Return = - 1.12 + 1.15 Accuracy

28 cases used, 2 cases contain missing values

Predictor      Coef  SE Coef      T      P
Constant     -1.124    1.293  -0.87  0.393
Accuracy      1.152    1.297   0.89  0.383

S = 0.0227900   R-Sq = 2.9%   R-Sq(adj) = 0.0%

Analysis of Variance

Source          DF         SS         MS     F      P
Regression       1  0.0004097  0.0004097  0.79  0.383
Residual Error  26  0.0135040  0.0005194
  Lack of Fit    6  0.0089677  0.0014946  6.59  0.001
  Pure Error    20  0.0045363  0.0002268
Total           27  0.0139137

Unusual Observations

Obs  Accuracy   Return      Fit   SE Fit  Residual  St Resid
  8      0.99  0.01000  0.01573  0.01100  -0.00573     -0.29 X
 24      0.99  0.01000  0.01573  0.01100  -0.00573     -0.29 X

X denotes an observation whose X value gives it large influence.
```

Regerssion equation

Adjusted coefficient of determination

Coefficient of determination

Standard error of estimate

Using Excel to conduct a regression analysis. From the Tools menu, select "Data Analysis..."

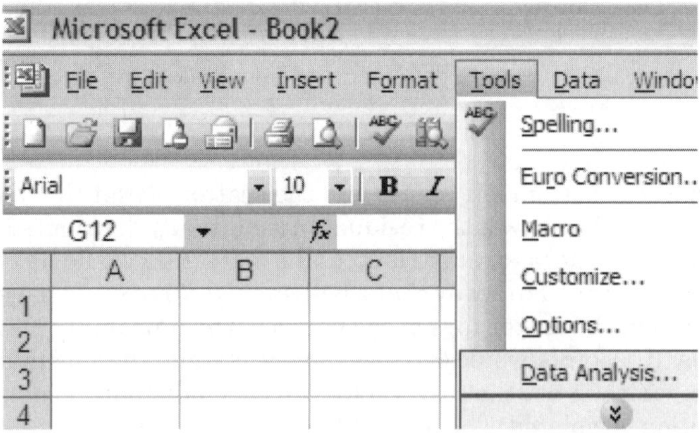

Microsoft Excel - Book2

File Edit View Insert Format Tools Data Windo

Arial 10 B I

G12 fx

Spelling...

Euro Conversion..

Macro

Customize...

Options...

Data Analysis...

In the "Data Analysis" dialog box, select "Regression."

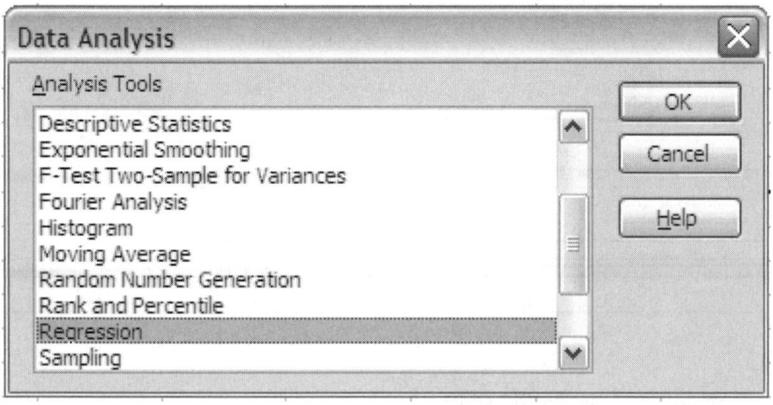

Insert the x and y columns into the appropriate fields and if we have inserted the titles of the columns, select the "Label" option.

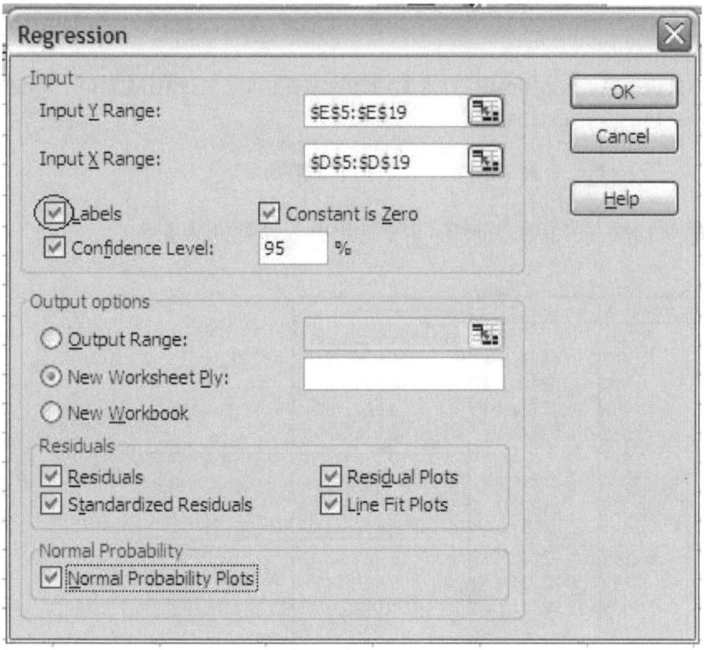

Then, select "OK."

	A	B
1	SUMMARY OUTPUT	
2		
3	*Regression Statistics*	
4	Multiple R	0.171598745
5	R Square	0.029446129
6	Adjusted R Square	-0.05143336
7	Standard Error	0.023720604
8	Observations	14
9		

10	ANOVA								
11		*df*	*SS*	*MS*	*F*	*Significance F*			
12	Regression	1	0.000204853	0.000204853	0.364074	0.557485382			
13	Residual	12	0.006752005	0.000562667					
14	Total	13	0.006956857						
15									
16		*Coefficients*	*Standard Error*	*t Stat*	*P-value*	*Lower 95%*	*Upper 95%*	*ower 95.0%*	*lpper 95.0%*
17	Intercept	-1.123750204	1.903378124	-0.590397772	0.565877	-5.270854876	3.023354	-5.27085	3.023354
18	Accuracy	1.152151374	1.909477896	0.603385552	0.557485	-3.008243559	5.312546	-3.00824	5.312546

10.1.4 How far are the results of our analysis from the true values: residual analysis

Now that we have the equation we were looking for, what makes us believe that we can use it to make predictions? How can we test it? Because we have the x and y values that were used to build the model, we can use them to see how far the regression is from its predicted values. We replace the xs that we had in the regression equation to obtain the predicted $\hat{y}$ s.

We will proceed by replacing the xs that we used to build the model into the regression equation:

$$\hat{y} = 1.152151x - 1.12375$$

Table 10.2 gives us the predicted values and the residuals.

TABLE 10.2

Return Y	Accuracy X	Predicted $\hat{Y}$	Residual $\hat{Y} - Y$
0.05	0.998	0.026096698	0.023903302
0.05	0.997	0.024944547	0.025055453
0.004	0.999	0.027248849	-0.023248849
0.003	0.999	0.027248849	-0.024248849
0.006	0.993	0.020335943	-0.014335943
0.06	0.997	0.024944547	0.035055453
0.009	0.9905	0.017455565	-0.008455565
0.01	0.989	0.015727339	-0.005727339
0.05	0.997	0.024944547	0.025055453
0.004	0.9999	0.028285785	-0.024285785
0.05	0.999	0.027248849	0.022751151
0.04	0.9989	0.027133634	0.012866366
0.005	0.999	0.027248849	-0.022248849
0.005	0.9989	0.027133634	-0.022133634

0.024944547 − 0.06

1.152151 * 0.04 − 1.12375

Minitab residual table:

↓	C1	C2	C3
	Return Y	Accuracy X	RESI1
1	0.050	0.9980	0.0239031
2	0.050	0.9970	0.0250553
3	0.004	0.9990	-0.0232490
4	0.003	0.9990	-0.0242490
5	0.006	0.9930	-0.0143361
6	0.060	0.9970	0.0350553
7	0.009	0.9905	-0.0084557
8	0.010	0.9890	-0.0057275
9	0.050	0.9970	0.0250553
10	0.004	0.9999	-0.0242860
11	0.050	0.9990	0.0227510
12	0.040	0.9989	0.0128662
13	0.005	0.9990	-0.0222490
14	0.005	0.9989	-0.0221338

Excel residual table:

22	RESIDUAL OUTPUT			
23				
24	*Observation*	*Predicted Return*	*Residuals*	*Standard Residuals*
25	1	0.026096867	0.023903133	1.048842149
26	2	0.024944716	0.025055284	1.099397235
27	3	0.027249019	-0.023249019	-1.020140379
28	4	0.027249019	-0.024249019	-1.064019236
29	5	0.02033611	-0.01433611	-0.629052149
30	6	0.024944716	0.035055284	1.53818581
31	7	0.017455732	-0.008455732	-0.371027862
32	8	0.015727505	-0.005727505	-0.251316375
33	9	0.024944716	0.025055284	1.099397235
34	10	0.028285955	-0.024285955	-1.065639956
35	11	0.027249019	0.022750981	0.998287064
36	12	0.027133804	0.012866196	0.564553998
37	13	0.027249019	-0.022249019	-0.976261521
38	14	0.027133804	-0.022133804	-0.971206013

Knowing the residuals is very important because it shows how the regression line fits the original data and therefore helps the experimenter determine if the regression equation is fit to be used for prediction.

Residuals are errors of estimate because they are deviations from the regression line. Had the regression been so perfect that we could predict with 100 percent certainty what the value of every $\hat{y}$ for any given x would be, all the points would have resided on the regression line and all the residuals would have been zero, $\hat{y} - y = 0$. Because the residuals are vertical distances from the regression line, the sum of all the residuals $\hat{y} - y$ is zero.

10.1.5 Standard error of estimate

The experimenter would want to know how accurate he can be in making predictions of the y value for any given value of x based on the regression analysis results. The validity of his estimate will depend on how the errors of prediction will be obtained from his regression analysis, particularly on the average error of prediction. With so many single residuals, it is difficult to look at every one of them individually and make a conclusion for all the data. The experimenter would want to have a single number that reflects all the residuals.

If we add all the deviations of observation from the regression, we obtain zero. To avoid that hurdle (as we saw it when we defined the standard deviation), the deviations are squared to obtain the *sum of square of error* (SSE),

$$SSE = \sum (y - \hat{y})^2$$

To obtain the average deviation from the regression line, we use the square root of the SSE divided by $n - 2$. We use the square root because we had squared the residuals, and we subtract 2 from n because we lose two degrees of freedom from using two sample treatments.

The standard error of estimate (SEE) therefore becomes

$$SEE = \sqrt{\frac{SSE}{n-2}} = \sqrt{\frac{\sum (y - \hat{y})^2}{n-2}} = \sqrt{\frac{0.000563}{12}} = 0.02373$$

10.1.6 How strong is the relationship between *x* and *y*: correlation coefficient

The regression analysis helped us build a model that can help us make predictions on how the response variable y would react to changes in the input variable x. The reaction depends on the strength of the correlation between the two variables. The strength of the relationships between two sets of data is described in statistics by the correlation coefficient, usually noted with the letter r. The correlation coefficient is a number between -1 and $+1$. When it is equal to zero, we conclude that there is absolutely no relationship between the two sets of data. If it is equal to $+1$, there is a strong positive relationship between the two.

An increase in one value of the input variable will lead to an increase of the corresponding value in the exact same proportion; a decrease in the value of x will lead to a decrease in the value of the corresponding y in the same proportion. The two sets of data increase and decrease in the same directions and in the same proportions.

If r equals -1, then an increase in the value of x will lead to a decrease of the corresponding y in the exact same proportion. The two sets of data increase and decrease in opposite directions but in the same proportions. Any value of r between zero and $+1$ and between zero and -1 is interpreted according to how close it is to those numbers.

The formula for the correlation coefficient is given as

$$r = \frac{\sum (X-\overline{X})(Y-\overline{Y})}{\sqrt{\sum (X-\overline{X})^2 \sum (Y-\overline{Y})^2}}$$

	Return	Accuracy	$(x-\overline{x})$	$(y-\overline{y})$	$(x-\overline{x})(y-\overline{y})$	$(x-\overline{x})^2$	$(Y-\overline{Y})^2$
	0.05	0.998	0.0012	0.025285714	0.000030343	0.0000014	0.000639
	0.05	0.997	0.0002	0.025285714	0.000005057	0.0000000	0.000639
	0.004	0.999	0.0022	−0.02071429	−0.000045571	0.0000048	0.000429
	0.003	0.999	0.0022	−0.02171429	−0.000047771	0.0000048	0.000472
	0.006	0.993	−0.0038	−0.01871429	0.000071114	0.0000144	0.00035
	0.06	0.997	0.0002	0.035285714	0.000007057	0.0000000	0.001245
	0.009	0.9905	−0.0063	−0.01571429	0.000099000	0.0000397	0.000247
	0.01	0.989	−0.0078	−0.01471429	0.000114771	0.0000608	0.000217
	0.05	0.997	0.0002	0.025285714	0.000005057	0.0000000	0.000639
	0.004	0.9999	0.0031	−0.02071429	−0.000064214	0.0000096	0.000429
	0.05	0.999	0.0022	0.025285714	0.000055629	0.0000048	0.000639
	0.04	0.9989	0.0021	0.015285714	0.000032100	0.0000044	0.000234
	0.005	0.999	0.0022	−0.01971429	−0.000043371	0.0000048	0.000389
	0.005	0.9989	0.0021	−0.01971429	−0.000041400	0.0000044	0.000389
Mean	0.024714	0.9968		Totals	0.000177800	0.0001543	0.006957

$$r = \frac{\sum(X-\overline{X})(Y-\overline{Y})}{\sqrt{\sum(X-\overline{X})^2 \sum (Y-\overline{Y})^2}} = \frac{SS_{xy}}{\sqrt{(SS_{xx})(SS_{yy})}}$$

$$= \frac{0.000177800}{\sqrt{0.0001543 \times 0.006957}} = 0.171599$$

Correlation coefficient	Interpretation
−1.0	Strong negative correlation
−0.5	Moderate negative correlation
0.0	Absolutely no relationship between the two sets of data
+0.5	Moderate positive relationship
+1.0	Strong positive relationship

Using Minitab. Paste the data into a worksheet and from the Stat menu, select "Basic Statistics" and then "Correlation. . ."

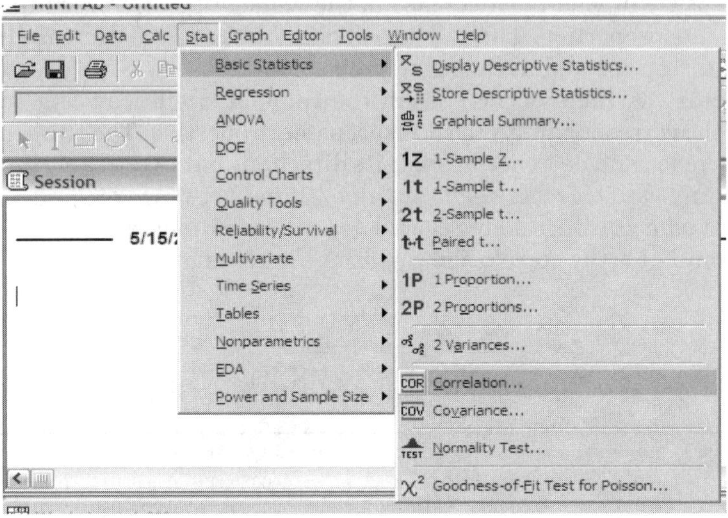

In the "Correlation" dialog box, select the appropriate columns, insert them into the *Variables* text box, and then select "OK."

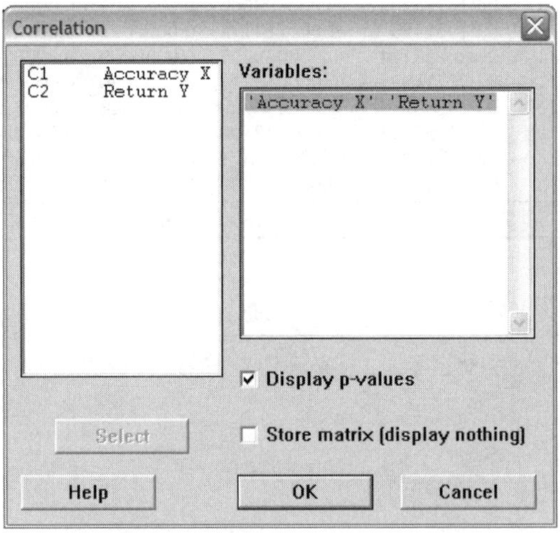

The results are given with the P-value.

Correlations: Accuracy X, Return Y

```
Pearson correlation of Accuracy X and Return Y = 0.172
P-Value = 0.557
```

Using Excel. We can find the correlation coefficient in several ways using Excel. One quick way is the following: Select a cell where we want to insert the result and then click on the "f_x" button to insert a function, as indicated in Figure 10.6.

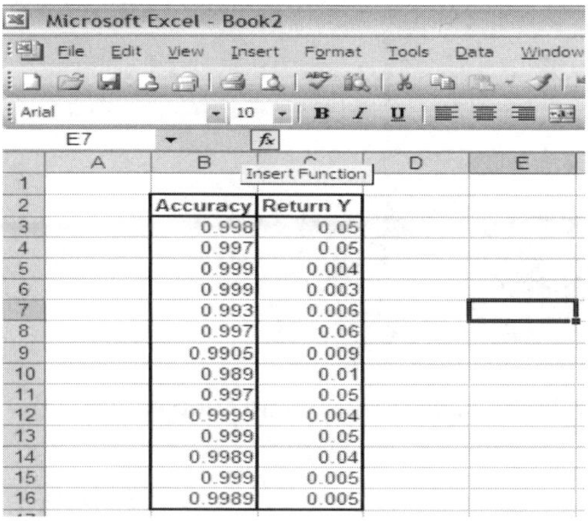

	A	B	C	D	E
1					
2		Accuracy	Return Y		
3		0.998	0.05		
4		0.997	0.05		
5		0.999	0.004		
6		0.999	0.003		
7		0.993	0.006		
8		0.997	0.06		
9		0.9905	0.009		
10		0.989	0.01		
11		0.997	0.05		
12		0.9999	0.004		
13		0.999	0.05		
14		0.9989	0.04		
15		0.999	0.005		
16		0.9989	0.005		

Figure 10.6

Then select "Statistical" from the drop-down list.

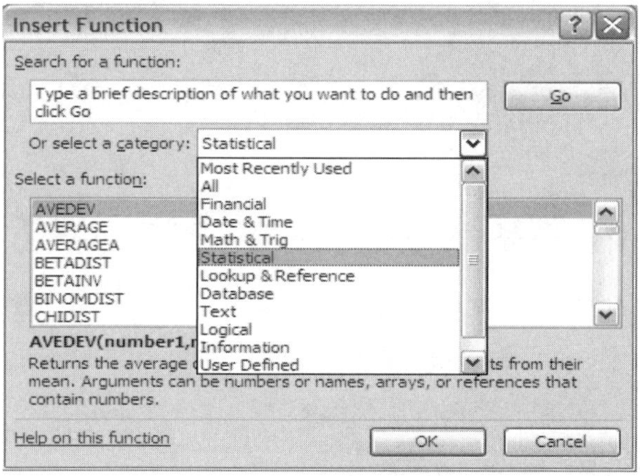

Select "CORREL" from the *Select a function:* text box, and then select "OK."

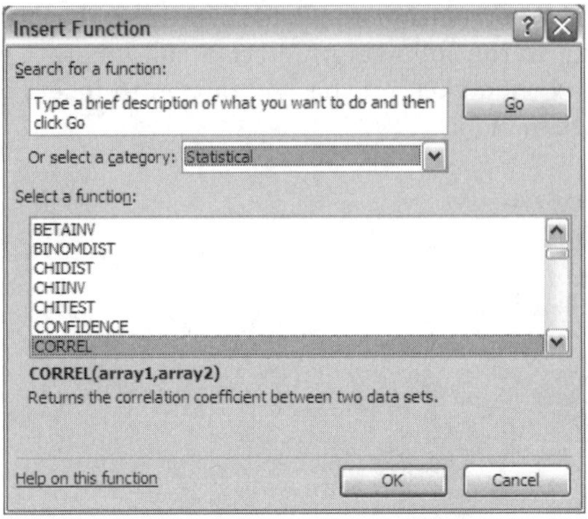

Insert Function

Search for a function:

Type a brief description of what you want to do and then click Go

Go

Or select a category: Statistical

Select a function:

BETAINV
BINOMDIST
CHIDIST
CHIINV
CHITEST
CONFIDENCE
CORREL

CORREL(array1,array2)

Returns the correlation coefficient between two data sets.

Help on this function

OK Cancel

Insert the fields in *Array1* and *Array2* accordingly and then select "OK."

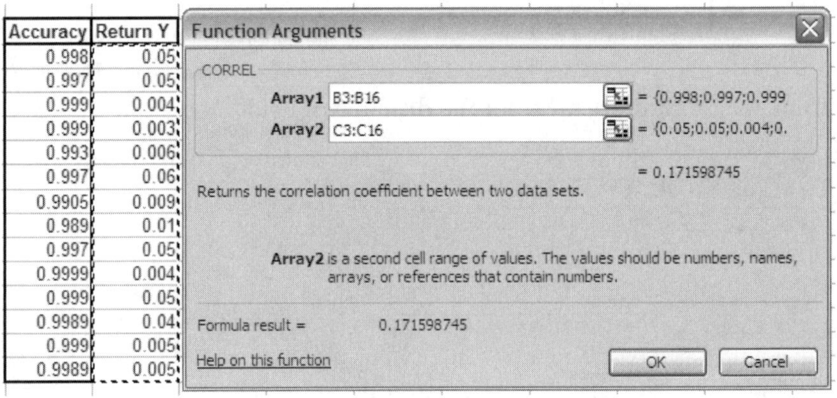

Accuracy	Return Y
0.998	0.05
0.997	0.05
0.999	0.004
0.999	0.003
0.993	0.006
0.997	0.06
0.9905	0.009
0.989	0.01
0.997	0.05
0.9999	0.004
0.999	0.05
0.9989	0.04
0.999	0.005
0.9989	0.005

Function Arguments

CORREL

Array1 B3:B16 = {0.998;0.997;0.999

Array2 C3:C16 = {0.05;0.05;0.004;0.

= 0.171598745

Returns the correlation coefficient between two data sets.

Array2 is a second cell range of values. The values should be numbers, names, arrays, or references that contain numbers.

Formula result = 0.171598745

Help on this function

OK Cancel

The result appears in the selected cell.

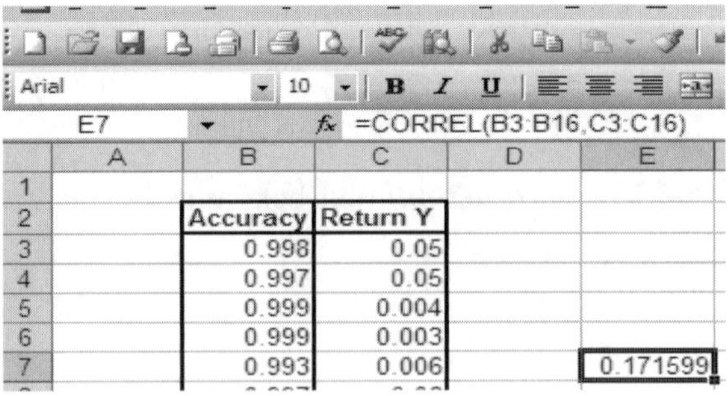

Arial 10 B I U

E7 =CORREL(B3:B16,C3:C16)

	A	B	C	D	E
1					
2		Accuracy	Return Y		
3		0.998	0.05		
4		0.997	0.05		
5		0.999	0.004		
6		0.999	0.003		
7		0.993	0.006		0.171599

10.1.7 Coefficient of determination, or what proportion in the variation of *y* is explained by the changes in *x*

The interpretation of the correlation coefficient is approximate and vague and does not give an accurate account of the changes in the y variable that are explained by changes in the x variable. The conclusions derived using the correlation coefficient were that there is a strong, moderate, or inexistent correlation between the changes in the values of the variables.

Whereas the correlation coefficient measures the strength of the relationship between the two sets of data, the coefficient of determination shows the proportion of variation in the variable y that is explained by the variations in x. The coefficient of determination is the square of the coefficient of correlation. In our case, the coefficient of determination would be

$$r^2 = 0.172^2 = 0.029584$$

or in terms of percentage, 2.96 percent. So 2.96 percent of the changes in the y variable are explained by changes in the x variables.

Note that even though the coefficient of determination is the square of the correlation coefficient, the correlation coefficient is not necessarily the square root of the coefficient of determination. This is because a square root is always positive and the correlation coefficient may be negative.

10.1.8 Testing the validity of the regression line: hypothesis testing for the slope of the regression model

The regression equation we obtained is based on a sample. If another sample were taken for testing, we may very well have ended up with a different equation. So for the equation we found to be valid, it must be reflective of the parameters of the population.

If the sample regression model is identical to the population regression model and the slope of the population equation is equal to zero, we should be able to predict with accuracy the value of the response variable for any value of x because it will be equal to the constant b_0 and β_0.

To test the validity of the regression line as a tool for predictions, we will need to test whether β_1 (which is the population slope) is equal to zero. What we are testing is the population slope using the slope of the sample. The null and alternate hypotheses for the test will be

$$H_0 : \beta_1 = 0$$

$$H_\alpha : \beta_1 \neq 0$$

The equality sign suggests that we are faced with a two-tailed curve. We will use the t test, which is obtained from the t-distribution with a degree of freedom of $n - 2$ to conduct the hypothesis testing. The formula for the t test is given as

$$t = \frac{b_1 - \beta_1}{S_{b_1}}$$

where

$$S_{b_1} = \frac{SSE}{\sqrt{SS_X}} = \frac{\sqrt{\sum(y - \hat{y})^2/(n-2)}}{\sqrt{(x - \overline{x})^2}} = \frac{0.023721}{\sqrt{0.0001543}}$$

$$= \frac{0.023721}{0.0123226} = 1.9095$$

The regression equation we obtained from our analysis was

$$\hat{y} = 1.152151x - 1.12375$$

The sample slope in this case is 1.152151 and $S_{b_1} = 1.9095$. Because we hypothesized that $\beta_1 = 0$,

$$t = \frac{1.152151 - 0}{1.9095} = 0.60338$$

Because we are faced with a two-tailed test, for $\alpha = 0.05$, the critical t would be

$$t_{\alpha/2,n-2} = t_{0.025,12} = 2.179$$

The calculated $t = 0.60338$ is lower than the critical t, therefore we cannot reject the null hypothesis. We must conclude that there is not a significant relationship between x and y that would justify using the regression model for predictions.

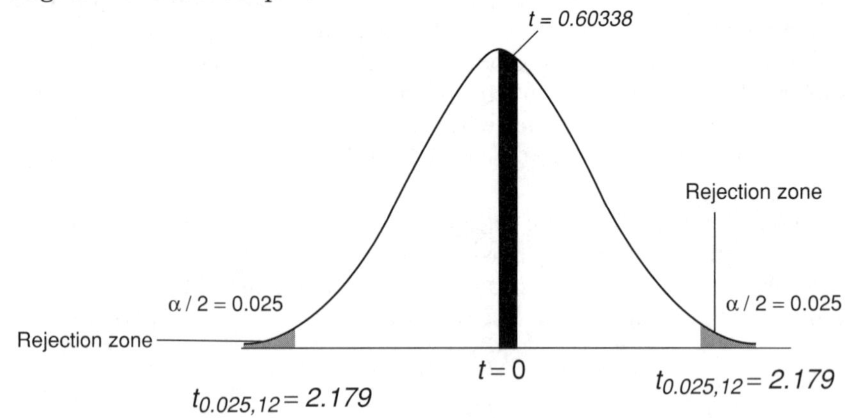

Minitab results

Regression Analysis: Return Y versus Accuracy X

```
The regression equation is
Return Y = - 1.12 + 1.15 Accuracy X


Predictor      Coef  SE Coef      T      P
Constant     -1.124    1.903  -0.59  0.566
Accuracy X    1.152    1.909   0.60  0.557
```

Notice that the t statistic is nothing but the ratio of the coefficients to the standard errors. The P-values are higher than 0.05, which suggests that the results are insignificant.

Excel results

		Coefficients	Standard Error	t Stat	P-value
17	Intercept	-1.123750204	1.903378124	-0.590397772	0.565877
18	Accuracy	1.152151374	1.909477896	0.603385552	0.557485

10.1.9 Using the confidence interval to estimate the mean

One of the main reasons why one would want to build a regression line is to use it to make predictions. For instance, based on the equation that we created we should be able to use a point of estimate and determine what the predicted y would be. What would happen if the in-house accuracy rate is, say, 0.9991? All we need to do is replace x by 0.9991 in the equation to obtain the predicted value of return.

$$\hat{y} = 1.15 \times 0.9991 - 1.12 = 0.028965$$

Yet the validity of the results will depend on the data used to build the regression equation. The equation was built based on a sample. If another sample were taken, we might have ended up with a different equation. So how confident can we be with the results that we have obtained?

We cannot be 100 percent confident about the projected values of y for every x, but we can find a confidence interval that would include the predicted y for a set confidence level. For a given value x_0, the confidence interval to estimate $\hat{y}$ will be

$$\hat{y} \pm t_{\alpha/2,n-2} S_e \sqrt{\frac{1}{n} + \frac{(x_0 - \overline{x})^2}{SS_{XX}}}$$

with

$$SS_{XX} = \frac{\sum (x - \overline{x})^2}{n}$$

$$S_e = SEE = \sqrt{\frac{SSE}{n-2}} = \sqrt{\frac{\sum (y - \overline{y})^2}{n-2}}$$

and $t_{\alpha/2,n-2}$ is found on the t table. For $n = 14$ and a confidence level of 95 percent, $t_{\alpha/2,n-2} = t_{0.025,12}$, which corresponds to 2.179 on the table. So for a point of estimate of 0.9999, the value of $\hat{y}$ will be 0.029885 and the confidence interval will be

$$\hat{y} \pm t_{\alpha/2,n-2} S_e \sqrt{\frac{1}{n} + \frac{(x_0 - \overline{x})^2}{SS_{XX}}} = 0.029885 \pm 2.179 \times 0.02373 \sqrt{\frac{1}{14} + \frac{(0.9999 - 0.9968)^2}{0.0001543}}$$

$$= 0.029885 \pm 0.018$$

10.1.10 Fitted line plot

Minitab offers a graphical method of depicting both the confidence interval and the predicted interval for a regression model. For the Accuracy-Return problem, we can generate a fitted line plot using Minitab.

After pasting the data into a worksheet, from the Stat menu, select "Regression" and then "Fitted Line Plot..."

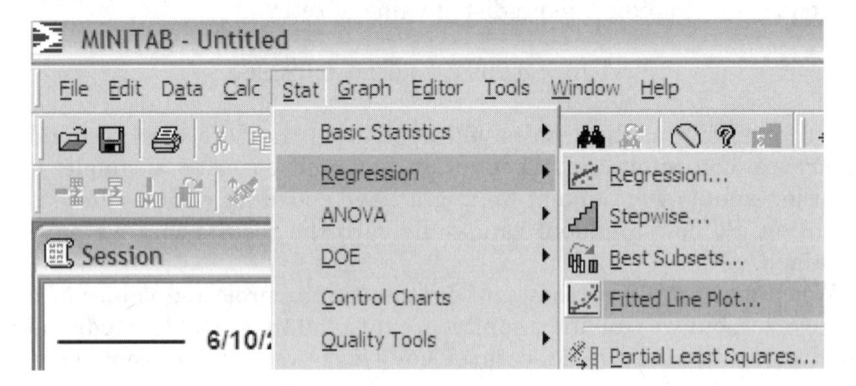

After the "Fitted Line Plot" dialog box pops up, select "Return" and "Accuracy" for *Response* and *Predictor*, respectively.

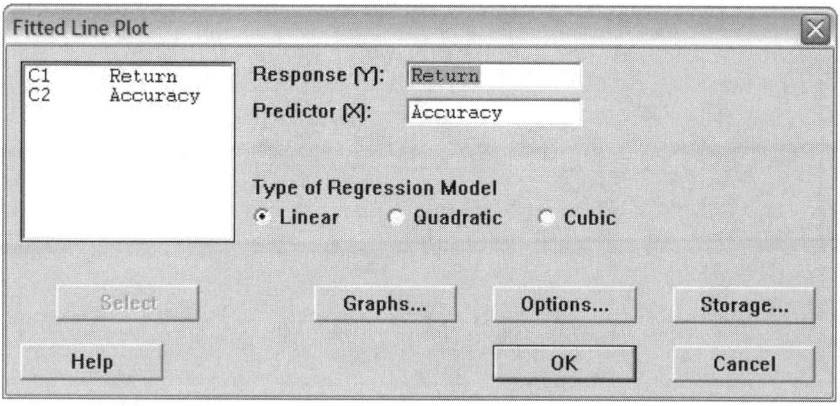

Select "Options. . . " to obtain the dialog box shown in Figure 10.7, then select the *Display confidence interval* and *Display prediction interval* options. By default, the *Confidence level* is set at "95.0."

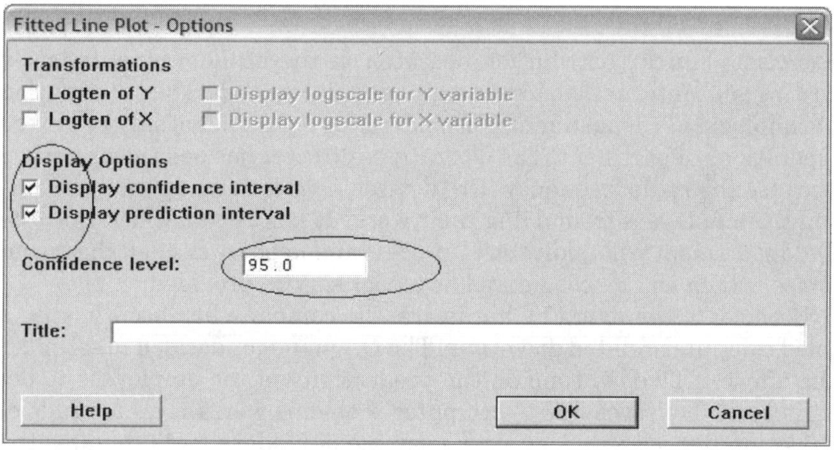

Figure 10.7

Then select "OK" and "OK" again to obtain the graph.

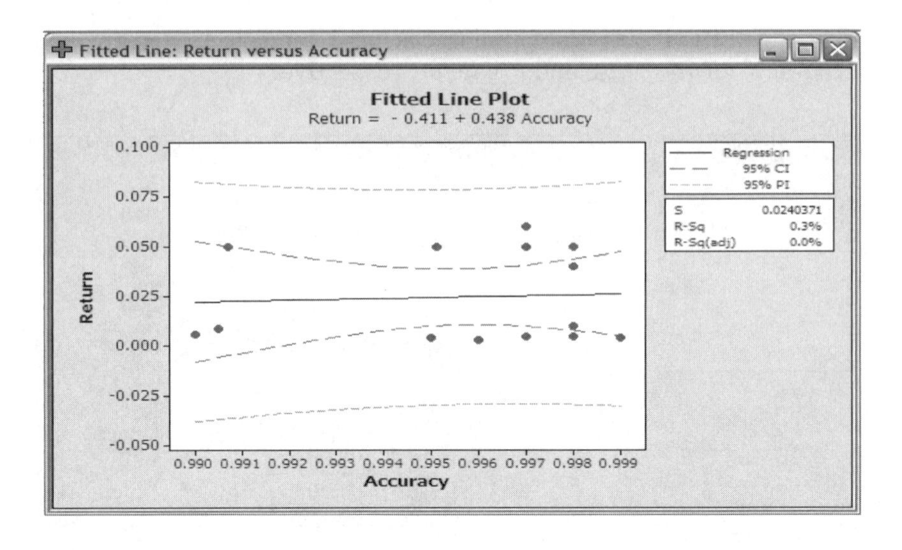

For a confidence level of 95 percent, we can see that most of the points are outside the range. The Black Belt must conclude that there is not any relationship between the accuracy rate and the volume of return; therefore, "Accuracy," as it is used at Dakar Electromotive, is an incorrect metric for Customer Services to predict the call volumes.

Exercise. Fatick Distribution operates on the basis of activity-based costing and internal customer-supplier relationship. The Information Technology (IT) department is considered as a supplier of services to the Operations department. The Operations director has been complaining because the radio frequency (RF) devices keep locking up and preventing the employees from doing their work. It has been decided that the QA department will audit the IT process and keep track of all the downtimes, which will be considered as a poor service provided by IT.

Because of the aging IT hardware, the company has been having a lot of computer-related downtime. The QA manager decided to estimate the effect of IT downtime on the productivity of the employees to determine if the losses due to computer problems warrant an upgrade of the computer system. He takes a sample of 25 days and tabulates the downtimes (in minutes) and the productivity of the associates for those days in Table 10.3. The table can be found in *DowntimeProductivity.mpj* and *Downtime.xls* on the included CD

TABLE 10.3

Downtime	35	29	15	14	0	32	18	16	16	10	32	15
Productivity	94	97	98	98	99	89	92	95	95	97	94	97

Downtime	9	8	14	0	10	9	8	6	5	4	3	8	7
Productivity	97	98	98	99	97	98	98	99	99	98	98	98	99

Using Excel then Minitab, run a Regression analysis to determine the effect of downtime on productivity.

a. Determine which is the x-variable and which is the y-variable.

b. Using Minitab and Excel, plot the residuals on one graph and interpret the results.

c. Are the residuals normally distributed?

d. Find the correlation coefficient between productivity and downtime and interpret the result.

e. What proportion in the variations in productivity is explained by the changes in downtime?

f. What would have been the predicted productivity if the downtime were 45 minutes?

g. What effect would changing the confidence level from 95 percent to 99 percent have on the significant F?

10.2 Building a Model with More than Two Variables: Multiple Regression Analysis

In our simple regression analysis, the variations of only one variable ("Accuracy") were used to explain the variations in the response variable ("Return"). After conducting the analysis, we concluded that the proportion of the variations in "Return" explained by the changes in "Accuracy" was insignificant. Therefore, changes in other variables must cause to the variations in "Return." Most likely, there are several variables with each contributing a certain proportion. When more than one independent variable explains the variations in the response variable, the model will be called a *multiple regression* model.

The underlying premises for the building of the model are similar to the ones for the simple regression with the difference that we have more than one x factor. To differentiate between population and sample, we will use

$$\hat{y} = \beta_1 x_1 + \beta_2 x_2 + \cdots + \beta_i x_i + \beta_n x_n$$

for the expected regression model for the population, and

$$\hat{y} = b_1 x_1 + b_2 x_2 + \cdots + b_i x_i + b_n x_n$$

for the samples.

Six Sigma case study. A Customer Service manager has been urged to reduce the operating costs in her department. The only way she can do it is through a reduction of staff. Reducing her staff while still responding to all her customers' calls within a specified time frame would require addressing the causes of the calls and improving on them. She initiates a Six Sigma project to reduce customer complaints. In the "Analyze" phase of the project, she categorizes the reasons for the calls into three groups: "damaged plastic," "machine overheating," and "ignorance" (for the customers who do not know how to use the product).

She believes that these are the main causes for the customers to keep her representatives too long over the phone. She tabulates a random sample of 15 days, shown in Table 10.5 (in minutes). The dependent (or response) variable is "Call Time" and the independent variables (or regressors) are "Overheat," "Plastic," and "Ignorance." The table can be found in *Calltime.xls* and *Calltime.mpj* on the included CD

TABLE 10.5

Call Time	Overheat	Plastic	Ignorance
69	15	17	7
76	17	23	6
89	20	24	10
79	18	23	7
76	17	23	9
76	17	25	5
78	18	26	8
56	14	24	9
87	20	28	11
65	17	26	3
89	23	27	9
67	19	26	10
76	17	25	8
67	15	26	7
71	16	27	9

She wants to build a model that explains how the time that the customer representatives spend talking to the customers relates to the different causes of the calls and help make predictions on how reducing the number of the calls based on the reasons for the calls can affect the call times.

In this case, we have one dependent (or response) variable, "Call Time," and three independent variables, "Overheat," "Plastic," and "Ignorance." The model we are looking for will be under the form of

$$Call\ Time = b_1 \times Overheat + b_2 \times Plastic + b_3 \times Ignorance + b_0$$

where b_1, b_2, and b_3 are coefficients of the dependent variables. They represent the proportions of the change in the dependent variable when the independent variables change by a factor of one. The term b_0 is a constant that would equal the response variable if all the independent variables were equal to zero.

Using Excel, the process used to find the multiple regression equation is the same as the one used for the simple linear regression. The output also will be similar and the interpretations that we make of the values obtained are the same. In that respect, Excel's scope is limited compared to Minitab.

Excel's output for the Customer Service data:

	A	B
1	SUMMARY OUTPUT	
2		
3	*Regression Statistics*	
4	Multiple R	0.836097793
5	R Square	0.69905952
6	Adjusted R Square	0.616984843
7	Standard Error	5.741501357
8	Observations	15

10.2.1 Hypothesis testing for the coefficients

When we conducted the simple regression analysis, we only had one coefficient for the slope, and we conducted hypothesis testing to determine if the population's regression slope was equal to zero. The null and alternate hypotheses were

$$H_0 : \beta_1 = 0$$

$$H_\alpha : \beta_1 \neq 0$$

We used the regression equation that we found to test the hypothesis.

In the case of a multiple regression analysis, we have several coefficients to conduct a hypothesis test; therefore, ANOVA would be more appropriate to test the null hypothesis. The null and alternate hypotheses would be:

$$H_0 : \beta_1 = \beta_2 = \beta_3 = 0$$

$$H_\alpha : \text{At least one coefficient is different from zero.}$$

Using Minitab. We will not be able to address all the capabilities of Minitab with regard to multiple regression analysis because it would take a voluminous book by itself, so we will be selective with the options offered to us.

After pasting the table into a Minitab worksheet, from the Stat menu, select "Regression" and then select "Regression" again. In the "Regression" dialog box, select "Options."

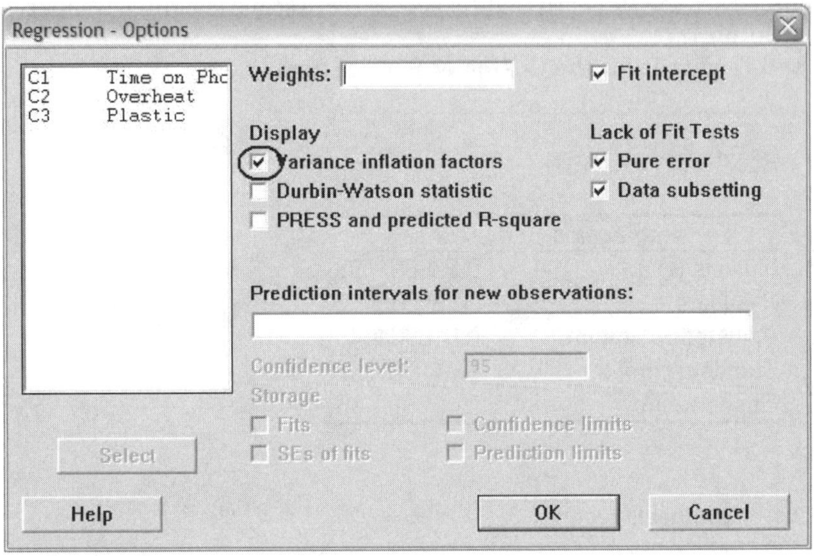

Regression - Options

C1 Time on Phc
C2 Overheat
C3 Plastic

Weights:
Fit intercept

Display
Variance inflation factors
Durbin-Watson statistic
PRESS and predicted R-square

Lack of Fit Tests
Pure error
Data subsetting

Prediction intervals for new observations:

Confidence level: 95
Storage
Fits
SEs of fits
Confidence limits
Prediction limits

Select
Help
OK
Cancel

Select the option *Variance inflation factor*, then select "OK" and "OK" again to get the output shown in Figure 10.8.

Using Excel. If we reject the null hypothesis, we would conclude that at least one independent variable is linearly related to the dependent variable. On the ANOVA table shown in Figure 10.9, the calculated F (8.5174) is much higher than the critical F (0.0033). Therefore, we must reject the null hypothesis and conclude that a least one independent variable is correlated to the dependent variable.

The circled coefficients represent the coefficients for each independent variable and the intercept represents the constant.

The residuals are interpreted in the same manner they were interpreted in the simple regression analysis. When calculating the residuals, we replace the values of "Overheat," "Plastics," and "Ignorance" for each line in the equation to obtain the predicted "Call Time;" then the predicted "Call Time" is subtracted from the actual "Call Time."

24	RESIDUAL OUTPUT			
25				
26	*Observation*	*Predicted Call Time*	*Residuals*	*Standard Residuals*
27	1	70.39937938	-1.399379384	-0.274965164
28	2	73.39366319	2.606336812	0.512121185
29	3	84.45305633	4.546943671	0.893432564
30	4	77.19996837	1.800031627	0.35368964
31	5	74.21406055	1.785939452	0.350920658
32	6	71.854221	4.145779001	0.8146074
33	7	75.57446956	2.425530444	0.476594398
34	8	62.98255415	-6.98255415	-1.372007596
35	9	82.19456931	4.805430689	0.944222883
36	10	70.67430122	-5.674301224	-1.114947942
37	11	92.87914413	-3.879144135	-0.762216104
38	12	79.65424053	-12.65424053	-2.486441745
39	13	72.67461836	3.325381641	0.653406873
40	14	64.70248557	2.297514427	0.45144043
41	15	68.14926834	2.850731656	0.560142521

Interpretation of the results. Note that on the ANOVA table, we do not have the F-critical value obtained from the F table, but we do have the P-value, which is 0.003, therefore less that the critical value of 0.05. So we must reject the null hypothesis and conclude that at least one independent variable is correlated with the dependent variable.

***P*-values for the coefficients.** The P-values are the results of hypotheses testing for every individual coefficient. The tests will help determine if the variable whose coefficient is being tested is significant in the model, i.e., if it must be kept or deleted from the model.

The P-value is compared to the α level, which in general is equal to 0.05. If the P-value is less that 0.05, we are in the rejection zone and we conclude that the variable is significant and reject the null hypothesis. Otherwise, we cannot reject the null hypothesis.

In our example, all the P-values are greater than the 0.05 except for "Overheat," which is 0.001. So "Overheat" is the only independent variable that is significantly correlated with the dependent factor "Call Time." Both "Plastic" and "Ignorance" are higher than 0.1 and are therefore insignificant and should be removed from the model.

Adjusted coefficient of determination. In the previous section on simple linear regression analysis, we defined the coefficient of determination as the proportion in the variation of the response variable that is explained by the independent factor. The same definition holds with multiple regression analysis.

Regression Analysis: Call Time versus Overheat, Plastic, Ignorance

```
The regression equation is
Call Time = 26.3 + 3.53 Overheat - 0.633 Plastic + 0.273 Ignorance

Predictor       Coef  SE Coef      T      P  VIF
Constant       26.25    15.69   1.67  0.122
Overheat      3.5328   0.7809   4.52  0.001  1.4
Plastic      -0.6330   0.6474  -0.98  0.349  1.2
Ignorance     0.2735   0.7893   0.35  0.736  1.2

S = 5.74150   R-Sq = 69.9%   R-Sq(adj) = 61.7%

Analysis of Variance

Source          DF       SS      MS     F      P
Regression       3   842.32  280.77  8.52  0.003
Residual Error  11   362.61   32.96
Total           14  1204.93

No replicates.
Cannot do pure error test.

Source      DF  Seq SS
Overheat     1  808.33
Plastic      1   30.03
```

Regression equation

Variance Inflation Factors

P- Values

$$\frac{842.32}{1204.93} = \text{R-Sq} = 69.9\%$$

Figure 10.8

But taking into account sample sizes and the degrees of freedom of independent factors is recommended to assure that the coefficient of determination is not inflated. The formula for the adjusted coefficient of determination is

$$AdjR^2 = 1 - \left|(1 - R^2)\frac{n-1}{n-1-k}\right|$$

where k is the number of independent factors, which is three.

10	ANOVA								
11		df	SS	MS	F	Significance F			
12	Regression	3	842.3201172	280.7733724	8.517359432	0.003297956			
13	Residual	11	362.6132161	32.96483783					
14	Total	14	1204.933333						
15									
16		Coefficients	Standard Error	t Stat	P-value	Lower 95%	Upper 95%	Lower 95.0%	Upper 95.
17	Intercept	26.25332732	15.68952885	1.673302466	0.122434986	-8.27911031	60.7857649	-8.27911031	60.78576
18	Overheat	3.532839399	0.780875666	4.524202191	0.000866041	1.814142777	5.25153602	1.81414278	5.25153
19	Plastic	-0.6329882	0.647388151	-0.97775685	0.349208449	-2.057880635	0.79190423	-2.05788063	0.791904
20	Ignorance	0.273465787	0.789311534	0.346461156	0.735532721	-1.463798065	2.01072964	-1.46379806	2.010729
21									

Figure 10.9

Multicollinearity. *Multicollinearity* refers to a situation where at least two independent factors are highly correlated. An increase in one of the factors would lead to an increase or a decrease of the other. When this happens, the interpretation that we make of the coefficient of determination may be inaccurate. The proportion of the changes in the dependent factor due to the variations in the independent factors might be overestimated.

One way to estimate multicollinearity is the use of the *variance inflation factor* (VIF). It consists in using one independent variable at a time as if it were a dependent variable and conducting a regression analysis with the other independent variables. This way, the experimenter will be able to determine if a correlation is present between the x factors.

The coefficient of determination for each independent variable can be used to estimate the VIF,

$$VIF = \left(1 - R_i^2\right)^{-1}$$

In most cases, multicollinearity is suspected when a VIF greater than 10 is present. In this case, the VIF are relatively small.

Exercise

a. Complete the missing fields in Table 10.6.

b. What is the coefficient of determination?

c. Interpret the results.

d. What does a P-value of zero suggest?

Exercise

a. Complete the missing items on this Minitab Output of Figure 10.10.

b. Based on the P-values, what can we conclude about the input factors?

c. What proportion of the input factors cause variations in the output factor?

d. What can we say about the VIF?

TABLE 10.6

ANOVA						
Source of Variation	*SS*	*df*	*MS*	*F*	*P-value*	*F Critical*
Between Groups	7680.519		3840.259		0.0000	3.402826
Within Groups		24				
Total	9368.963	26				

Regression Analysis: Light versus PSI, Roughness

```
The regression equation is
Light = 60.1 - 6.01 PSI + 5.06 Roughness


Predictor      Coef   SE Coef       T      P   VIF
Constant      60.08     90.58    0.66  0.532
PSI          -6.007     5.149   -1.17  0.288   1.0
Roughness     5.060     5.298    0.95  0.376   1.0


S = 14.4426    R-Sq =          R-Sq(adj) = 0.3%


Analysis of Variance

Source           DF       SS       MS      F       P
Regression        2                     1.01   0.419
Residual Error        1251.5    208.6
Total             8   1672.9
```

Figure 10.10

10.2.2 Stepwise regression

Six Sigma case study. Senegal Electric is a company that manufactures and sells robots. It markets the robots through five channels: websites, TV, magazine, radio, and billboards. A Six Sigma project was initiated to reduce marketing cost and improve its impact on sales. In the "Analyze" phase of the project, a Black Belt decides to build a regression model that will determine the significant factors that affect sales and eliminate the insignificant ones from the model. She randomly selects 15 days of a month and tabulates the number of hits from the website, the number of coupons from the magazine, the numbers of radio and TV broadcasts, and the number of people who bought the robots because they saw the billboards.

She wants to build a model that uses all the variables whose variations significantly explain the variation in sales. At the same time, she wants the model to be simple and relevant. This data is presented in Table 10.7 and can be found in the files *Senegal electric.xls* and *Senegal electric.mpj* on the included CD.

In this case, "Sales" is the response factor and "TV," "Radio," "Magazine," "Website," and "Billboard" are the regressors (or independent factors).

After using Minitab to run a multiple regression analysis, we obtain the output shown in Figure 10.11.

TABLE 10.7

Sales	TV	Radio	Magazine	Website	Billboard
1292	12	17	220	11987	60
1234	11	17	298	8976	58
1254	12	16	276	8720	20
1983	16	21	190	19876	729
1678	13	17	276	2342	305
1876	14	19	230	1456	198
2387	18	19	311	2135	511
1234	10	16	325	1238	1153
1365	11	18	368	1243	131
2354	18	22	143	2313	989
1243	9	11	109	1215	1111
2345	18	25	215	163	1102
1342	11	18	160	9808	1003
1235	11	16	543	706	107
1243	11	17	24	973	8
1293	12	15	19	169	50

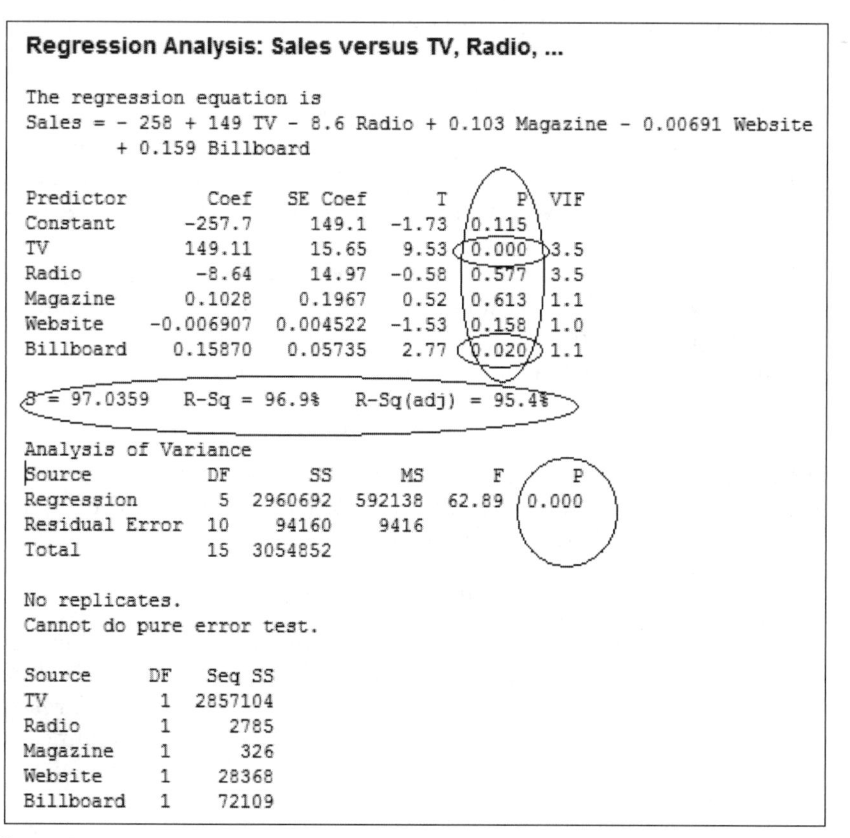

Regression Analysis: Sales versus TV, Radio, ...

```
The regression equation is
Sales = - 258 + 149 TV - 8.6 Radio + 0.103 Magazine - 0.00691 Website
        + 0.159 Billboard

Predictor        Coef    SE Coef      T      P   VIF
Constant       -257.7      149.1  -1.73  0.115
TV             149.11      15.65   9.53  0.000   3.5
Radio           -8.64      14.97  -0.58  0.577   3.5
Magazine       0.1028     0.1967   0.52  0.613   1.1
Website     -0.006907   0.004522  -1.53  0.158   1.0
Billboard     0.15870    0.05735   2.77  0.020   1.1

S = 97.0359   R-Sq = 96.9%   R-Sq(adj) = 95.4%

Analysis of Variance
Source          DF       SS      MS      F      P
Regression       5  2960692  592138  62.89  0.000
Residual Error  10    94160    9416
Total           15  3054852

No replicates.
Cannot do pure error test.

Source     DF   Seq SS
TV          1  2857104
Radio       1     2785
Magazine    1      326
Website     1    28368
Billboard   1    72109
```

Figure 10.11

The reading we can make of the results is that because the P-values for "TV" is zero and the one for "Billboard" is 0.02, these two factors are significant in the model at an α level of 0.05. The website ads have an insignificant negative impact on the sales.

Both the coefficient of determination and the adjusted coefficient of determination show that a significant proportion in the variation of the sales is due to the variations in the regressors. Because "Website," "Radio," and "Magazine" are insignificant for the model, the Black Belt has decided to drop them from the model, but she is wondering what would happen to the model if the insignificant factors are taken off. What impact will that action have on R^2?

One way she can find out is to use stepwise regression. All the independent variables in the model contribute to the value of R^2, so if one of the variables is removed from or added to the model, this will change its value. *Stepwise regression* is a process that helps determine a model with only independent variables that are significant at a given α level. Three types of stepwise regression are generally used: the standard stepwise regression, the forward selection, and the backward elimination.

Standard stepwise. The *standard stepwise regression* is a selection method that starts with building a model with only one regressor, then adding regressors one at a time, keeping the significant ones and rejecting the insignificant ones, until there is no more significant regressor out of the model.

Simple linear regression models for each regressor are initially built to predict the response variable. The initial model will be the simple regression model with the highest absolute value of t at that α level. The first model will therefore be under the form of

$$\hat{y} = b_1x_1 + b_0$$

Then the next step will consist in finding the next regressor whose absolute value of t combined with an initial one would provide the highest t value in the model. The model becomes

$$\hat{y} = b_1x_1 + b_2x_2 + b_0$$

Then the whole model is examined to determine if all the t values are still significant at that α level. If they are, both of them are kept and the process is started again to add another regressor. Every time a regressor is added, the model is tested to see if the t values are still significant. When one has become insignificant, it is removed from the model.

After adding x_3 to the model, it becomes

$$\hat{y} = b_1x_1 + b_2x_2 + b_3x_3 + b_0$$

If after testing all the t values, the t value for x_2 has proved insignificant, x_2 is removed from the model and it becomes

$$\hat{y} = b_1x_1 + b_3x_3 + b_0$$

Use Minitab to run a stepwise regression at $\alpha = 0.05$. After pasting the data into a Minitab Worksheet, from the Stat menu select "Regression" and then select "Stepwise ..."

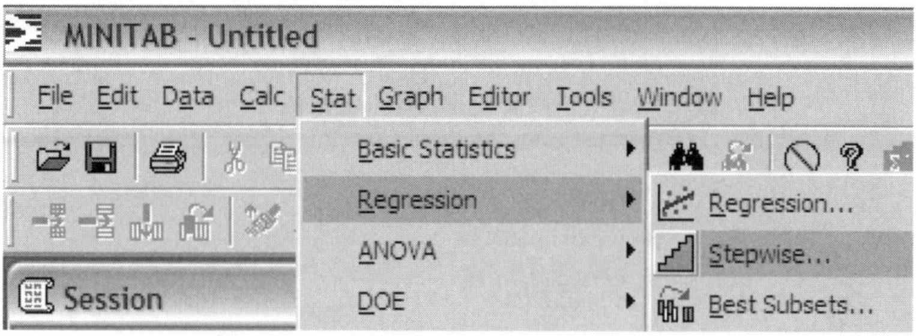

After the "Stepwise Regression" dialog box pops up, enter the variables into the appropriate fields as shown in Figure 10.12; then select "Methods..."

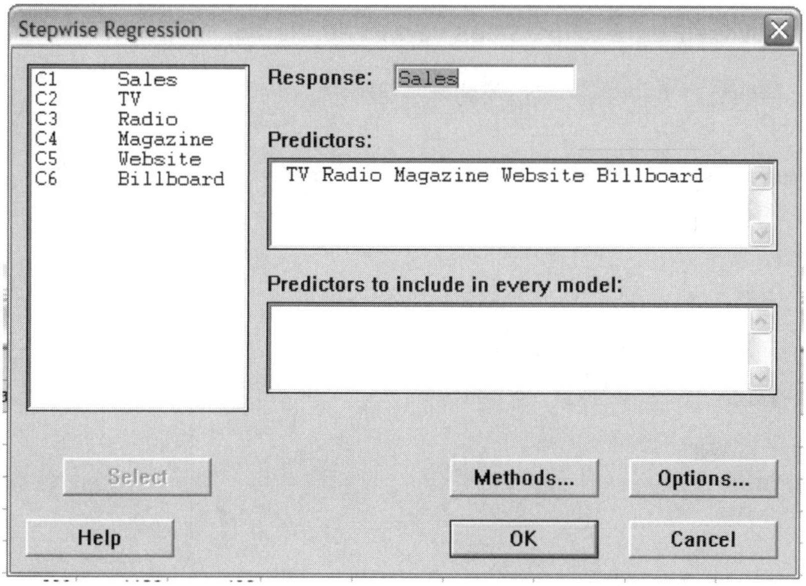

Figure 10.12

In the "Stepwise Method" box, the options *Use alpha values* and *Stepwise (forward and backward)* should be selected by default. The α levels are "0.15," so change them to "0.05."

Stepwise - Methods

Use alpha values   Use F values

Stepwise (forward and backward)
Predictors in initial model:

Alpha to enter: 0.05   F to enter: 4
Alpha to remove: 0.05   F to remove: 4

Forward selection
Alpha to enter: 0.25   F to enter: 4

Backward elimination
Alpha to remove: 0.1   F to remove: 4

Select

Help   OK   Cancel

Then select "OK" and "OK" again to get the output shown in Figure 10.13.

Stepwise Regression: Sales versus TV, Radio, ..

```
  Alpha-to-Enter: 0.05  Alpha-to-Remove: 0.05


Response is Sales on 5 predictors, with N = 16


Step              1       2
Constant     -311.8  -306.0

TV            146.6   140.5
T-Value       14.22   15.94
P-Value       0.000   0.000

Billboard             0.156
T-Value                2.73
P-Value               0.017

S               119    98.3
R-Sq          93.53   95.89
R-Sq(adj)     93.06   95.26
Mallows C-p     9.0     3.3
```

Figure 10.13

An examination of the output shows that the only two regressors that are kept are "TV" and "Billboard," and their *P*-values show that they are significant in the model. After removing the other regressors, R^2 and $adjR^2$ have not drastically changed; they went from 96.9 and 95.4 to 95.89 and 95.26, respectively. This means that the variations in the variables that were dropped from the model did not significantly account for the variations in sales.

Forward selection. As in the standard stepwise regression, the *forward selection* begins with building a model with the single largest coefficient of determination R^2 to predict the dependent variable. After that, it selects the second variable that produces the highest absolute value of t. Once that regressor is added to the model, the forward selection does not examine the model to see if the t values have changed and one needs to be removed. Forward selection is therefore similar to the standard stepwise regression with the difference that it does not remove regressors once they are added. With the data that we have, if we run a forward selection we would have the exact same results as in the case of the standard stepwise regression.

Backward elimination. The *backward elimination* begins the opposite to forward selection. It builds a complete model using all the regressors and then it looks for the insignificant regressors. If it finds one, it permanently deletes it from the model; if it does not find any, it keeps the model as is. The least significant regressors are deleted first. If we had ran a backward elimination using Minitab with data that we have at the same α level, we would have the same results.

Exercise. The rotations per minute (RPM) is critical to the quality of a wind generator. Several components affect the RPM of a particular generator. Among them, the weight of the fans, the speed of the wind, and the pressure. After having designed the Conakry model of a wind generator, the reliability engineer wants to build a model that will show how the "Rotation" variable relates to the "Wind," "Pressure," and "Weight" variables. After testing the generator under different settings, he tabulates the results shown in Table 10.8. The table can also be found in *Windmill.xls* and *Windmill.mpj* on the included CD.

Using the information from Table 10.8:

a. Show that "Wind" and "Pressure" are highly correlated.

b. Show that "Rotation" is highly dependent on the input factors.

c. Show that only "Weight" is significant in the equation.

d. Show that the VIF is too high for "Wind" and "Pressure."

TABLE 10.8

Rotation	Weight	Pressure	Wind
2345	69	98	75
2365	70	100	77
2500	74	114	88
2467	73	113	87
2345	69	127	98
2347	69	129	99
2134	63	101	78
2368	70	124	95
2435	72	113	87
2654	78	112	86
2345	69	104	80
2346	69	116	89
2435	72	127	98
2436	72	113	87
2435	72	116	89
2543	75	112	86
2435	72	113	87

e. What would happen to the coefficient of determination if the insignificant factors are removed from the equation?

f. Interpret the probability plot for the residuals.

g. Use stepwise regression, the forward selection, and backward elimination to determine the effect of removing some factors from the model.

Chapter 11

Design of Experiment

Incorrect business decisions can have very serious consequences for a company. The decisions that a company makes can be as simple as choosing on what side of a building to install bathrooms, or as complex as what should be the layout of a manufacturing plant, or as serious as whether it should invest millions of dollars in the acquisition of a failing company or not.

In any event, when we are confronted with a decision we must choose between at least two alternatives. The choice we make depends on many factors but in quality driven operations, the most important factor is the satisfaction that the customers derive from using the products or services delivered to them. The quality of a product is the result of a combination of factors. If that combination is suboptimal, quality will suffer and the company will lose as a result of rework and repair.

The best operations' decisions are the result of a strategic thinking that consists in conducting several experiments, combining relevant factors in different ways to determine which combination is the best. This process is known in statistics as *Design Of Experiment* (DOE).

Because several factors affect the quality levels of products and services (from now on, we will refer to the generated products or services as *response factors* or *response variables*) and they affect them differently, it is necessary to know how the input factors and their interactions affect the response variable.

One part of statistics that helps determine if different inputs affect the response variables differently is the ANOVA. ANOVA is a basic step in the DOE that is a formidable tool for decision-making based on data analysis. The types of ANOVA that are more commonly used are:

- The completely randomized experimental design, or one-way ANOVA. One-way ANOVA compares several (usually more than two) samples' means to determine if there is a significant difference between them.
- The factorial design, or two-way ANOVA, which takes into account the effect of noise factors.

Because one-way ANOVA has been extensively dealt with in the previous chapters, we will concentrate on two-way ANOVA in this chapter. Through an example, we will show the mathematical reasoning to better understand the rationale behind the results generated by Minitab and Excel. We will conduct a test in two ways: we will first use the math formulas and then we will use Minitab or Excel and extensively interpret the generated results after the test.

11.1 The Factorial Design with Two Factors

In our one-way ANOVA examples, all we did was determine if there was a significant difference between the means of the three treatments. We did not consider the treatment levels in those examples.

The *factorial designs* experiments conducted in a way that several treatments are tested simultaneously. The difference from the one-way ANOVA is that in factorial design, the level of every treatment is tested for all treatments.

Consider that the heat generated by a type of electricity generator depends on its RPM and on the time it is operating. Samples taken while two generators are running for four hours are summarized in Table 11.1. For the first hour, the generators ran at 500 RPM and the heat generated was 65 degrees Celsius for both generators.

In this example, the variations in the level of heat produced by the generators can be due to the time they have been running, or to the fluctuations in the RPM, or the interaction of the time and RPM variations. Had we been running a one-way ANOVA, we would have only considered one treatment (either the time or the RPM).

With the two-way ANOVA, we consider all the RPMs for every timeframe and we also consider the timeframes (hours) for every RPM. The row effects and the column effects are called the *main effects*, and the combined effects of the rows and columns is called the *interaction effect*. In this example, we will call a "cell" the intersection between an RPM (column) and a length of time (row). Every cell has two observations. Cell 1 is comprised of observations (65, 65). So we have three cells per row and four rows, which make a total of 12 cells.

TABLE 11.1

Hours	500 RPM	550 RPM	600 RPM
1	65	80	84
	65	81	85
2	75	83	85
	80	85	86
3	80	86	90
	85	87	90
4	85	89	92
	88	90	92

As in the case of the one-way ANOVA, the two-way ANOVA is also a hypothesis test and in this case, we are faced with three hypotheses: The first hypothesis will stipulate that there is no difference between the means of the RPM treatments.

$$H_0 : \mu_{r1} = \mu_{r2} = \mu_{r3}$$

where μ_1, μ_2, and μ_3 are the means of the RPM treatments.

The second hypothesis will stipulate that the number of hours that the generators operate does not make any difference on the heat.

$$H_0 : \mu_{h1} = \mu_{h2} = \mu_{h3}$$

where μ_{h1}, μ_{h2}, and μ_{h3} are the means of the hours that generators were operating.

The third stipulation will be that the effect of the interaction of the two main effects (RPM and time) is zero. If the interaction effect is significant, a change in one treatment will have an effect on the other treatment. If the interaction is very important, we say that the two treatments are *confounded.*

Conduct the ANOVA for the data we gathered. At first, we will use the formulas and mathematically solve the problem, and then verify the results we obtained using Excel. We will show how to use Excel step by step to conduct a factorial design, two-way ANOVA.

11.1.1 How does ANOVA determine if the null hypothesis should be rejected or not?

The way ANOVA determines if the null hypothesis should be rejected or not is by assessing the sources of the variations from the means. In Table 11.1, all the observations are not identical; they range between 65 and 92 degrees. The means of the different main factors (the different RPMs and the different timeframes) are not identical, either. For a confidence level of 95 percent (an α level of 0.05), ANOVA seeks to determine the sources of the variations between the main factors. If the sources of the variations are solely within the treatments (in this case, within the columns or rows), we would not be able to reject the null hypothesis. If the sources of variations are between the treatments, we reject the null hypothesis.

Table 11.2 summarizes what we have just discussed:

TABLE 11.2

Sources of variation	Sums of squares	Degrees of freedom	Mean square	F-statistics
RPM	tSS	$a-1$	MSt = tSS / $(a-1)$	F = MSt / MSE
TIME	lSS	$b-1$	MSL = lSS / $(b-1)$	F = MSL / MSE
Interaction Time RPM	ISS	$(a-1)(b-1)$	MSI = ISS / $[(a-1)(b-1)]$	F = MSI / MSE
Error	SSE	$ab(n-1)$	MSE = SSE / $[ab(n-1)]$	
Total	TSS	$N-1$		

The formulas for the sums of squares to solve a two-way ANOVA with interaction are given as follows.

The sums of squares are nothing but deviations from means,

$$lSS = nt\sum_{i=1}^{l}(\overline{X}_i - \overline{X})^2$$

$$tSS = nl\sum_{j=1}^{t}(\overline{X}_j - \overline{X})^2$$

$$ISS = n\sum_{i=1}^{l}\sum_{j=1}^{t}(\overline{X}_{ij} - \overline{X}_i - \overline{X}_j + \overline{X})^2$$

$$SSE = \sum_{i=1}^{l}\sum_{j=1}^{t}\sum_{k=1}^{n}(\overline{X}_{ijk} - \overline{X}_{ij})^2$$

$$TSS = \sum_{i=1}^{l}\sum_{j=1}^{t}\sum_{k=1}^{n}(\overline{X}_{ijk} - \overline{X})^2$$

where lSS is the sum of squares for the rows, tSS is the sum of squares for the treatments, ISS is the sum of squares for the interactions, SSE is the error of the sum of squares, TSS is the total sum of squares, n is the number of observed data in a cell ($n = 2$), t is the number of treatments, l is the number of row treatments, i is the number of treatment levels, j is the column treatment levels, k is the number of cells, X_{ijk} is any observation, $\overline{X}_{ij}$ is the cell mean, $\overline{X}_i$ is the level mean, $\overline{X}_j$ is the treatment mean, and $\overline{X}$ is the mean of all the observations.

11.1.2 A mathematical approach

The extra cells to the right in Table 11.3 are the means of the six numbers to their left. The extra cells at the very bottom are the means of the columns Plug in the numbers to the equations:

TABLE 11.3

Hours	500 RPM	550 RPM	600 RPM	
1	65	80	84	
	65	81	85	76.66667
Mean	65	80.5	84.5	
2	75	83	85	
	80	85	86	82.33333
Mean	77.5	84	85.5	
3	80	86	90	
	85	87	90	86.33333
Mean	82.5	86.5	90	
4	85	89	92	
	88	90	92	89.33333
Mean	86.5	89.5	92	
	77.875	85.125	88	

$$lSS = nt\sum_{i=1}^{l}(\overline{X}_i - \overline{X})^2 = 2 \times 3((76.66 - 83.667)^2 + (82.333 - 83.667)^2 + (86.33 - 86.667)^2 + (89.333 - 86.667)^2)$$

Therefore

$$lSS = nt\sum_{i=1}^{l}(\overline{X}_i - \overline{X})^2 = 2 \times 3(49 + 1.778 + 7.1111 + 32.111) = 540$$

$$tSS = nl\sum_{j=1}^{t}(\overline{X}_j - \overline{X})^2 = 2 \times 4(33.5434 + 2.126736 + 18.77778)$$

$$= 435.5833$$

Add a row at the bottom of each cell to visualize the means for the cells. The mean of the first cell is 65, because (65 + 65) / 2 = 65.

$$ISS = n\sum_{i=1}^{l}\sum_{j=1}^{t}(\overline{X}_{ij} - \overline{X}_i - \overline{X}_j + \overline{X})^2$$

$$= 2\left[(65 - 76.6667 - 77.875 + 83.667)^2 + (80.5 - 76.6667 - 85.125 + 83.667)^2\right] + \cdots + (92 - 89.333 - 88 + 83.667)^2 = 147.75$$

The SSE is obtained from

$$SSE = \sum_{i=1}^{l}\sum_{j=1}^{t}\sum_{k=1}^{n}(X_{ijk} - \overline{X}_{ij})^2$$

$$= (65-65)^2 + (65-65)^2 + (80-80.5)^2 + \cdots + (90-89)^2 + (92-92)^2 + (92-92)^2 = 34$$

The TSS will be the sum of SSE, ISS, tSS, and lSS,

$$TSS = 34 + 147.75 + 540 + 435.5833 = 1157.3333$$

We can now insert the numbers in our table. Note that what we were looking for was the F-statistic.

TABLE 11.4

Sources of variation	Sums of squares	Degrees of freedom	Mean square	F-statistic	F-critical
RPM	435.45	2	217.79	76.86	3.89
TIME	540	3	180	63.53	3.49
Interaction Time RPM	147.75	6	24	8.69	3
Error	34	12	2.833		
Total	1157.333	23			

To determine if we must reject the null hypothesis, we must compare the F-statistic to the F-critical value found on the F table. If the F-critical value (the one on the F-table) is greater than the F-statistic (the one we calculated), we would not reject the null hypothesis; otherwise, we do. In this case, the F-statistics for all the main factors and interaction are greater than their corresponding F-critical values, so we must reject the null hypotheses. The length of time the generators are operating, the RPM variations, and the interaction of RPM and time have an impact on the heat that the generators produce. But once we determine that the interaction between the two main factors is significant, it is unnecessary to investigate the main factors.

Using Minitab. Open the file *Generator.mpj*. From the Stat menu, select "ANOVA" and then select "Two-Way." In the "Two-Way Analysis of Variance" dialog box, select "Effect" for the *Response* field. For *Row Factor*, select "Time" and for *Column Factor*, select "RPM." Select "OK" to obtain the results shown in Figure 11.1.

Using Excel. We must have Data Analysis installed in Excel to perform ANOVA. If you do not have it installed, from the Tools menu select "Add

Two-way ANOVA: Effect versus Time, RPM

```
Source        DF       SS       MS      F      P
Time           3   540.00  180.000  63.53  0.000
RPM            2   435.58  217.792  76.87  0.000
Interaction    6   147.75   24.625   8.69  0.001
Error         12    34.00    2.833
Total         23  1157.33

S = 1.683   R-Sq = 97.06%   R-Sq(adj) = 94.37%
```

Figure 11.1

Ins." A dialog box should pop up, select all the options, and then select "OK."

Now that we have Data Analysis, open the file *Generator.xls* from the included CD. From the Tools menu, select "Data Analysis . . . "

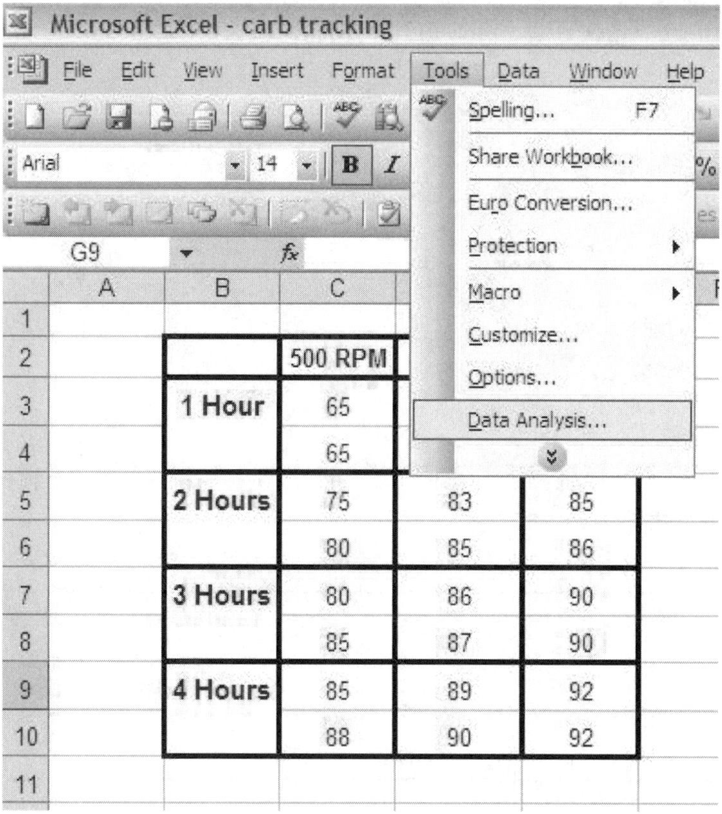

Microsoft Excel - carb tracking

File Edit View Insert Format Tools Data Window Help

Spelling... F7
Share Workbook...
Euro Conversion...
Protection
Macro
Customize...
Options...
Data Analysis...

	500 RPM		
1 Hour	65		
	65		
2 Hours	75	83	85
	80	85	86
3 Hours	80	86	90
	85	87	90
4 Hours	85	89	92
	88	90	92

Select "ANOVA: Two-Factor With Replication."

Data Analysis

Analysis Tools

Anova: Single Factor
Anova: Two-Factor With Replication
Anova: Two-Factor Without Replication
Correlation
Covariance
Descriptive Statistics
Exponential Smoothing
F-Test Two-Sample for Variances
Fourier Analysis
Histogram

OK
Cancel
Help

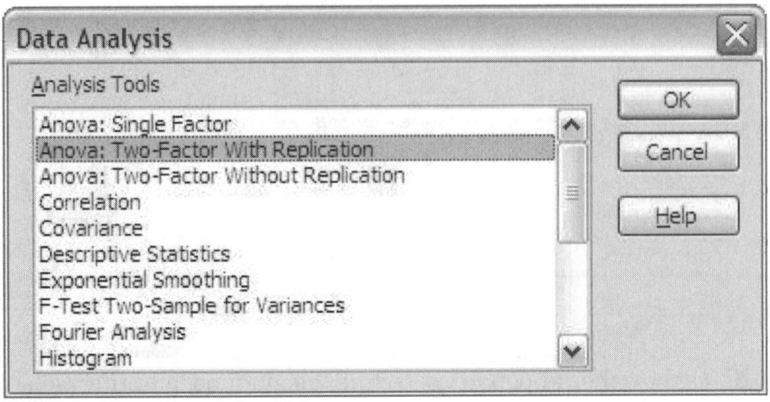

When selecting the *Input Range*, we include the labels (the titles of the rows and columns).

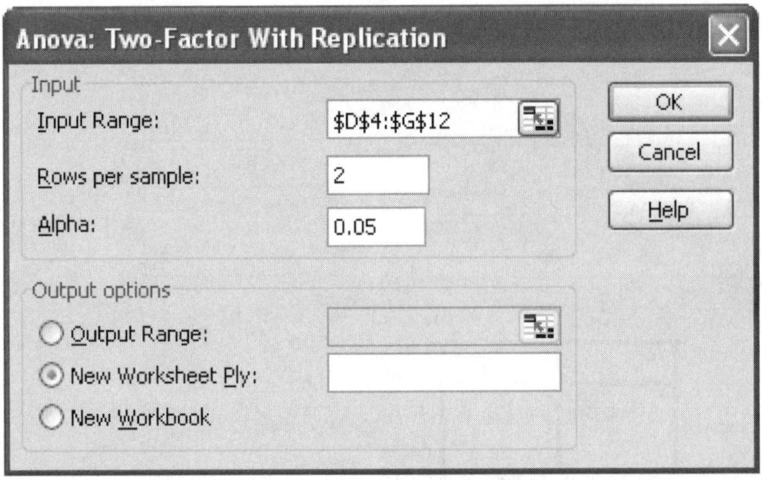

Anova: Two-Factor With Replication

Input

Input Range: $D$4:$G$12
Rows per sample: 2
Alpha: 0.05

Output options

Output Range:
New Worksheet Ply:
New Workbook

OK
Cancel
Help

Then select "OK."

35	ANOVA						
36	*Source of Variation*	*SS*	*df*	*MS*	*F*	*P-value*	*F crit*
37	Sample	540	3	180	63.52941	1.23E-07	3.490295
38	Columns	435.5833	2	217.7917	76.86765	1.44E-07	3.885294
39	Interaction	147.75	6	24.625	8.691176	0.000847	2.99612
40	Within	34	12	2.833333			
41							
42	Total	1157.333	23				
43							

We can also determine the significance of our results based on the P-value. In this case, all the P-values are infinitesimal, much lower than 0.05, which confirms the conclusion we previously made: the RPMs, the time, and interactions thereof have an effect on the heat produced by the generators. But once we determine that the interaction between the two main factors is significant, it is unnecessary to investigate the main factors.

Example Sanghomar is a company that manufactures leather products. It has four lines that use that same types of machines. The quality engineer has noticed variations in the thickness of the sheets of leather that come from the different lines. He thinks that only two factors, the machines and the operators, can affect the quality of the products. He wants to run a two-way ANOVA and takes samples of the products generated by five operators and the four machines. The data obtained are summarized in Table 11.5. The same data are in the files *Leatherthickness.xls* and *Leatherthickness.mpj* on the included CD.

TABLE 11.5

Employee	Machine 1	Machine 2	Machine 3	Machine 4
1	9.01	9.01	9.06	9.03
1	9.20	9.20	9.24	9.20
2	9.06	9.07	9.06	9.06
2	9.00	9.00	8.98	9.03
3	9.01	9.01	9.21	8.97
3	8.90	8.99	9.09	8.90
4	9.02	9.02	9.02	9.02
4	9.40	8.98	9.40	9.40
5	9.89	9.89	9.05	9.89
5	9.60	9.07	9.05	8.90

Using Excel and Minitab:

a. Determine if there is a difference between the performances of the machines.

b. Determine if there is a difference between the performances of the employees.

c. What can we say about the interaction between machines and employees?

Solution Minitab output:

Two-way ANOVA: Thickness versus Employees, Machines

```
Source       DF       SS        MS     F      P
Employees     4  0.84766  0.211915  3.60  0.023
Machines      3  0.05383  0.017942  0.31  0.821
Interaction  12  0.56586  0.047155  0.80  0.646
Error        20  1.17625  0.058813
Total        39  2.64360

S = 0.2425   R-Sq = 55.51%   R-Sq(adj) = 13.24%
```

```
                        Individual 95% CIs For Mean Based on
                        Pooled StDev
```

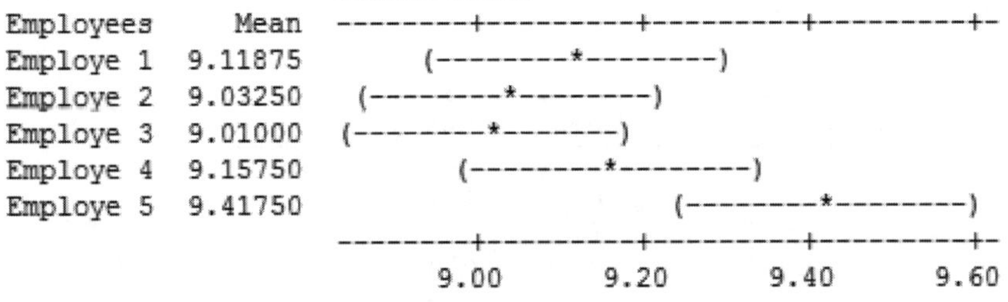

```
Employees      Mean  --------+---------+---------+---------+-
Employe 1   9.11875       (--------*--------)
Employe 2   9.03250   (--------*--------)
Employe 3   9.01000  (--------*-------)
Employe 4   9.15750         (--------*--------)
Employe 5   9.41750                     (--------*--------)
                     --------+---------+---------+---------+-
                          9.00      9.20      9.40      9.60
```

```
                     Individual 95% CIs For Mean Based on
                     Pooled StDev
Machines   Mean  ----+---------+---------+---------+-----
Machine1  9.209          (------------*-------------)
Machine2  9.124   (------------*-------------)
Machine3  9.116  (-------------*------------)
Machine4  9.140   (-------------*------------)
                 ----+---------+---------+---------+-----
```

Excel output:

ANOVA						
Source of Variation	*SS*	*df*	*MS*	*F*	*P-value*	*F crit*
Sample	0.84766	4	0.211915	3.603231	0.022809	2.866081
Columns	0.0538275	3	0.017942	0.30508	0.821384	3.098391
Interaction	0.56586	12	0.047155	0.801785	0.645565	2.277581
Within	1.17625	20	0.058813			
Total	2.6435975	39				

a. The P-value for the machines is 0.821, which means that we cannot reject the null hypothesis because there is not a significant difference between the performances of the machines.

b. The P-value for the employees is 0.023, which means that we must reject the null hypothesis and conclude that there is a significant difference between the performances the employees.

c. The P-value for the interaction between machines and employees is 0.646, which means that we cannot reject the null hypothesis. The interaction of employees and machines does not make a significant difference in the quality of the products.

11.2 Factorial Design with More than Two Factors (2^k)

In the previous examples, only two factors were considered to have an effect on the response variable. In most situations, more than two independent variables affect the response factor. When the factors affecting the response variable are too many, collecting many samples for analysis that would reflect multiple factorial levels may become time-consuming and costly, and the analysis itself may become complex. An alternative to that approach would be the use of 2^k, a two levels with k factors design. This approach simplifies the analysis because it only considers two levels, high (+1) and low (−1) for each factor, resulting in 2^k trials. The simplest form of a 2^k design with more than two factors is the 2^3. In this case, the number of trials would be $2^3 = 2 \times 2 \times 2 = 8$.

Example New barcode scanners are being tested for use at a Memphis Distribution Center to scan laminated barcodes. The speed at which the scanners perform is critical to the productivity of the employees in the warehouse. The quality engineer has determined that three factors can contribute to the speed of the scanners: the distance between the operator and the barcodes, the ambient light in the warehouse, and the reflections from the laminated barcodes. She conducts an experiment based on a time study that measured the time it took a scanner to read a barcode. In this study, eight trials are conducted, and the results of the trials are summarized in Table 11.6. The first trial—conducted at a close distance, in a dark environment, and without any reflection from the laminated barcode—yielded a high response of 2.01 seconds and a low response of 1.95 seconds.

The objective of the experimenter is to determine if the different factors or their interactions have an impact on the response time of the Barcode scanner. In other words:

- Does the distance from which the operator scans the barcode have an impact on the time it takes the scanner to read the barcode?
- Is the scanner going to take longer to read the barcode if the warehouse is dark?

- Do the reflections due to the plastic on the laminated barcode reduce the speed of reading of the scanner?
- Do the interactions between these factors have an influence on the performance of the scanner?

To answer these questions, the experimenter postulates hypotheses. The null hypothesis would suggest that the different factors do not have any impact on the time the scanner takes to read the barcode, and the alternate hypothesis would suggest the opposite.

$$H_0 : \mu_{low} = \mu_{high}$$
$$H_a : \mu_{low} \neq \mu_{high}$$

At the end of the study, the experimenter will either reject the null hypothesis or she will fail to reject them.

Then the experimenter collects samples of observations and tabulates the different response levels as shown in Table 11.6.

TABLE 11.6

Distance	Light	Reflection	High Response	Low Response	Mean Response
Close	Dark	None	2.01	1.95	1.980
Close	Bright	Glary	2.08	2.04	2.060
Close	Bright	None	2.07	1.90	1.985
Close	Dark	Glare	3.00	1.98	2.490
Far	Dark	Glare	4.00	3.90	4.250
Far	Dark	None	4.02	4.10	4.360
Far	Bright	None	4.02	3.80	3.910
Far	Bright	Glare	4.05	4.30	4.175

The test that the experimenter conducts is a *balanced ANOVA*. A balanced ANOVA requires all the treatments to have the same number of observations, which we have in this case.

Solution Open the file *RFscanner.mpj* from the included CD. From the Stat menu, select "ANOVA" and from the drop-down list, select "Balanced ANOVA." Fill in the "Balanced Analysis of Variance" as indicated in Figure 11.2:

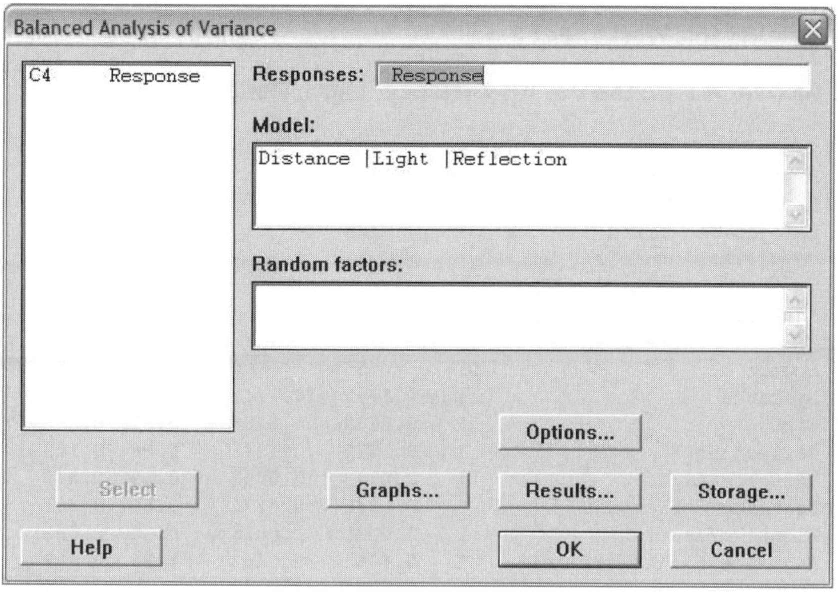

Figure 11.2

Select "Options..." and select the *Use the restricted form of model* option. Select "OK" and select "OK" again to get the output shown in Figure 11.3.

Interpretation of the results. What we are seeking to determine is if the three factors (distance, light, and reflection) separately have an impact on the time it takes to scan a laminated barcode. As in the case of the one-way ANOVA, what will help us make a determination is the *P*-value. For a confidence level of 95 percent, if the *P*-value is less that 0.05, we must reject the null hypothesis and conclude that the factor has an impact on the response time; otherwise, we will fail to reject the null hypothesis.

In our example, all the main factors except for the distance and all the interactions have a *P*-value greater than 0.05. Therefore, we must conclude that only the null hypothesis for the distance will be rejected; the other ones should not be rejected. Distance is the only main factor affecting the time it takes the scanner to read the barcode.

Results for: Worksheet 2

ANOVA: Response versus Distance, Light, Reflection

Factor	Type	Levels	Values
Distance	fixed	2	Close, Far
Light	fixed	2	Bright, Dark
Reflection	fixed	2	Glary, None

Analysis of Variance for Response

Source	DF	SS	MS	F	P
Distance	1	14.1941	14.1941	189.97	0.000
Light	1	0.0233	0.0233	0.31	0.592
Reflection	1	0.1541	0.1541	2.06	0.189
Distance*Light	1	0.0743	0.0743	0.99	0.348
Distance*Reflection	1	0.0371	0.0371	0.50	0.501
Light*Reflection	1	0.0028	0.0028	0.04	0.852
Distance*Light*Reflection	1	0.1463	0.1463	1.96	0.199
Error	8	0.5977	0.0747		
Total	15	15.2295			

S = 0.273347   R-Sq = 96.08%   R-Sq(adj) = 92.64%

Residual Plots for Response

Figure 11.3

Chapter

12 The Taguchi Method

I recently had a very unpleasant experience with a notebook computer that I bought about six months ago. At first, I was having some irritating problems with the LCD (liquid Crystal display). It would be very dim for about five minutes when I turned the computer on; the system had to warm up before the LCD would display correctly. I did bear with that situation until it started to black out while I was in the middle of my work. Because it was still under warranty, I sent it back to the manufacturer for repair and decided that I would never buy a product from that manufacturer again.

12.1 Assessing the Cost of Quality

The quality of a product is one of the most important factors that determine a company's sales and profit. Quality is measured in relation with the characteristics of the products that customers' expect to find on it, so the quality level of the products is ultimately determined by the customers. The customers' expectations about a product's performance, reliability, and attributes are translated into CTQ characteristics and integrated in the products' design by the design engineers.

While designing the products, the design engineers must also take into account the resources' capabilities (machines, people, materials, and so on), i.e., their ability to produce products that meet the customers' expectations. They specify with precision the quality targets for every aspect of the products.

But quality comes with a cost. The definition of the cost of quality is contentious. Some authors define it as the cost of nonconformance, i.e., how much producing nonconforming products would cost a company. This is a one-sided approach because it does not consider the cost incurred to prevent nonconformance and, above all in a competitive market, the cost of improving the quality targets.

For instance, in the case of an LCD display manufacturer, if the market standard for a 15-inch LCD with a resolution of 1024 × 768 is 786,432 pixels and a higher resolution requires more pixels, improving the quality of the 15-inch LCD and pushing the company's specifications beyond the market standards would require the engineering of LCDs with more pixels, which would require extra cost.

The cost of quality is traditionally measured in terms of the cost of conformance and the cost of nonconformance, to which we will add the cost of innovation. The cost of conformance includes the appraisal and preventive costs whereas the cost of nonconformance includes the costs of internal and external defects.

12.1.1 Cost of conformance

Preventive costs. Preventive costs are the costs incurred by the company to prevent nonconformance. It includes the costs of:

- Process capability assessment and improvement
- The planning of new quality initiatives (process changes, quality improvement projects, and so on)
- Employee training

Appraisal costs. Appraisal costs are the costs incurred while assessing, auditing, and inspecting, products and procedures to assure conformance of products and services to specifications. It is intended to detect quality related failures. It includes:

- Cost of process audits
- Inspection of products received from suppliers
- Final inspection audit
- Design review
- Prerelease testing

12.1.2 Cost of nonconformance

The cost of nonconformance is, in fact, the cost of having to rework products and the loss of customers that results from selling poor quality products.

Internal failure. Internal failures are failures that occur before the products reach the customers.

- Cost of reworking products that failed audit
- Cost of bad marketing
- Scrap

External failure. External failures are reported by the customers.

- Cost of customer support
- Cost of shipping returned products
- Cost of reworking products returned from customers
- Cost of refunds
- Loss of customer goodwill
- Cost of discounts to recapture customers

In the short term, there is a positive correlation between quality improvement and the cost of conformance and a negative correlation between quality improvement and the cost of nonconformance. In other words, an improvement in the quality of the products will lead to an increase in the cost of conformance that generated it. This is because an improvement in the quality level of a product might require extra investment in R&D, more spending in appraisal cost, more investment in failure prevention, and so on.

But a quality improvement will lead to a decrease in the cost of nonconformance because fewer products will be returned from the customers, therefore less operating costs of customer support, and there will be less internal rework.

For instance, one of the CTQs for an LCD is the number of pixels it contains. The brightness of each pixel is controlled by individual transistors that switch the backlights on and off. The manufacturing of LCDs is very complex and very expensive, and it is very difficult to determine the number of dead pixels on an LCD before the end of the manufacturing process. So to reduce the number of scrapped units, if the number of dead pixels is infinitesimal or the dead pixels are almost invisible, the manufacturer would consider the LCDs as "good enough" to be sold. Otherwise, the cost of scrap or internal rework would be so prohibitive that it would jeopardize the cost of production. Improving the quality level of the LCDs to zero dead pixels would therefore increase the cost of conformance. On the other hand, not improving the quality level of the LCDs will lead to an increase in the probability of

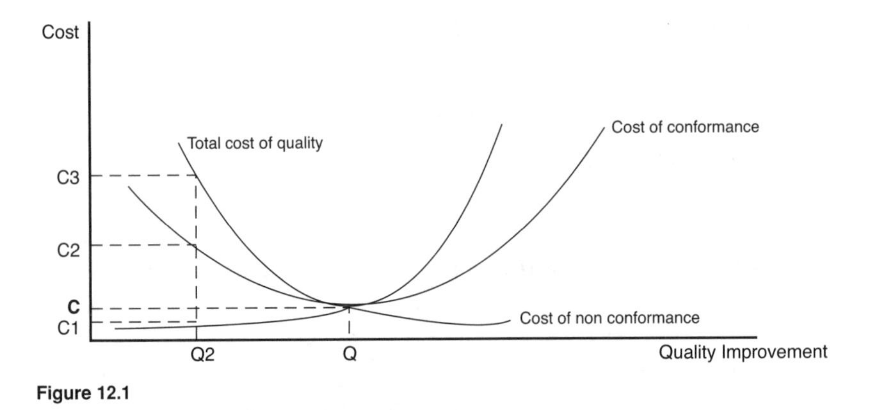

Figure 12.1

having returned products from customers and internal rework, therefore increasing the cost of nonconformance.

The graph in Figure 12.1 plots the relationship between quality improvement and the cost of conformance on one hand and the cost of nonconformance on the other hand.

If the manufacturer determines the quality level at Q2, the cost of conformance would be low (C1), but the cost of nonconformance would be high (C2) because the probability for customer dissatisfaction will be high and more products will be returned for rework, therefore increasing the cost of rework, the cost of customers services, and shipping and handling.

The total cost of quality would be the sum the cost of conformance and the cost of nonconformance. That cost would be C3 for a quality level of Q2.

$$C3 = C1 + C2$$

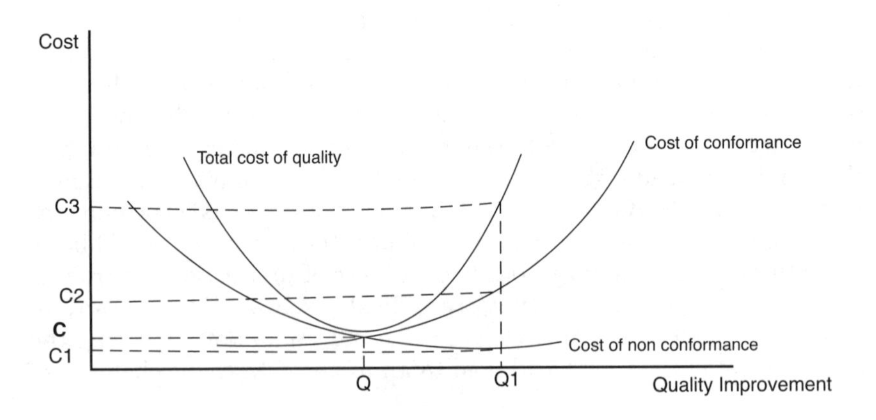

Should the manufacturer decide that the quality level should be at Q1, the cost of conformance (C2) would be higher than the cost of nonconformance (C1) and the total cost of quality would be at C3. The total cost of quality is minimized only when the cost of conformance and the cost of nonconformance are equal.

It is worthy to note that currently, the most frequently used graph to represent the throughput yield in manufacturing is the normal curve. For a given target and specified limits, the normal curve helps estimate the volume of defects that should be expected. Whereas the normal curve estimates the volume of defects, the "U" curve estimates the cost incurred as a result of producing parts that do not match the target.

The graph of Figure 12.2 represents both the volume of expected conforming and nonconforming parts and the costs associated to them at every level.

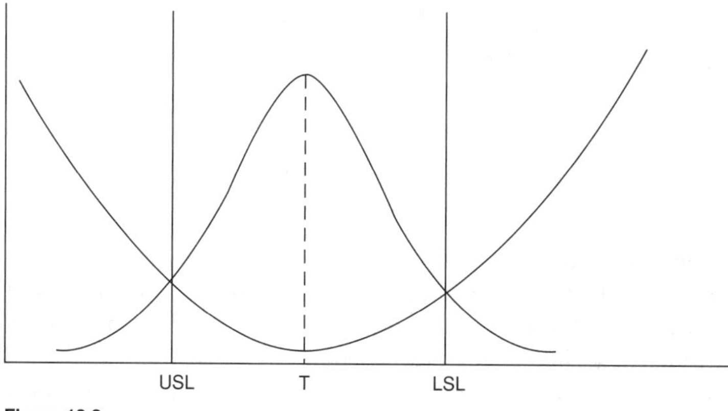

Figure 12.2

12.2 Taguchi's Loss Function

In the now-traditional quality management acceptance, the engineers integrate all the CTQs in the design of their new products and clearly specify the target for their production processes as they define the characteristics of the products to be sent to the customers. But because of unavoidable common causes of variation (variations that are inherent to the production process and that are hard to eliminate) and the high costs of conformance, they are obliged to allow some variation or tolerance around the target. Any product that falls within the specified tolerance is considered as meeting the customers' expectations, and any product outside the specified limits would be considered as nonconforming.

But according to Taguchi, the products that do not match the target—even if they are within the specified limits—do not operate as intended and any deviation from the target, be it within the specified limits or not, will generate financial loss to the customers, the company, and to society, and the loss is proportional to the deviation from the target.

Suppose that a design engineer specifies the length and diameter of a certain bolt that needs to fit a given part of a machine. Even if the customers do not notice it, any deviation from the specified target will cause the machine to wear out faster, causing the company a financial loss under the form of repairs of the products under warranty or a loss of customers if the warranty has expired.

Taguchi constructed a *loss function equation* to determine how much society loses every time the parts produced do not match the specified target. The loss function determines the financial loss that occurs every time a CTQ of a product deviates from its target. The loss function is the square of the deviation multiplied by a constant k, with k being the ratio of the cost of defective products and the square of the tolerance.

The loss function quantifies the deviation from the target and assigns a financial value to the deviation,

$$l(y) = k(y - T)^2$$

with

$$k = \frac{\Delta}{m^2}$$

where Δ is the cost of a defective product, T is the engineered target and m is a measure of the deviation from the target. and $m = LSL - T$ or $m = T - USL$. According to Taguchi, the cost of quality in relation to the deviation from the target is not linear because the customers' frustration increases (at a faster rate) as more defects are found on a product. That's why the loss function is quadratic.

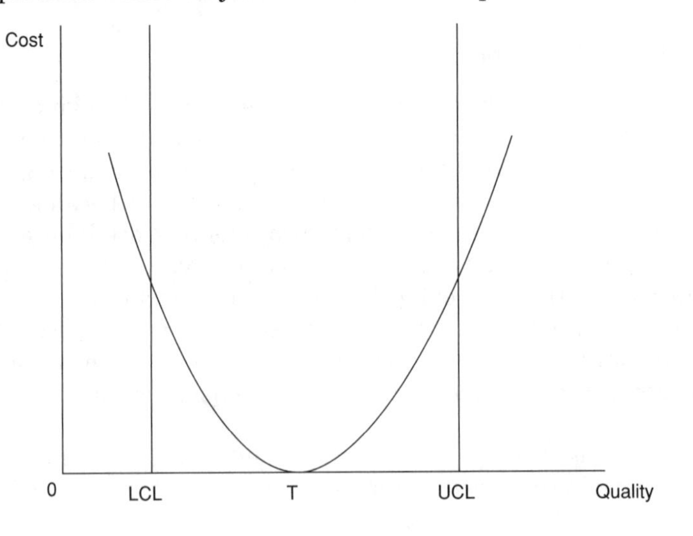

The graph that depicts the financial loss to society that results from a deviation from the target resembles the total cost of quality "U" graph that we built earlier, but the premises that helped build them are not the same. While the total cost curve was built based on the costs of conformance and nonconformance, Taguchi's loss function is primarily based on the deviation from the target and measures the loss from the perspective of the customers' expectations.

Example Suppose a machine manufacturer specifies the target for the diameter of a given rivet to be 6 inches and the upper and lower limits to be 5.98 and 6.02 inches, respectively. A bolt measuring 5.99 inches is inserted in its intended hole of a machine. Five months after the machine was sold, it breaks down as a result of loose parts. The cost of repair is estimated at \$95. Find the loss to society incurred as a result of the part not matching its target.

Solution We must first determine the value of the constant k:

$$l(y) = k(y - T)^2$$

with

$$k = \frac{\Delta}{m^2}$$

$$T = 6$$

$$USL = 6.02$$

$$m = (USL - T) = 6.02 - 6 = 0.02$$

$$\Delta = 95$$

$$k = (95/0.0004) = 237{,}500$$

Therefore,

$$l(y) = 0.0001 \times \$237{,}500 = \$23.75$$

Not producing a bolt that matches the target would have resulted in a financial loss to society that amounted to \$23.75.

12.3 Variability Reduction

Because the deviation from the target is the source of financial loss to society, what needs to be done to prevent any deviation from the set target? The first thought might be to reduce the specification range and improve the online quality control—to bring the specified limits closer to the target and inspect more samples during the production process to find the defective products before they reach the customers. But this would not be a good option because it would only address the symptoms and not the root causes of the problem. It would be an expensive alternative because it would require more inspection, which would at best help detect nonconforming parts early enough to prevent them from reaching the customers. The root of the problem is, in fact, the variation within the production process, i.e., the value of σ, the standard deviation from the mean.

Suppose that the length of a screw is a CTQ characteristic and the target is determined to be 15 inches with an LCL of 14.96 and a UCL of 15.04 inches. The sample data of Table 12.1 was taken for testing.

TABLE 12.1

15.02
14.99
14.96
15.03
14.98
14.99
15.03
15.01
14.99

All the observed items in this sample fall within the control limits even though all of them do not match the target. The mean is 15 and the standard deviation is 0.023979. Should the manufacturer decide to improve the quality of the output by reducing the range of the control limits to 14.98 and 15.02, three of the items in the sample would have failed audit and would have to be reworked or discarded.

Suppose that the manufacturer decides instead to reduce the variability (the standard deviation) around the target and leave the control limits untouched. After process improvement, the sample data of Table 12.2 is taken.

TABLE 12.2

15.01
15.00
14.99
15.01
14.99
14.99
15.00
15.01
15.00

The mean is still 15 but the standard deviation has been reduced to 0.00866, and all the observed items are closer to the target. Reducing the variability around the target has resulted in improving quality in the production process at a lower cost. This is not to suggest that the tolerance around the target should never be reduced; addressing the tolerance limits should be done under specific conditions and only after the variability around the target has been reduced.

Because variability is a source of financial loss to producers, customers, and society at large, it is necessary to determine what the sources of variation are so that actions can be taken to reduce them. According to Taguchi, these sources of variation that he calls *noise factors* can be reduced to three:

- The *inner noise*. Inner noises are deteriorations due to time. Product wear, metal rust or fading colors, material shrinkage, and product waning are among the inner noise factors.
- The *outer noise*, which are environmental effects on the products. They are factors such as heat, humidity, operating conditions, or pressure. These factors have negative effects on products or processes. In the case of the notebook computer, at first the LCD would not display until it heated up, so humidity was the noise factor that was preventing it from operating properly. The manufacturer has no control over these factors.
- The *product noise*, or manufacturing imperfections. Product noise is due to production malfunctions, and can come from bad materials, inexperienced operators, or incorrect machine settings.

But if the online quality control is not the appropriate way to reduce production variations, what must be done to prevent deviations from the target?

According to Taguchi, a preemptive approach must be taken to thwart the variations in the production processes. That preemptive approach that he calls *offline quality control* consists in creating a robust design—in other words, designing products that are insensitive to the noise factors.

12.3.1 Concept design

The production of a product begins with the *concept design*, which consists in choosing the product or service to be produced and defining its structural design and the production process that will be used to generate it. These factors are contingent upon, among other factors, the cost of production, the company's strategy, the current technology, and the market demand. So the concept design will consist of:

- Determining the intended use of the product and its basic functions
- Determining the materials needed to produce the selected product
- Determining the production process needed to produce the product

12.3.2 Parameter design

The next step in the production process is the *parameter design*. After the design architecture has been selected, the producer will need to set the parameter design. The parameter design consists in selecting the best combination of control factors that will optimize the quality level of the product by reducing the product's sensitivity to noise factors. *Control factors* are parameters over which the designer has control. When an engineer designs a computer, he has control over factors such as the CPU, system board, LCD, memory, cables, and so on. He determines what CPU best fits a motherboard, what memory and what wireless network card to use, and how to design the system board that would make it easier for the parts to fit. The way he combines these factors will impact the quality level of the computer.

The producer wants to design products at the lowest possible cost and, at the same time, have the best quality result under current technology. To do so, the combination of the control factors must be optimal while the effect of the noise factors must be so minimal that they will not have any negative impact on the functionality of the products. The experiment that leads to the optimal results will require the identification of the noise factors because they are part of the process and their effects must be controlled.

One of the first steps the designer will take is to determine what the optimal quality level is. He will need to determine what the functional requirements are, assess the CTQ characteristics of the product, and specify their targets. The determination of the CTQs and their targets depends, among other criteria, on the customer requirements, the cost of production, and current technology. The engineer is seeking to produce the optimal design,a product that is insensitive to noise factors.

The quality level of the CTQ characteristics of the product under optimal conditions depends on whether the response experiment is static or dynamic. The response experiment (or output of the experiment) is said to be *dynamic* when the product has a signal factor that steers the output. For instance, when I switch on the power button on my computer, I am sending a signal to the computer to load my operating system. It should power up and display within five seconds and it should do so exactly the same way every time I switch it on. As in the case of my notebook computer, if it fails to display because of the humidity, I conclude that the computer is sensitive to humidity and that humidity is a noise factor that negatively impacts the performance of my computer.

The response experiment is said to be *static* when the quality level of the CTQ characteristic is fixed. In that case, the optimization process will seek to determine the optimal combination of factors that enables the process to reach the targeted value. This happens in the absence of

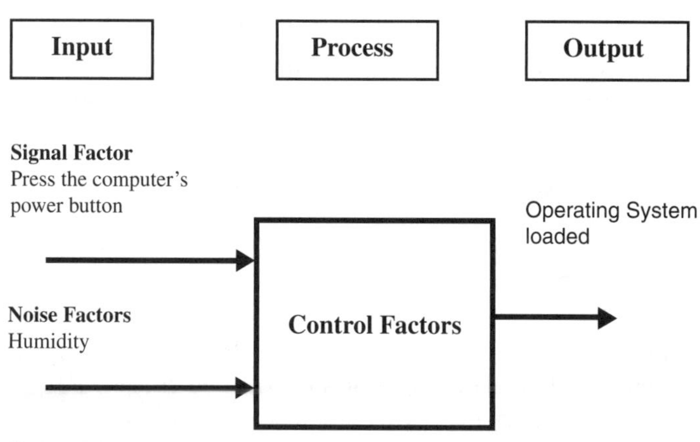

Figure 12.3

a signal factor, where the only input factors are the control factors and the noise factors. When we build a product, we determine all the CTQ targets and we want to produce a balanced product with all the parts matching the targets.

The optimal quality level of a product depends on the nature of the product itself. In some cases, the more a CTQ characteristic is found on a product, the happier the customers are; in other cases, the less the CTQ characteristic is present, the better it is. Some products require the CTQs to match their specified targets.

According to Taguchi, to optimize the quality level of products, the producer must seek to minimize the noise factors and maximize the *signal-to-noise* (S/N) ratio. Taguchi uses logarithmic functions to determine the signal-to-noise ratios that optimize the desired output.

The bigger, the better. If the number of minutes per dollar customers get from their cellular phone service provider is critical to quality, the customers will want to get the maximum number of minutes they can for every dollar they spend on their phone bills.

If the lifetime of a battery is critical to quality, the customers will want their batteries to last forever. The longer the battery lasts, the better it is. The signal-to-noise ratio for the bigger-the-better is

$$S/N = -10 \times \log_{10} \ (\textit{mean square of the inverse of the response})$$

$$S/N = 10 \log_{10} \left(\frac{1}{n} \sum \frac{1}{y^2} \right)$$

The smaller, the better. Impurity in drinking water is critical to quality. The fewer impurities customers find in their drinking water, the better it is. Vibrations are critical to quality for a car—the fewer vibrations the customers feel while driving their cars, the better, and the more attractive the cars are. The signal-to-noise ratio for the smaller -the better is

$$S/N = -10 \times \log_{10} \ (\textit{mean square of the response})$$

$$S/N = -10 \log_{10} \left(\frac{\sum y^2}{n} \right)$$

The nominal, the best. When a manufacturer is building mating parts, he would want every part to match the predetermined target. For instance, when he is creating pistons that must be anchored on a given part of a machine, failure to have the length of the piston match a predetermined size will result in it being either too short or too long, resulting in a reduction of the quality of the machine. In this case, the manufacturer wants all the parts to match their target.

When a customer buys ceramic tiles to decorate their bathroom, the size of the tiles is critical to quality; having tiles that do not match the predetermined target will result in them not being correctly lined up against the bathroom walls. The S/N equation for the nominal-the-best is

$$S/N = -10 \times \log_{10} \ (\textit{the square of the mean divided by the variance})$$

$$S/N = 10 \log_{10} \left(\frac{y^2}{s^2} \right)$$

12.3.3 Tolerance design

Parameter design may not completely eliminate variations from the target. This is why tolerance design must be used for all parts of a product to limit the possibility of producing defective products. The tolerance around the target is usually set by the design engineers; it is defined as the range within which variation may take place. The tolerance limits are set after testing and experimentation. The setting of the tolerance must be determined by criteria such as the set target, the safety factors, the functional limits, the expected quality level, and the financial cost of any deviation from the target.

The safety limits measure the loss incurred when products that are outside the specified limits are produced,

$$\theta = \sqrt{\frac{A_D}{A}}$$

with A_0 being the loss incurred when the functional limits are exceeded, and A being the loss when the tolerance limits are exceeded. Tolerance specifications for the response factor will be

$$\Delta = \frac{\Delta_D}{\theta}$$

with Δ_0 being the functional limit.

Example The functional limits of a conveyor motor are ±0.05 (or 5 percent) of the response RPM. The adjustments made at the audit station before a motor leaves the company costs $2.50 and the cost associated with defective motors once they have been sold is on average $180. Find the tolerance specification for a 2500 RPM motor.

Solution First, we must find the economical factor, which is determined by the loss incurred when the functional limits or the tolerance limits are exceeded.

$$\theta = \sqrt{\frac{A_D}{A}} = \sqrt{\frac{180}{2.6}} = 8.486$$

Now we can determine the tolerance specification. The tolerance specification will be the value of the response factor plus or minus the allowed variation from the target. Tolerance specification for the response factor is

$$\Delta = \frac{\Delta_D}{\theta} = \frac{0.06}{8.486} = 0.0069$$

The variation from the target is

$$2500 \times 0.0059 = 14.73$$

Thus, the tolerance specification will be 2500 ± 14.73.

Chapter

13

Measurement Systems Analysis –MSA: Is Your Measurement Process Lying to You?

In our SPC study, we showed how to build control charts. Control charts are used to monitor production processes and help make adjustments when necessary to keep the processes under control. We have also noticed that no matter how well-controlled a process is, there is always variation in the quality level of the output. All the points on a control chart are never on the same line. In fact, when we think about a control chart, what comes to mind is a zig-zag line with points at the edges. If all the tested parts were identical and the testing processes were precise and accurate, then they would all have been aligned on one line.

To improve the quality of a production system, it is necessary to determine the sources of the variations, whether they are common or special. The variations in a production process are due either to the actual differences between the parts produced or to the process used to assess the quality of the parts, or a combination of these two factors.

For instance, when we test the weight of some parts produced by the same machine using the same process and we notice a weight variation in the results of the test, that variation can only be due to either an actual difference in weight between the parts themselves or to the testing process (the device we use to test the parts and the people who perform the testing). If the testing process is faulty, we might think that there are differences between the parts when, in actuality, there is not any. A faulty measurement system will necessarily lead to wrong conclusions.

If we measure the same part repeatedly, chances are that there will be a variation in the results that we obtain. The measurement process is never perfect but there is always a possibility to reduce the measurement process variations:

$$\sigma_T^2 = \sigma_p^2 + \sigma_m^2$$

where σ_T^2 is the total variation, σ_p^2 is the part-to-part variation, and σ_m^2 is the variation due to the measurement process.

The variations due to the measurement system can be broken down into the variations due to the operator and those due to the instrument used for the measurement,

$$\sigma_m^2 = \sigma_o^2 + \sigma_d^2$$

Before collecting data and testing a process output, it is necessary to analyze the measurement system to assure that the procedures used are not faulty and that it would therefore not lead to erroneous conclusions —rejecting the null hypothesis when in actuality it is correct. The errors due to the measurement system can be traced to two factors: precision and accuracy.

The *accuracy* is measured in terms of the deviation of the measurement system from the actual value of the part being measured. If the actual weight of an engine is 500 pounds and a measurement results in 500 pounds, we conclude that the measurement is accurate. If the measurement results in 502 pounds, we conclude that the measurement deviates from the actual value by 2 pounds and that it is not accurate. *Precision* refers to variations observed when the same instrument is used repeatedly.

13.1 Variation Due to Precision: Assessing the Spread of the Measurement

Precision refers to the variability observed from a repeated measurement process in an experiment under control. If an experiment that consists of repeating the same test using the same process is conducted and the results of the test show the same pattern of variability, we can conclude that there is a reproducibility of the process.

Precision, repeatability, and reproducibility. If the very same part is tested repeatedly with the very same instrument, we expect to find the exact same result if the measurement is precise. Suppose the length of a crankshaft is critical to the quality of an electric motor. We use an electronic meter to measure a randomly selected shaft for testing. In this case, at first what we are testing is not the crankshaft itself but the electronic meter. No matter how many times we test the same crankshaft, its actual dimensions will not change. If there are variations in the results of the testing, they are more likely to come from the electronic meter or the person performing the test. We repeat the test several times and we expect to reproduce the same result if the electronic meter is precise.

If the same operator tests the crankshaft repeatedly, they are very likely to do it the same way. In that case, if there is any variation it is likely to come from the device (electronic meter) used to test the crankshaft. If several operators test the same crankshaft repeatedly, they may do so in different ways. In that case, failure to reproduce the same results may come from either the device or the process used for the testing.

When we talk about precision, what is being addressed is *repeatability* and *reproducibility*. To determine the sources of variations when we fail to reproduce the same results after repeated testing, we can use several methods including the ANOVA, the $\overline{X}R$ chart, the gage R&R, and gage run charts. A *gage* in this context can be software, a physical instrument, a standard operating procedure (SOP), or any system or process used to measure CTQs.

We have seen how ANOVA and DOE can help determine sources of variations in a production process. ANOVA is based on the formulation of a null hypothesis and running a test that will result in rejecting or failing to reject that hypothesis. The rejection or the failure to reject the null hypothesis is determined by the sources of the variations. If the sources of variations are within treatment, the null hypothesis is not rejected; if the sources of variations are between treatment, the null hypothesis is rejected.

A *gage run chart* is a graphical representation of the observations by part and by operator. It enables the experimenter to make an assessment based on how close the observations are about their means and the presence of outliers. A *gage R&R* experiment is conducted to describe the performance of a measurement system through the quantification of the variations in the measurement process.

13.1.1 Gage repeatability & reproducibility crossed

$\overline{X}R$ chart. A quality control manager wants to tests new wattmeters used to measure the active electric power generated by a newly designed generator. He takes a sample of 20 units of wattmeters labeled from "A" to “T” and designates Macarlos, one of the auditors to test each one of them three times. Macarlos takes the measurements to construct $\overline{X}R$ control charts to assess the variations in the measurement system. He tabulates the results, shown in Table 13.1.

TABLE 13.1

Part #	Measurement 1	Measurement 2	Measurement 3	$\overline{X}$	Range
A	9.9895	10.0115	9.9887	9.996567	0.0228
B	9.9925	10.0726	9.9574	10.007500	0.1152
C	9.9959	9.9154	10.0827	9.998000	0.1673
D	9.9284	9.9035	10.0765	9.969467	0.1730
E	9.9933	9.9894	10.0050	9.995900	0.0156
F	10.0927	9.9390	10.0219	10.017870	0.1537
G	9.9571	10.0341	10.0680	10.019730	0.1109
H	10.0198	10.0222	9.9482	9.996733	0.0740
I	10.0316	9.9924	10.0352	10.019730	0.0428
J	10.0341	9.9509	9.9808	9.988600	0.0832
K	9.8503	10.0326	9.9435	9.942133	0.1823
L	9.9776	9.9956	10.0235	9.998900	0.0459
M	10.0148	10.0133	10.0025	10.010200	0.0123
N	10.0687	9.9999	9.9950	10.021200	0.0737
O	10.0166	9.9708	9.9324	9.973267	0.0842
P	10.0061	10.0189	10.0827	10.035900	0.0766
Q	9.9488	9.9544	9.9968	9.966667	0.0480
R	10.1188	9.9999	9.9608	10.026500	0.1580
S	10.0349	10.0932	9.9212	10.016430	0.1720
T	10.1071	10.0088	9.9233	10.013070	0.1838
			Mean	10.000720	0.099765

We recall from our discussion of SPC that the control limits and the center line for an $\overline{X}$-chart are obtained from the following equations:

$$UCL = \overline{\overline{X}} + A_2\overline{R}$$

$$CL = \overline{\overline{X}}$$

$$LCL = \overline{\overline{X}} - A_2\overline{R}$$

We obtain A_2 from the control charts constant table, Table 13.2.

The value $n = 3$, therefore $A_2 = 1.023$ and

$$UCL = 10.00071 + 1.023 \times 0.099765 = 10.1028$$

$$CL = 10.00071$$

$$UCL = 10.00071 - 1.023 \times 0.099765 = 9.8987$$

The interpretation we make of this $\overline{X}$-chart shown in Figure 13.1 is different from the ones we had in earlier chapters. In this example, each wattmeter is considered as a sample and each measurement is an

TABLE 13.2

Sample size	A_2	A_3	B_3	B_4	d_2	d_3
2	1.880	2.659	0.000	3.267	1.128	0.853
3	1.023	1.954	0.000	2.568	1.693	0.888
4	0.729	1.628	0.000	2.266	2.059	0.880
5	0.577	1.427	0.000	2.089	2.326	0.864
6	0.483	1.287	0.030	1.970	2.534	0.848
7	0.419	1.182	0.118	1.882	2.704	0.833
8	0.373	1.099	0.185	1.815	2.847	0.820
9	0.337	1.032	0.239	1.761	2.970	0.808
10	0.308	0.975	0.284	1.716	3.078	0.797
11	0.285	0.927	0.321	1.679	3.173	0.787
12	0.266	0.886	0.354	1.646	3.258	0.778
13	0.249	0.850	0.382	1.618	3.336	0.770
14	0.235	0.817	0.406	1.594	3.407	0.763
15	0.223	0.789	0.428	1.572	3.472	0.756
16	0.212	0.763	0.448	1.552	3.532	0.750
17	0.203	0.739	0.466	1.534	3.588	0.744
18	0.194	0.718	0.482	1.518	3.640	0.739
19	0.187	0.698	0.497	1.503	3.689	0.734
20	0.180	0.680	0.510	1.490	3.735	0.729
21	0.173	0.663	0.523	1.477	3.778	0.724
22	0.167	0.647	0.534	1.466	3.819	0.720
23	0.162	0.633	0.545	1.455	3.858	0.716
24	0.157	0.619	0.555	1.455	3.895	0.712
25	0.153	0.606	0.565	1.435	3.031	0.708

element in the sample, which translates to $n = 3$. Therefore, each point on the control chart represents the mean measurement of a wattmeter.

The $\overline{X}$ control chart shows that the points follow a normal pattern and are all within the control limits, which suggests that the variability around the mean is due to common causes and the auditor is not having problems getting accurate results. If the part-to-part variations between the wattmeters is under control, the conclusion that must be

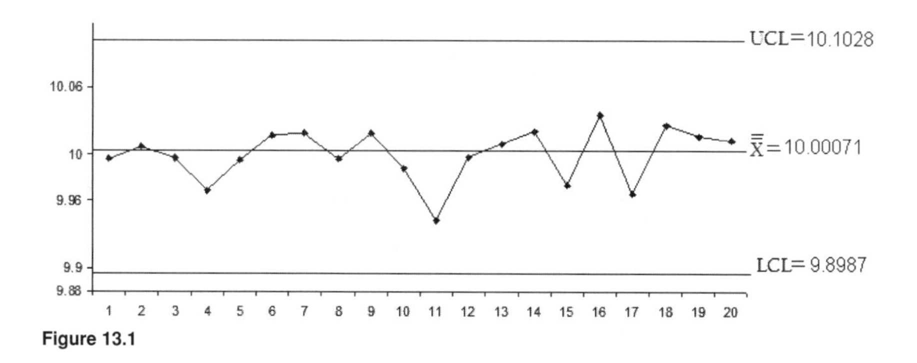

Figure 13.1

drawn would be that the measurement process is in-control and that the gage used in the measurement process is precise.

The R-chart will show the extent of the variations because it measures the difference between the observations collected from the same wattmeter. Here again, we can use the equations obtained from the chapter about SPC:

$$UCL = D_4\overline{R} = 2.575 \times 0.099765 = 0.2569$$

$$CL = \overline{R} = 0.099765$$

$$LCL = D_3\overline{R} = 0.099765 \times 0 = 0$$

with a standard deviation of

$$\sigma_{gage} = \frac{\overline{R}}{d_2} = \frac{0.099765}{1.693} = 0.05893$$

The R-chart of Figure 13.2 shows the variations in measurement for each wattmeter. It measures the difference between the highest and the lowest measurements for each unit. Therefore, each point on the chart represents a range, the difference between the highest and lowest measurements for each unit.

The R-chart shows a random pattern with all the points being inside the control limits; this confirms that the gage used in the measurement process is generating consistent results.

Repeatability and reproducibility. Because the true, actual value of the measurements of the wattmeters are known to be consistent and only exhibit random variations, had the results of the experiment that Macarlos conducted shown inconsistency, that inconsistency could have only come from errors in the measurement process. Measurement errors occur when either the operator did not perform the test properly or the instrument s/he is using is not consistent, or both.

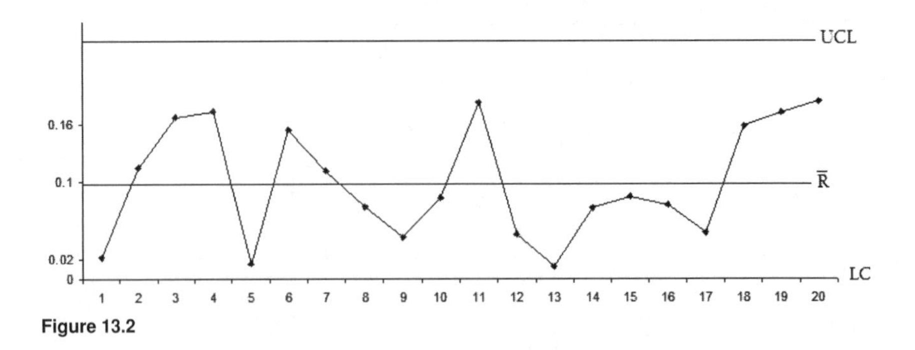

Figure 13.2

If several operators are testing the gage for consistency, we should be concerned about the repeatability and the reproducibility of the gage. *Repeatability* refers to a special cause of variation that can be traced to the measuring device—the variations occur when the same part is measured multiple times by the same operator.

Reproducibility refers to the special causes of variation due to the process used—when variations are observed after several operators test the same parts with the same device. The device in this context can be anything from software to a work instruction to a physical object.

Precision-to-tolerance ratio (P/T). Not only does the quality engineer want the measurement process to be consistent, stable, and in-control and the variation to be only due to common causes, but he also wants the process to be within preset specified limits. He is not only concerned about the process consistency but is also concerned about the measurement process capability.

The $\overline{X}$-chart of Figure 13.1 showed how the data are scattered about the measurement process mean $\overline{\overline{X}}$ and how the variations are patterned about the mean. But the quantification of the variations is better assessed by the $\overline{R}$-chart. As seen in previous chapters, $\overline{R}$ can be expressed in terms of the standard deviation,

$$\overline{R} = d_2 \sigma_{gage}$$

This equation can be rewritten as

$$\sigma = \frac{\overline{R}}{d_2}$$

$$C_r = \frac{UCL - LCL}{USL - LSL} = \frac{6\sigma_{gage}}{USL - LSL}$$

Because the standard deviation of the gage is also the measure of the precision of the gage, and the denominator measures the spread of the tolerance, this equation can be rewritten as

$$P/T = \frac{6\sigma_{gage}}{USL - LSL}$$

P/T is called the *precision-to-tolerance ratio*. In general, a unit is considered calibrated when the variations from the target are less than one tenth of the unit's actual CTQ value. Therefore if the P/T is less than 0.1, the measurement process is considered precise. The total measurement error of the gage can be divided into two parts: the error due to repeatability and the error due to reproducibility.

Let's get Back to our example, the QA manager looks at the results presented by Macarlos and decides to push the test a little further to validate the findings. He invites Jarrett and Tim to perform the same test as Macarlos with the same parts. In other words, they are asked to repeat the test to see if they will produce results that are consistent with the previous test. The results of their tests are tabulated in Table 13.3 with the mean measurements and the range for each part.

Because we have three operators measuring 20 parts, the variations can come from the gage or from the process used by the different operators, or from both the gage and the process. Because we have defined "repeatability" as the variations traced to the device and "reproducibility" as variations traced to the process, we can summarize the total variation in the following equation:

$$\sigma^2_{Total} = \sigma^2_{repeatability} + \sigma^2_{reproducibility}$$

Repeatability is the variations traced to the device. $\overline{R}$ measures the variations observed for each part for each operator using the gage and helps estimate the gage repeatability. The mean range is obtained by adding the three $\overline{R}$ values and dividing the result by 3,

$$\overline{\overline{R}} = \frac{\overline{R}_{Macarlos} + \overline{R}_{Jarrett} + \overline{R}_{Tim}}{3} = \frac{0.1 + 0.1 + 0.186}{3} = 0.129$$

$$\sigma_{gage} = \frac{\overline{\overline{R}}}{d_2}$$

because, in this case, $n = 3$, we obtain

$$d_2 = 1.693$$

from the control charts constant table, Table 13.4.

The gage repeatability is therefore

$$\sigma_{repeatability} = \frac{\overline{\overline{R}}}{d_2} = \frac{0.129}{1.693} = 0.0762$$

The gage reproducibility measures the measurement error due to the process used by the different operators. $\overline{\overline{X}}$ measures the overall mean measurement found by an operator for all the parts measured. In this case again, we will be interested in the standard deviation obtained for all the operators,

$$\sigma_{gage} = \frac{R_{\overline{\overline{X}}}}{d_2}$$

$$R_{\overline{\overline{X}}} = \overline{\overline{X}}_{\max} - \overline{\overline{X}}_{\min} = 10.003 - 9.997 = 0.006$$

TABLE 13.3

	Macarlos					Jarrett					Tim				
Part	M1	M2	M3	$\overline{X}$	Range	M1	M2	M3	$\overline{X}$	Range	M1	M2	M3	$\overline{X}$	Range
A	9.990	10.012	9.989	9.997	0.023	9.939	10.042	10.079	10.020	0.140	10.054	9.874	9.969	9.966	0.181
B	9.993	10.073	9.957	10.008	0.115	10.052	9.957	9.952	9.987	0.100	10.051	9.897	9.994	9.981	0.154
C	9.996	9.915	10.083	9.998	0.167	10.024	10.078	9.991	10.031	0.087	10.060	9.893	9.942	9.965	0.167
D	9.928	9.904	10.077	9.969	0.173	10.069	10.102	9.956	10.042	0.146	9.903	9.936	10.055	9.965	0.152
E	9.993	9.989	10.005	9.996	0.016	10.125	9.938	10.033	10.032	0.187	10.002	9.881	9.956	9.946	0.120
F	10.093	9.939	10.022	10.018	0.154	9.998	9.954	10.009	9.987	0.054	10.137	10.085	9.941	10.054	0.196
G	9.957	10.034	10.068	10.020	0.111	10.001	9.900	10.016	9.972	0.116	9.920	10.048	9.886	9.951	0.163
H	10.020	10.022	9.948	9.997	0.074	10.033	10.049	10.035	10.039	0.016	9.874	10.106	9.902	9.960	0.233
I	10.032	9.992	10.035	10.020	0.043	9.926	10.028	9.965	9.973	0.102	9.903	10.016	9.873	9.931	0.144
J	10.034	9.951	9.981	9.989	0.083	10.021	9.946	9.885	9.951	0.136	9.975	10.179	10.181	10.111	0.206
K	9.850	10.033	9.944	9.942	0.182	9.923	9.912	9.992	9.942	0.080	10.078	10.239	9.889	10.068	0.350
L	9.978	9.996	10.024	9.999	0.046	9.980	9.988	9.975	9.981	0.013	9.884	10.197	10.144	10.075	0.313
M	10.015	10.013	10.003	10.010	0.012	9.997	9.954	9.951	9.967	0.046	9.823	10.062	10.008	9.964	0.239
N	10.069	10.000	9.995	10.021	0.074	10.046	9.964	10.044	10.018	0.082	9.916	9.975	10.014	9.968	0.098
O	10.017	9.971	9.932	9.973	0.084	10.016	9.983	9.928	9.976	0.088	9.851	10.110	10.084	10.015	0.260
P	10.006	10.019	10.083	10.036	0.077	10.033	10.079	10.040	10.051	0.046	10.050	10.079	10.014	10.048	0.065
Q	9.949	9.954	9.997	9.967	0.048	9.869	9.997	9.947	9.938	0.128	9.926	9.986	10.146	10.019	0.220
R	10.119	10.000	9.961	10.027	0.158	9.973	10.009	10.072	10.018	0.099	10.142	9.968	10.102	10.071	0.174
S	10.035	10.093	9.921	10.016	0.172	10.074	10.012	9.977	10.021	0.097	10.057	9.898	10.083	10.013	0.185
T	10.107	10.009	9.923	10.013	0.184	10.117	9.874	9.961	9.984	0.243	10.033	9.978	9.932	9.981	0.102
				10.001	0.100				9.997	0.100				10.003	0.186
				$\overline{\overline{X}}$	$\overline{R}$				$\overline{\overline{X}}$	$\overline{R}$				$\overline{\overline{X}}$	$\overline{R}$

TABLE 13.4

Sample size	A_2	A_3	B_3	B_4	d_2	d_3
2	1.880	2.659	0.000	3.267	1.128	0.853
3	1.023	1.954	0.000	2.568	1.693	0.888
4	0.729	1.628	0.000	2.266	2.059	0.880
5	0.577	1.427	0.000	2.089	2.326	0.864
6	0.483	1.287	0.030	1.970	2.534	0.848
7	0.419	1.182	0.118	1.882	2.704	0.833
8	0.373	1.099	0.185	1.815	2.847	0.820
9	0.337	1.032	0.239	1.761	2.97	0.808
10	0.308	0.975	0.284	1.716	3.078	0.797
11	0.285	0.927	0.321	1.679	3.173	0.787
12	0.266	0.886	0.354	1.646	3.258	0.778
13	0.249	0.850	0.382	1.618	3.336	0.770
14	0.235	0.817	0.406	1.594	3.407	0.763
15	0.223	0.789	0.428	1.572	3.472	0.756
16	0.212	0.763	0.448	1.552	3.532	0.750
17	0.203	0.739	0.466	1.534	3.588	0.744
18	0.194	0.718	0.482	1.518	3.640	0.739
19	0.187	0.698	0.497	1.503	3.689	0.734
20	0.180	0.680	0.510	1.490	3.735	0.729
21	0.173	0.663	0.523	1.477	3.778	0.724
22	0.167	0.647	0.534	1.466	3.819	0.720
23	0.162	0.633	0.545	1.455	3.858	0.716
24	0.157	0.619	0.555	1.455	3.895	0.712
25	0.153	0.606	0.565	1.435	3.031	0.708

Therefore,

$$\sigma_{reproducibility} = \frac{R_{\overline{\overline{X}}}}{d_2} = \frac{0.006}{1.693} = 0.0035$$

The total variance will be the sum of the variance for the measure of reproducibility and the one for the repeatability,

$$\sigma^2_{Total} = \sigma^2_{repeatability} + \sigma^2_{reproducibility}$$

$$\sigma^2_{Total} = 0.0762^2 + 0.0035^2 = 0.00582$$

Example The diameter of the pistons produced at Joal Mechanics is critical to the quality of the products. Many products are being returned from customers due to variations in their diameters. The Quality Control manager decided to investigate the causes of the variations; he starts the task with an open mind and believes that the variations can be due to the operators, to the measurement devices, or to a variation in the actual sizes of the parts that have gone unnoticed. He selects 10 parts and three operators whose assignment it is to test the parts, and the results of their findings are tested using Minitab. Each operator should test every part. The results of the test can be found in the file *Joal.mpj* on the included CD.

Solution Open the file *Joal.mpj* on the included CD. From the Stat menu, select "Quality Tools" and from the drop-down list, select "Gage Study" and then select "Gage R&R Study (Crossed)." Fill out the fields as indicated in Figure 13.3. To obtain *Xbar and R* charts, select that option.

Figure 13.3

Click on the "OK" to get the results.

The first part of the output in Figure 13.4 shows the sources of the variations and the proportion of the contribution of each aspect of the measurement to the total variation. The total gage R&R measures the variations due to the measurement process. In this case, it is 88.74 percent, which is very high. The proportion due to reproducibility is very close to zero and the one due to repeatability in 88.74 percent; therefore, the measuring device is most likely to be blamed for the variations. The variations between parts are relatively small (11.26 percent).

The graphs in Figure 13.5 are a representation of the output in the session window. The top left graph shows how the contributions of the part-to-part reproducibility and repeatability are distributed. In this case, we clearly see that the part-to-part variation is relatively small, and that the repeatability carries the bulk of the variations.

The graphs at the center left and at the bottom left show how the different operators managed to measure the parts. It shows that Al and Sira have had trouble using their measuring device. Their measuring processes are unstable and out-of-control, whereas Ken has been able to produce an in-control and stable process.

Had we selected the ANOVA option, we would have ended up with the results shown in Figure 13.6. The ANOVA shows that for an α level of 0.05, neither the parts, the operators, nor the interaction between the parts and the operators have a statistical significance; their P-values

Gage R&R Study - XBar/R Method

Source	VarComp	%Contribution (of VarComp)
Total Gage R&R	0.676776	88.74
Repeatability	0.676776	88.74
Reproducibility	0.000000	0.00
Part-To-Part	0.085848	11.26
Total Variation	0.762624	100.00

Source	StdDev (SD)	Study Var (6 * SD)	%Study Var (%SV)
Total Gage R&R	0.822664	4.93598	94.20
Repeatability	0.822664	4.93598	94.20
Reproducibility	0.000000	0.00000	0.00
Part-To-Part	0.292998	1.75799	33.55
Total Variation	0.873283	5.23970	100.00

Number of Distinct Categories = 1

Gage R&R for Measurement

Figure 13.4

are all much greater than 0.05. The gage R&R shows that the repeatability contributes up to 88.84 percent to the variations.

13.1.2 Gage R&R nested

Not all products can be tested multiple times. Some products lose their value after a test; in this case, the test done on a product cannot be replicated. This type of testing is generally referred to as *destructive testing*. For instance, if we must apply pressure over a metal shaft to measure the strength it takes to bend it 90 degrees, after using a shaft for testing and it is bent, it would not be possible to replicate the test on the same part. The nature of the test conducted under these circumstances makes it essential to be very selective about the parts being tested because their lack of homogeneity can lead to incorrect conclusions. If the parts being tested are not identical and the part-to-part variations are too high, the results of the test would be misleading because one of the assumptions of the destructive test for measurement systems analysis is that only common causes of part-to-part variations are present.

In some nondestructive tests, several identical parts are tested with each part being tested by only one operator multiple times. The results of such tests can distinguish the proportions of the sources of variations

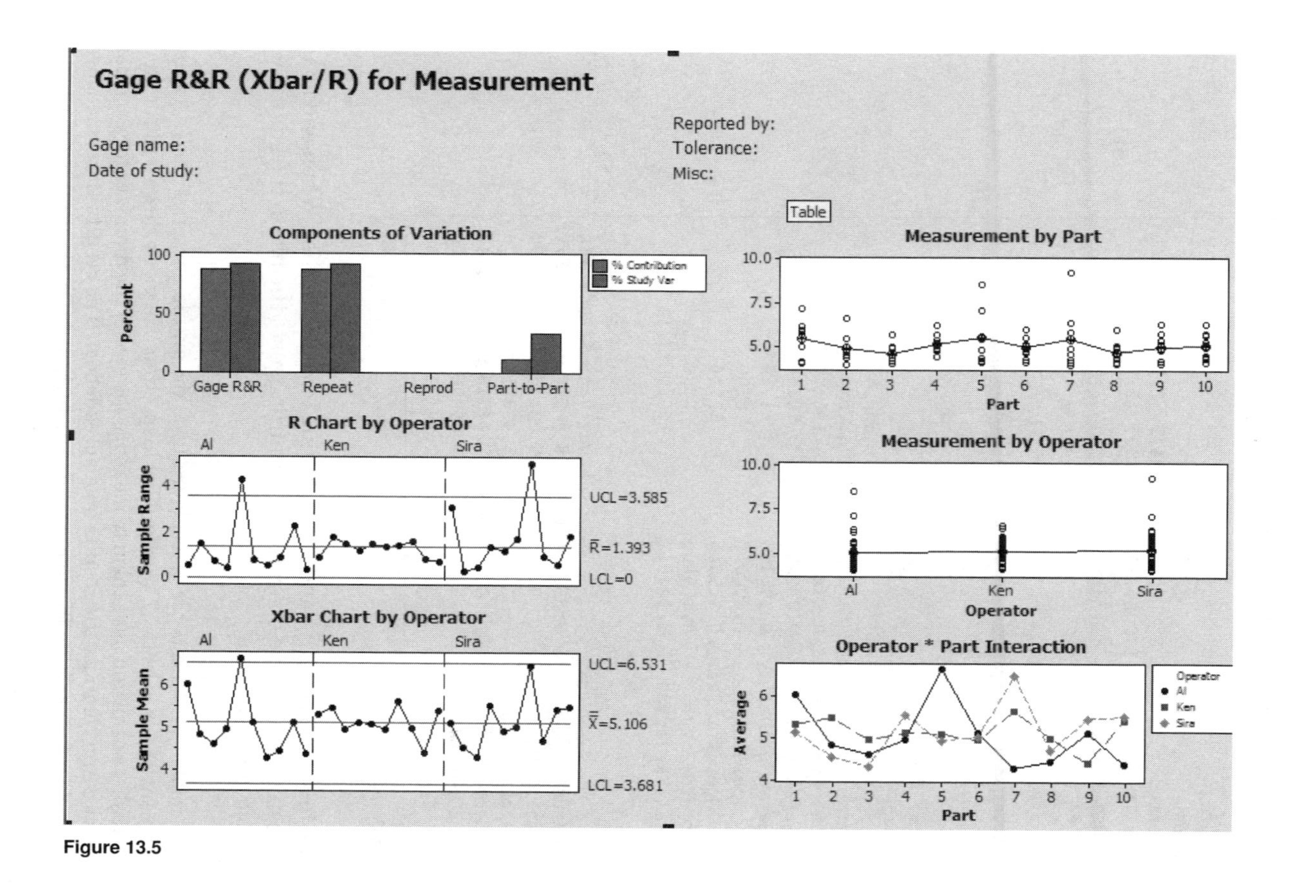

Gage R&R (Xbar/R) for Measurement
Gage name:
Date of study:
Reported by:
Tolerance:
Misc:
Components of Variation
Percent
100
50
0
Gage R&R
Repeat
Reprod
Part-to-Part
% Contribution
% Study Var
R Chart by Operator
Al
Ken
Sira
Sample Range
4
2
0
UCL=3.585
R̄=1.393
LCL=0
Xbar Chart by Operator
Al
Ken
Sira
Sample Mean
6
5
4
UCL=6.531
X̿=5.106
LCL=3.681
Table
Measurement by Part
10.0
7.5
5.0
1
2
3
4
5
6
7
8
9
10
Part
Measurement by Operator
10.0
7.5
5.0
Al
Ken
Sira
Operator
Operator * Part Interaction
Average
6
5
4
1
2
3
4
5
6
7
8
9
10
Part
Operator
Al
Ken
Sira

Figure 13.5

Gage R&R Study - ANOVA Method

Two-Way ANOVA Table With Interaction

Source	DF	SS	MS	F	P
Part	9	8.2518	0.91687	0.79357	0.627
Operator	2	0.2694	0.13471	0.11660	0.891
Part * Operator	18	20.7966	1.15537	1.37673	0.177
Repeatability	60	50.3527	0.83921		
Total	89	79.6706			

Alpha to remove interaction term = 0.25

Gage R&R

Source	VarComp	%Contribution (of VarComp)
Total Gage R&R	0.944597	100.00
Repeatability	0.839212	88.84
Reproducibility	0.105386	11.16
Operator	0.000000	0.00
Operator*Part	0.105385	11.16
Part-To-Part	0.000000	0.00
Total Variation	0.944597	100.00

Source	StdDev (SD)	Study Var (6 * SD)	%Study Var (%SV)
Total Gage R&R	0.971904	5.83142	100.00
Repeatability	0.916085	5.49651	94.26
Reproducibility	0.324631	1.94779	33.40
Operator	0.000000	0.00000	0.00
Operator*Part	0.324631	1.94779	33.40

Figure 13.6

due to reproducibility and the ones due to repeatability. When each part is tested by only one operator, Minitab suggests the use of the *gage R&R nested* method.

Example The diameter of washers used on an alternator is CTQ. A quality inspector selects 12 washers for inspection. He gives three auditors four washers each and asks them to measure the size of the diameter for each part twice and compute the results on a spreadsheet. The results of the measures are found on the file *Washers.mpj* on the included CD. Determine the sources of the variations in the sizes of the diameters.

Solution Open the file *Washers.mpj* on the included CD. From the Stat menu, select "Quality Tools" and from the drop-down list, select "Gage Study" and

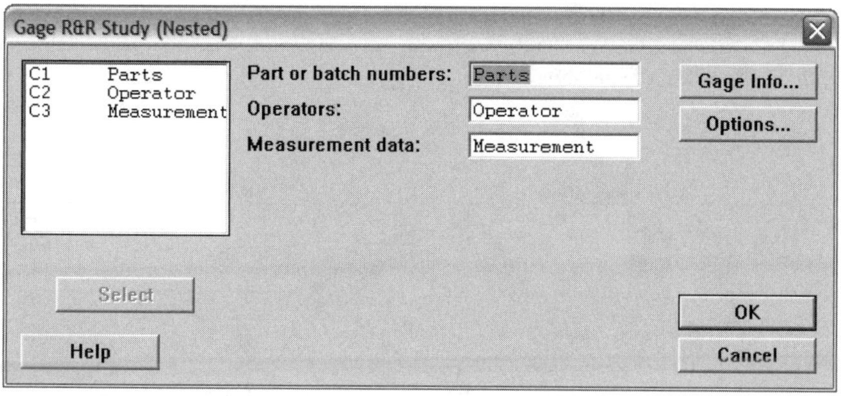

Gage R&R Study (Nested)

C1 Parts
C2 Operator
C3 Measurement

Part or batch numbers: Parts
Operators: Operator
Measurement data: Measurement

Gage Info...
Options...
Select
Help
OK
Cancel

Figure 13.7

then select "Gage R&R Study (Nested)." Fill out the dialog box as indicated in Figure 13.7 and then select "OK."

The output of Figure 13.8 should appear on the session window.

Gage R&R (Nested) for Measurement

Source	DF	SS	MS	F	P
Operator	2	0.70589	0.352944	1.31617	0.315
Parts (Operator)	9	2.41344	0.268161	2.14943	0.108
Repeatability	12	1.49711	0.124759		
Total	23	4.61644			

Gage R&R

Source	VarComp	%Contribution (of VarComp)
Total Gage R&R	0.135357	65.37
Repeatability	0.124759	60.25
Reproducibility	0.010598	5.12
Part-To-Part	0.071701	34.63
Total Variation	0.207058	100.00

Source	StdDev (SD)	Study Var (6 * SD)	%Study Var (%SV)
Total Gage R&R	0.367909	2.20745	80.85
Repeatability	0.353212	2.11927	77.62
Reproducibility	0.102946	0.61768	22.62
Part-To-Part	0.267770	1.60662	58.85
Total Variation	0.455036	2.73022	100.00

Number of Distinct Categories = 1

Figure 13.8

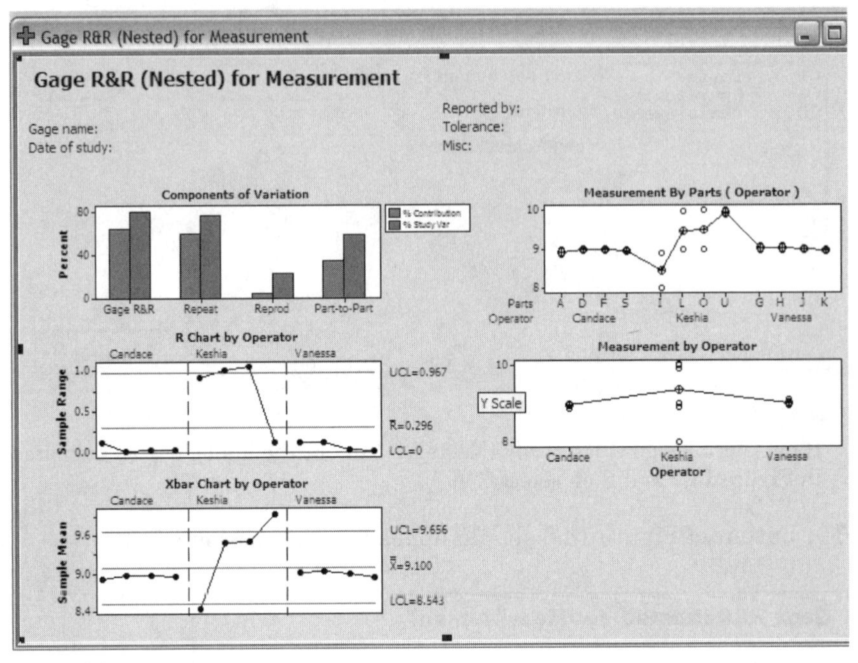

Figure 13.9

The results show that the proportion of the variations due to the gage R&R (65.37 percent) is about twice as high as the ones due to part-to-part variation (34.63 percent). The proportion due to repeatability is 60.25 percent compared to reproducibility at 5.12 percent. The measurement system is the primary source of the variations.

The graphs in Figure 13.9 give a pictorial account of the sources of variation.

The top left graph shows that the gage R&R is the source of variations and in it, repeatability accounts for a significant portion of the variations. The sample range and the sample mean graphs show that Keshia has had problems with the measuring device.

13.2 Gage Run Chart

A *gage run chart* is a graphical expression of the observations by part and by operator. It helps graphically assess the variations due to the operators or the parts. In the following example, the length of a shaft is critical to the quality of an electric motor, so five parts are tested by three operators. Each part is tested by every one of the three operators and the results are tabulated in the Table 13.5.

TABLE 13.5

Part ID	Operator	Measurement
A	Macarlos	12.00
B	Macarlos	11.90
C	Macarlos	12.01
D	Macarlos	11.98
E	Macarlos	12.01
A	Bob	12.05
B	Bob	11.90
C	Bob	12.00
D	Bob	11.99
E	Bob	12.01
A	Joe	12.01
B	Joe	11.91
C	Joe	12.00
D	Joe	12.40
E	Joe	12.00

Using Minitab. Minitab is very practical for generating a run gage chart. Open the file *Motorshaft.mpj* on the included CD. From the Stat menu, select "Quality Tools," from the drop-down list select "Gage Study," and then select "Gage Run Chart." In the "Gage Run Chart" dialog box, enter "Part numbers," the "Operator," and "Measurement" in their appropriate fields. Then, select "OK."

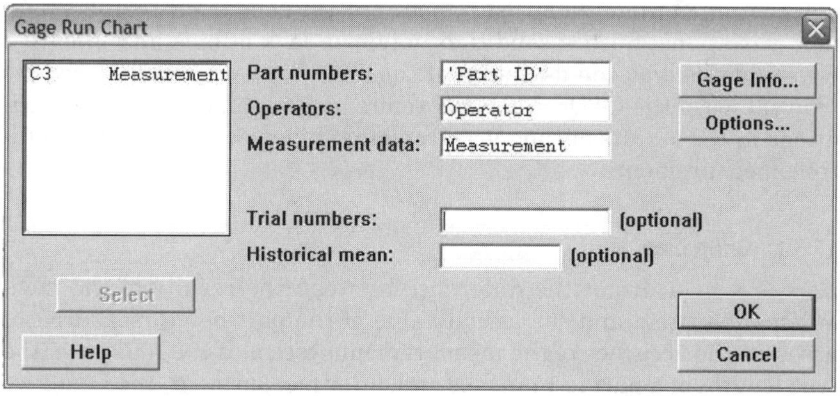

Gage Run Chart

C3 Measurement

Part numbers: 'Part ID'

Operators: Operator

Measurement data: Measurement

Trial numbers: (optional)

Historical mean: (optional)

Gage Info...

Options...

Select

Help

OK

Cancel

The gage run chart output of Figure 13.10 shows that Joe did not do a good job measuring part D. Part D is an outlier, very far away from the average. Part B is far from the mean for all three operators, and the measurements taken by all the operators are fairly close, so in this case the variation is most likely due to the part.

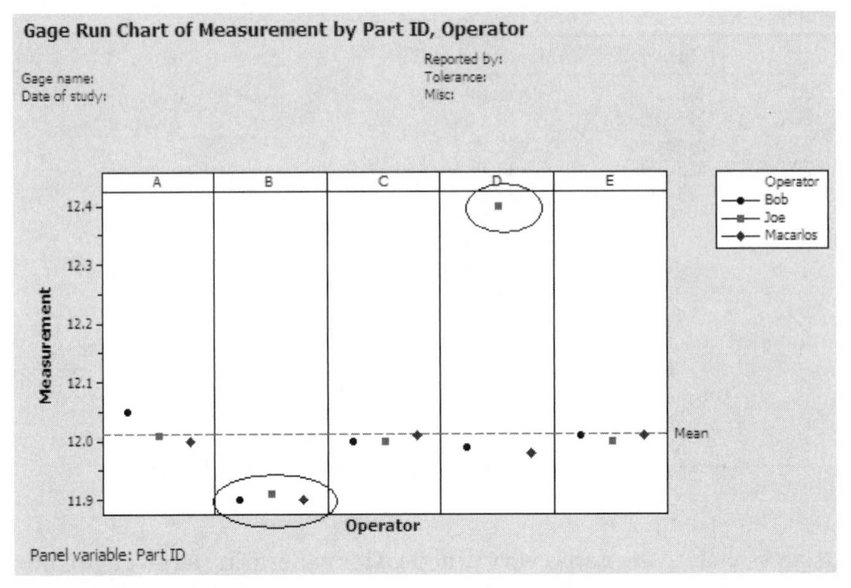

Figure 13.10

13.3 Variations Due to Accuracy

The *accuracy* of a measurement process refers to how close the results of the measurement are to the true values of the CTQ characteristics of products or services being measured. For an accurate process, the results obtained from the measurement process should only exhibit common variations. Such a condition implies a lack of bias and linearity. *Bias* is defined as the deviation of the measurement results from the true values of the CTQs. *Linearity* refers to gradual, proportional variations in the results of the measurements. Linearity also implies multiple measurements.

13.3.1 Gage bias

Bias is a measure of the difference between the results of the measurement system and the actual value of the part being measured; it assesses the accuracy of the measurement system. If the reference (the true length) of a part is 15 inches and, after measuring it, the result we obtain is 13 inches, we would conclude that the measurement system is biased by 2 inches.

Therefore, we can use the following formula to estimate the gage bias:

$$Bias = \frac{\sum_{i=1}^{n} x_i}{n} - \theta$$

with θ being the true value of the part being measured and $\frac{\sum_{i=1}^{n} x_i}{n}$ being the mean measurement observed. The equation is read as the difference between the average measurement result and the true value of the part.

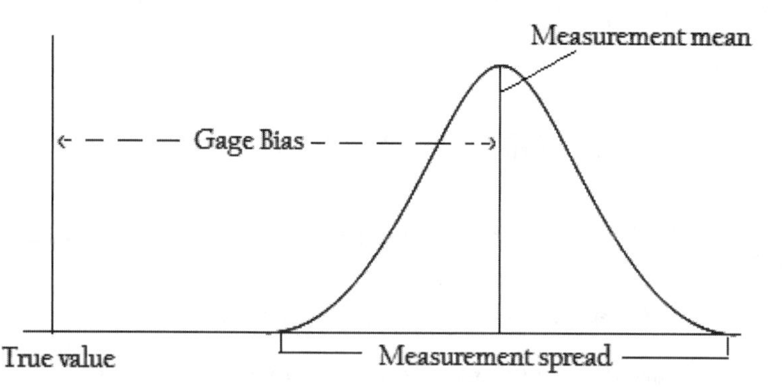

A gage bias assesses the extent to which the mean measurement deviates from the actual value of the product being measured. Because all the measurements taken are just a sample of the infinite number of the possible measurements, one way of measuring the statistical significance of the difference would be the use of hypothesis testing. The null hypothesis would consist of stating that there is not a difference between the measurements' mean and the actual value of the part being measured,

$$H_0 : \overline{X} = \theta$$

and the alternate hypothesis would state the opposite,

$$H_1 : \overline{X} \neq \theta$$

with $\overline{X}$ being the sample's mean and θ being the true value of the measurement. If the number of measurements taken is relatively small, we can use the t-test to test the hypothesis. In this case,

$$t = \frac{\overline{X} - \theta}{s/\sqrt{n}}$$

$$df = n - 1$$

Example The production process for the manufacturing of liners is know to have a standard deviation of 0.02 ounces, and with the upper and lower specified limits set at three standard deviations from the mean, the process mean is 15 ounces. A liner known to weigh 15 ounces is selected for a test and the measurements obtained are summarized in Table 13.6.

TABLE 13.6

15.0809	15.1299	15.0483	15.0307
15.0873	15.0414	15.0515	14.9630
14.9679	15.0351	15.0962	15.1002
15.0423	15.0559	15.0654	15.0731
15.1029	15.0793	14.9759	15.0363
14.9803	15.0753	15.0507	14.9833

Find the gage bias and determine if the bias is significant at an α level of 0.05.

Solution We find the measurements' mean to be 15.04805, and because the true value is 15,

$$\textit{Gage bias} = 15.04805 - 15 = 0.04805$$

The null and alternate hypotheses to test the significance of the difference would be

$$H_0 : \overline{X} = 15$$

$$H_1 : \overline{X} \neq 15$$

The standard deviation is 0.045884 and the measurement mean is 15.048. The t-test value would be

$$t = \frac{\overline{X} - \mu}{\sigma/\sqrt{n}} = \frac{15.0480 - 15}{\dfrac{0.045884}{\sqrt{24}}} = 5.1248$$

Minitab output:

One-Sample T: C2

```
Test of mu = 15 vs not = 15


Variable   N     Mean   StDev  SE Mean          95% CI              T      P
C2        24  15.0480  0.0459   0.0094  (15.0287, 15.0674)   5.13  0.000
```

The P-value is equal to zero, therefore we must reject the null hypothesis and conclude that there is a statistically significant difference between the true value and the measurements' mean.

13.3.2 Gage linearity

In the previous example, only one part was measured against a known actual value of a part. Because a gage is supposed to give dimensions or attributes of different parts of a same nature, the same gage can be used to measure different parts of the same nature but of different sizes. In this case, every part would be measured against a known actual value of a part. If the gage is accurate, an increase in the dimensions of the parts being measured should result in a proportional increase of the measurements taken. Even if the gage is biased, if the gage exhibits linearity we should expect the same proportional variations.

Suppose that we are using a voltmeter to measure the voltage of the current that flows through an electrical line. If the actual voltage applied is 120 volts and that voltage is doubled and then tripled, if the voltmeter is accurate we should expect it to read 120 volts, then 240 volts, and finally 360 volts.

If the first reading of the voltmeter was not exact and was off by 5 volts, we should expect the readings to be 125 volts for the for the first reading, 250 Volts for the second reading, and 375 volts for the third reading. If these results are obtained, we can conclude that the gage exhibits linearity.

If all the known actual values of different parts of dimensions that increase proportionally at a constant rate are plotted on a graph, we should obtain a straight line, and the equation of that line should be of the form

$$Y = aX + b$$

If measurements of the same parts are taken using an accurate and precise gage, the measurements obtained should be on the same line as the previous one if plotted on the same graph. Otherwise, the points representing the measurements would be scattered around the regression line (the reference line).

To run a gage linearity test, we can use regression analysis to determine the regression line and observe the spread of the data plots of the gage measurements about the line. The regression analysis would be a simple linear one with the independent variable being the known actual values and the dependent variables being the gage bias. If the equation of the regression line is under the form of

$$Y = X$$

in other words if $a = 1$ and $b = 0$, we would conclude that the gage is a perfect instrument to measure the parts because every gage measurement would be equal to the true value of the part being measured. Therefore, the bias would be equal to zero and the regression plot would look like that of Figure 13.11.

To have a good estimate of the measurements, each part should be measured several times—at least four times—and the bias would be the

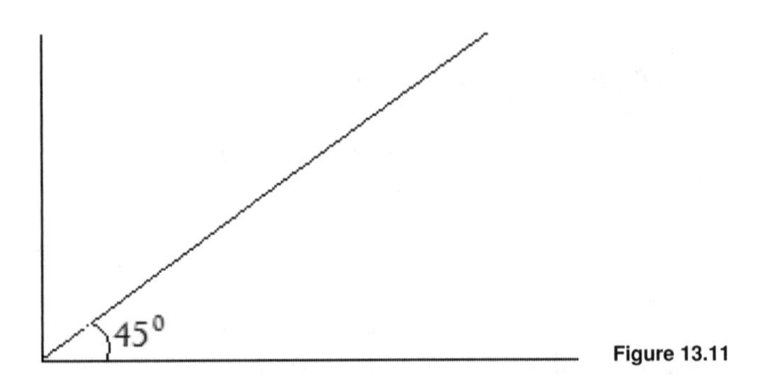

Figure 13.11

difference between the actual known value and the mean measurement for each part.

Example A scale is used to measure the weight of pistons. The true values of the pistons are known and five measurements for each piston are taken using the same scale. The results of the measurements are given in Table 13.7. Find the equation of the regression line to estimate the bias at any value and determine if the scale is a good gage to measure the weight of the parts.

TABLE 13.7

True value	M1	M2	M3	M4	M5	Mean	Bias
5	4.70	4.90	4.89	4.98	5.03	4.900	−0.100
10	9.98	9.78	10.03	10.09	10.70	10.116	0.116
15	14.99	14.66	14.98	14.93	14.98	14.908	−0.092
20	20.04	19.54	19.84	19.99	19.98	19.878	−0.122
25	25.08	24.42	24.76	24.10	24.44	24.560	−0.440
30	30.03	29.30	29.57	29.84	29.11	29.570	−0.430
35	34.78	34.18	33.58	32.98	32.38	33.580	−1.420
40	39.89	39.06	39.23	39.40	39.57	39.430	−0.570
45	44.08	43.94	43.80	43.66	43.52	43.800	−1.200
50	50.09	49.82	49.55	49.28	49.01	49.550	−0.450
55	54.01	54.70	54.39	54.08	54.77	54.39	− 0.610
60	60.00	59.58	59.16	59.74	59.32	59.560	−0.440
65	65.04	64.46	64.88	64.30	65.72	64.880	−0.120
70	70.01	69.34	69.67	69.00	69.33	69.470	−0.530
75	74.04	74.22	74.40	74.58	74.76	74.400	−0.600
80	80.02	80.10	80.18	80.26	80.34	80.180	0.180
85	85.04	84.98	84.92	85.86	84.80	85.120	0.120
90	90.20	89.86	89.52	89.18	90.84	89.920	−0.080
95	95.02	95.74	94.46	95.18	94.90	95.060	0.060
100	100.03	99.62	99.21	99.80	99.39	99.610	−0.390

Solution Minitab output for the equation of the regression line is given in Figure 13.12.

Regression Analysis: Bias versus Value

```
The regression equation is
Bias = - 0.457 + 0.00192 Value

S = 0.424562   R-Sq = 1.8%   R-Sq(adj) = 0.0%
```

Figure 13.12

The slope of the regression line—which represents the percent linearity—is 0.00192, and the y-intercept is -0.457. The gage would have been considered linear if the slope were equal to one; in this case, it is very far from one. The coefficient of determination is equal to 1.8 percent, which is very low and therefore suggests that the proportion in the variations in the bias explained by the true values is insignificant.

The determination of the gage bias is based on whether the y-intercept is equal to zero. If it is equal to zero, the gage would be considered as unbiased; otherwise, it is. In this case, it is equal to -0.457. We must conclude that the scale used for measuring the weight of the pistons is not fit and would lead to wrong conclusions.

The scatter plot of bias versus value is shown in Figure 13.13.
The vertical distances between the regression line and every point represent the errors of measurement.

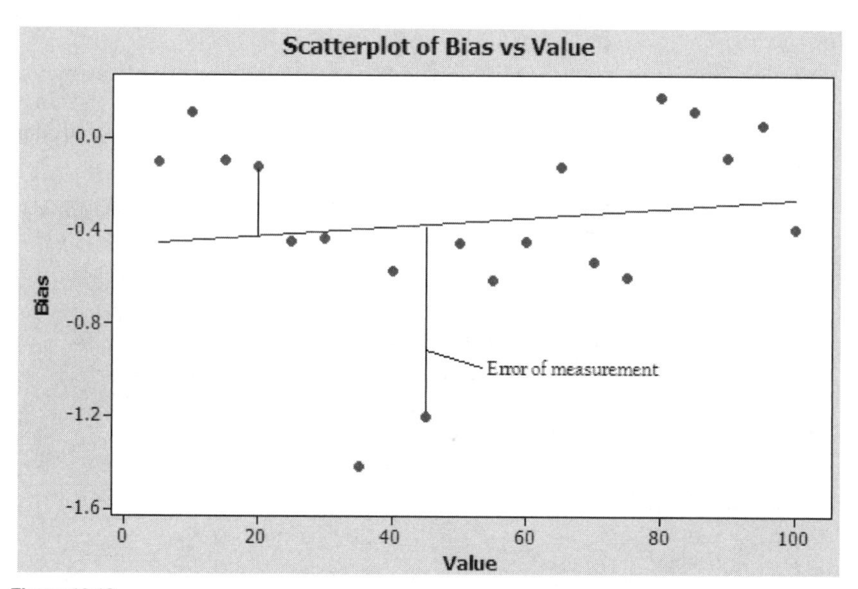

Figure 13.13

The spread of the points shows no correlation between the values of the bias and the true values. The y-intercept is far from zero and the regression line is almost horizontal. These two results suggest bias and a lack of linearity.

Example A thermometer is used to measure the heat generated by motors used on conveyors at Boal Mechanicals. The heat is known to depend on the type of motor used. Six motors labeled A, B, C, D, E, and F with known heat levels are selected for testing. The results of the tests are summarized in Table 13.8, which can also be found on the included CD in the file *Baol.mpj.*

Test the thermometer for its fitness to be used as a measuring tool for the business.

TABLE 13.8

Motor	True value	Gage measurement	Motor	True value	Gage measurement
A	15	15.0000	D	30	30.0028
A	15	15.0400	D	30	29.5911
A	15	15.0613	D	30	30.0003
A	15	15.0092	D	30	29.9907
A	15	15.0048	D	30	29.9992
B	20	20.0026	E	35	34.8712
B	20	19.9667	E	35	35.3279
B	20	20.0001	E	35	35.0023
B	20	19.9527	E	35	35.0023
B	20	20.0002	E	35	34.9994
C	25	24.9894	F	40	40.0050
C	25	25.0628	F	40	40.0021
C	25	24.7427	F	40	39.9978
C	25	25.0081	F	40	39.9995
C	25	25.0774	F	40	40.0397

Solution Open the file *Boal.mpj* on the included CD. From the Stat menu, select "Quality Tools," then select "Gage Study," and then select "Gage Linearity and Bias Study." fill out the "Gage Linearity and Bias Study" dialog box as indicated in Figure 13.14.

Figure 13.14

Then select "OK" to obtain the output of Figure 13.15.

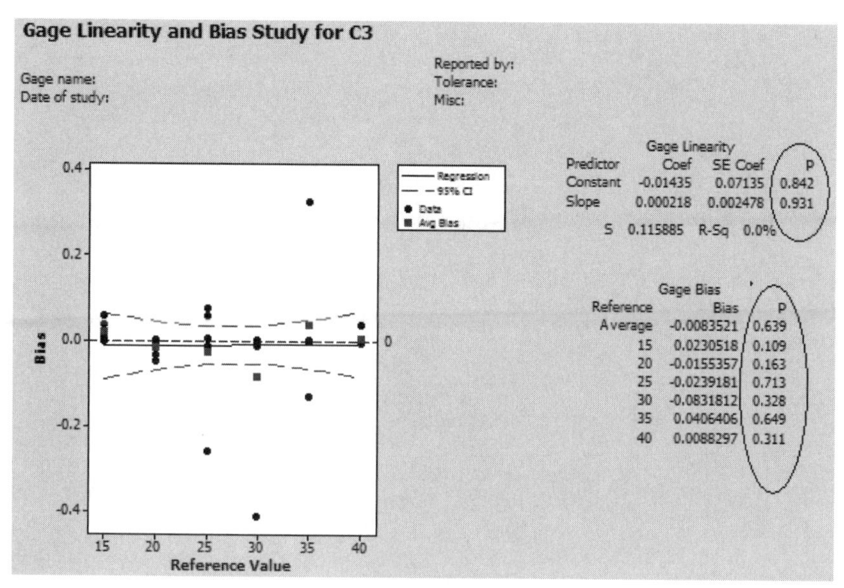

Figure 13.15

The slope which represents linearity is equal to 0.000218. The slope is very low and therefore suggests a good measure of linearity.

Chapter

14

Nonparametric Statistics

So far, all the probability distributions that we have seen are based on means, variances, standard deviations, and proportions. For instance, the Poisson distribution is based on the mean, and the normal and log-normal distributions are based on the mean and standard deviation.

The hypothesis tests and the estimations that we conducted were based on assumptions about the distributions that the data being analyzed follow, and that those distributions depended on means and standard deviations. The standard error-based t-test was founded on the assumption that the samples were randomly taken from populations that were normally distributed, and the analyses done were contingent upon the standard deviation, the mean, and the sample size. This also applied to ANOVA and ANOM.

In these contexts, statisticians call the mean and the standard deviation *parameters*. If we use probability distributions that involve these parameters to estimate the probability of an event to occur or to determine if there is a difference between samples' statistics, we are conducting a *parametric procedure* to derive an estimation or to determine if a hypothesis must be rejected or not.

But what if the data being analyzed do not follow any probability distribution? What if we cannot derive a mean or a standard deviation from the data? What if the data are qualitative, ranked, nominal, ordinal, or nonadditive? In these cases, distribution-free or *nonparametric* techniques will be used to analyze the data. In this chapter, we will discuss a few nonparametric tests.

14.1 The Mann-Whitney U test

14.1.1 The Mann-Whitney U test for small samples

The *Mann-Whitney U* test is better explained through an example.

Example A business owner has two plants of unequal sizes that use several types of vehicles that use unleaded fuel. The daily consumption of fuel is not normally distributed. He wants to compare the amount of fuel that the two plants use a day. He takes a sample of seven days from Plant A and a sample of five days from Plant B. Table 14.1 shows the two samples.measured in gallons.

We can make several observations from this table. First, the sample sizes are small and we only have two samples, so the first thing that comes to mind would be to use the standard error-based t-test. But the t-test assumes that the populations from which the samples are taken should be normally distributed—which is not the case in this example—therefore, the t-test cannot be used. Instead, the Mann-Whitney U test will be.

The Mann-Whitney U test assumes that the samples are independent and from dissimilar populations.

Step 1: Define the null hypothesis Just as in the case of the t-test, the Mann-Whitney U test is a hypothesis test. The null and alternate hypotheses are

H_0 : The daily consumption of fuel is the same in the two plants.

H_1 : The daily consumption of fuel in the two plants is different.

The result of the test will lead to the rejection of the null hypothesis or a failure to reject the null hypothesis.

Step 2: Analyze the data The first step in the analysis of the data will consist in naming the groups. In our case, they are already named A and B. The next step will consist in grouping the two columns into one and sorting the observations in ascending order and ranked from 1 to n. Each observation will be paired with the name of the original group to which it belonged. We obtain the columns shown in Table 14.2.

TABLE 14.1

A	B
15	17
24	23
19	10
9	11
12	18
13	
16	

TABLE 14.2

Observation	Group	Rank
9	A	1
10	B	2
11	B	3
12	A	4
13	A	5
15	A	6
16	A	7
17	B	8
18	B	9
19	A	10
23	B	11
24	A	12

We will call ω_1 the sum of the ranks of the observations for group A and ω_2 the sum of the ranks of the observations for group B.

$$\omega_1 = 1 + 4 + 5 + 6 + 7 + 10 + 12 = 45$$
$$\omega_2 = 2 + 3 + 8 + 9 + 11 = 33$$

Step 3: Determine the values of the U statistic The computation of the U statistic will depend on the samples' sizes.

The samples are small when n_1 and n_2 are both smaller than 10. In that case,

$$U_1 = n_1 n_2 + \frac{n_1 (n_1 + 1)}{2} - \varpi_1$$
$$U_2 = n_1 n_2 + \frac{n_2 (n_2 + 1)}{2} - \varpi_2$$

The test statistic U will be the smallest of U_1 and U_2.

If any or both of the sample sizes is greater than 10, then U will be approximately normally distributed and we could use the Z transformation with

$$\mu = \frac{n_1 \cdot n_2}{2}$$
$$\sigma = \sqrt{\frac{n_1 n_2 (n_1 + n_2 + 1)}{12}}$$

And

$$Z = \frac{U - \mu}{\sigma}$$

In our case, both sample sizes are less than 10, therefore

$$U_1 = 7 \times 5 + \frac{7(7+1)}{2} - 45 = 35 + 28 - 45 = 18$$
$$U_2 = 7 \times 5 + \frac{5(5+1)}{2} - 33 = 35 + 15 - 33 = 17$$

Because the calculated test statistic is the smaller of the two, we will have to consider $U_2 = 17$, so we will use $U = 17$ with $n_2 = 7$ and $n_1 = 5$. From the Mann-Whitney table below, we obtain a P-value equal to 0.5 for a one-tailed graph. Because we are dealing with a two-tailed graph, we must double the P-value and obtain 1.

Mann-Whitney Table

U	4	5	6	7
13	0.4636	0.2652	0.1474	0.825
14	0.5364	0.3194	0.1830	0.1588
15		0.3775	0.2226	0.1297
16		0.4381	0.2669	0.1588
17		0.5000	0.3141	0.1914
18			0.3654	0.2279

Using Minitab Open the file *Fuel.mpj* on the included CD. From the Stat menu, select "Nonparametrics" and from the drop-down menu, select "Mann-Whitney." Fill out the Mann-Whitney dialog box as indicated in Figure 14.1.

Figure 14.1

```
Mann-Whitney Test and CI: A, B

   N  Median
A  7   15.00
B  5   17.00     ——Medians

                                                   Confidence Interval

Point estimate for ETA1-ETA2 is -1.00
96.5 Percent CI for ETA1-ETA2 is (-8.00,7.00)                        P - Value
W = 45.0
Test of ETA1 = ETA2 vs ETA1 not = ETA2 is significant at 1.0000
```

Figure 14.2

Select "OK" to obtain the output of Figure 14.2.

The P-value is the highest it can be, therefore we cannot reject the null hypothesis. We must conclude that there is not enough statistical evidence to say that the two sets of data are not identical.

14.1.2 The Mann-Whitney U test for large samples

Example In the previous example, we used small samples; in this one, we will use large samples.

Tambacounda Savon is a soap manufacturing company located in Senegal. It operates two shifts and the quality manager wants to compare the quality level of the output of the two shifts. He takes a sample of 12 days from the first shift and 11 days from the second shift and obtains the following errors per 10,000 units. At a confidence level of 95 percent, can we say that the two shifts produce the same quality level of output?

TABLE 14.4

First shift	Second shift
2	14
4	5
7	1
9	7
6	15
3	4
12	9
13	10
10	17
0	16
11	8
5	

Solution

Step 1: Define the hypotheses The null hypothesis in this case will suggest that there is not any difference between the quality level of the output of the two shifts, and the alternate hypothesis will suggest the opposite.

H_0 : The quality level of the first shift is the same as the one for the second shift.

H_1 : The quality level of the first shift is different from the one of the second shift.

Step 2: Analyze the data Here again, we pool all the data in one column or line and we rank them from the smallest to the highest while still maintaining the original groups to which they belonged.

TABLE 14.5

Defects	Shift	Rank
0	First	1
1	Second	2
2	First	3
3	First	4
4	First	5.5
4	Second	5.5
5	First	7.5
5	Second	7.5
6	First	9
7	First	10.5
7	Second	10.5
8	Second	12
9	First	13.5
9	Second	13.5
10	First	15.5
10	Second	15.5
11	First	17
12	First	18
13	First	19
14	Second	20
15	Second	21
16	Second	22
17	Second	23

$$\omega_{first} = 1 + 3 + 4 + 5.5 + 7.5 + 9 + 10.5 + 13.5 + 15.5 + 17 + 18 + 19 = 123.5$$

$$\omega_{second} = 2 + 5.5 + 7.5 + 10.5 + 13.5 + 15.5 + 20 + 21 + 22 + 23 = 152.5$$

We can now find the value of U:

$$U_{first} = 12 \times 11 + \frac{12(12+1)}{2} - 123.5 = 86.5$$

$$U_{second} = 12 \times 11 + \frac{11(11+1)}{2} - 152.5 = 45.5$$

$$\mu = \frac{n_1 \times n_2}{2} = \frac{12 \times 11}{2} = 66$$

$$\sigma = \sqrt{\frac{n_1 n_2 (n_1 + n_2 + 1)}{12}} = \sqrt{\frac{132(12+11+1)}{12}} = \sqrt{264} = 16.25$$

The next step will consist in finding the Z score. We use U_{second}

$$Z = \frac{U - \mu}{\sigma} = \frac{45.5 - 66}{16.25} = -1.262$$

What would have happened if we had used U_{first} instead of U_{second}?

$$Z = \frac{U - \mu}{\sigma} = \frac{86.5 - 66}{16.25} = 1.262$$

We would have obtained the same result with the opposite sign.

At a confidence level of 95 percent, we would reject the null hypothesis if the value of Z is outside the interval $[-1.96, +1.96]$. In this case, $Z = -1.262$ is well within that interval; therefore, we should not reject the null hypothesis.

Minitab output

Mann-Whitney Test and CI: First Shift, Second Shift

```
              N  Median
First Shift   12  6.500
Second Shift  11  9.000


Point estimate for ETA1-ETA2 is -3.000
95.5 Percent CI for ETA1-ETA2 is (-7.001,1.999)
W = 123.5
Test of ETA1 = ETA2 vs ETA1 not = ETA2 is significant at 0.2184
The test is significant at 0.2178 (adjusted for ties)
```

The P-value of 0.2184 is greater than 0.05, which suggests that for an α level of 0.05, we cannot reject the null hypothesis.

Minitab does not offer the Z value but because we obtained it from our calculations, we can find the value of P. On the Z-score table, a value of $Z = 1.262$ corresponds to about 0.3962. Because we are faced with a two-tailed graph, we must double that value and subtract the result from one to obtain 0.2076. The difference between what we obtained and the Minitab output is attributed to rounding errors.

14.2 The Chi-Square Tests

In our hypothesis testing examples, we used means and variances to determine if there were statistically significant differences between samples. When comparing two or more samples' means, we expect their values to be identical for the samples to be considered similar. Yet while using the standard error-based t-test to analyze and interpret hypotheses, we have seen that even when samples do not have the exact same means, we sometimes cannot reject the null hypothesis and must conclude that the samples' means are not significantly statistically different.

The chi-square test compares the observed values to the expected values to determine if they are statistically different when the data being analyzed do not satisfy the t-test assumptions.

14.2.1 The chi-square goodness-of-fit test

Suppose that a molding machine has historically produced metal bars with varying strength (measured in PSI) and the strengths of the bars are categorized in Table 14.5. The ideal strength is 1998 PSI.

TABLE 14.6

Strength (PSI)	Proportion
2000	5%
1999	9%
1998	65%
1997	10%
1996	6%
1995	5%

After the most important parts of the machine have been changed, a shift supervisor wants to know if the changes made have made a difference to the production. She takes a sample of 300 bars and finds that their strengths in PSI are as shown in Table 14.6.

TABLE 14.7

Strength (PSI)	Bars
2000	22
1999	45
1998	198
1997	30
1996	9
1995	1

Based on the sample that she took, can we say that the changes made to the machine have made a difference?

In this case, we cannot use a hypothesis testing based on the mean because we cannot simply add the percentages, divide them by six, and conclude that we have or do not have a mean strength, nor can we add the number of bars and divide them by six to determine the mean.

Because the data that we have is not additive, we will use a nonparametric test called the *chi-square goodness-of-fit test*. The chi square goodness-of-fit test compares the expected frequencies (Table 14.5) to the actual or observed frequencies (Table 14.6). The formula for the test is

$$\chi^2 = \sum \frac{(f_a - f_e)^2}{f_e}$$

with f_e as the expected frequency and f_a as the actual frequency. The degree of freedom will be given as

$$df = k - 1$$

Chi-square cannot be negative because it is the square of a number. If it is equal to zero, all the compared categories would be identical, therefore chi-square is a one-tailed distribution.

The null and alternate hypotheses will be

H_0 : The distribution of quality of the products after the parts were changed is the same as before the parts were changed.

H_1 : The distribution of the quality of the products after the parts were changed is different than it was before they were changed.

We will first transform Table 14.5 to obtain the absolute values of the number of products that would have been obtained had we chosen a sample of 300 products before the parts were changed.

TABLE 14.8

Strength (PSI)	Proportion	
2000	5% × 300	15
1999	9% × 300	27
1998	65% × 300	195
1997	10% × 300	30
1996	6% × 300	18
1995	5% × 300	15
Total		300

Now we can use the formula to determine the value of the calculated chi-square,

$$\chi^2 = \sum \frac{(f_a - f_e)^2}{f_e} = \frac{(15-22)^2}{15} + \frac{(27-45)^2}{27} + \frac{(195-198)^2}{195} + \frac{(30-30)^2}{30}$$

$$+ \frac{(18-9)^2}{18} + \frac{(15-1)^2}{15} = 32.88$$

With a confidence level of 95 percent, $\alpha = 0.05$ and a degree of freedom of 5 ($df = 6 - 1$), the critical value of $\chi^2_{0.05,5}$ is equal to 11.0705.

The next step will be to compare the calculated χ^2 with the critical $\chi^2_{0.05,5}$ found on the table Chi square. If the critical $\chi^2_{0.05,5}$ critical value is greater than the calculated χ^2, we cannot reject the null hypothesis; otherwise, we reject it.

Because the calculated χ^2 value (32.88) is much higher that the critical value (11.0705), we must reject the null hypothesis. The changes made on the machine have indeed resulted in changes in the quality of the output.

Example Konakry Motor Company owns five warehouses in Bignona. The five plants are being audited for ISO-9000 compliance. The audit is performed to test the employees' understanding and conformance with the companies standardized processes. The employees at the different plants are expected to have the same probability to be selected for audit.

The random samples taken from the different plants were:

Plant 1	76 employees
Plant 2	136 employees
Plant 3	89 employees
Plant 4	95 employees
Plant 5	93 employees

Can we conclude at a significance level of 0.05 that the employees at the five plants had the same probability of being selected?

Solution

Step 1: Define the hypotheses In this case, the ratios of the number of employees audited at each plant to the overall number of employees audited are expected to be the same if there is not any statistical difference between them at a confidence level of 95 percent. So the null hypothesis will be

$$H_0 : p_1 = p_2 = p_3 = p_4 = p_5$$

with p_1, p_2, p_3, p_4, and p_5, being the ratios of the employees selected from each plant to the overall number of employees selected. The alternate hypothesis would be that at least one ratio is different form the rest of the ratios.

Step 2: Determine when the null hypothesis should be rejected The rejection or nonrejection of the null hypothesis is based on whether the calculated χ^2 that we will obtain from our analysis is greater or smaller than the expected $\chi^2_{0.05,4}$ found on the chi-square table.

A value of 0.05 represents the α level for 95 percent confidence while 4 represents the degree of freedom $(5 - 1)$. $\chi^2_{0.05,4}$ happens to be equal to 9.48773. If the calculated $\chi^2 > 9.48773$, we would have to reject the null hypothesis; otherwise, we should not.

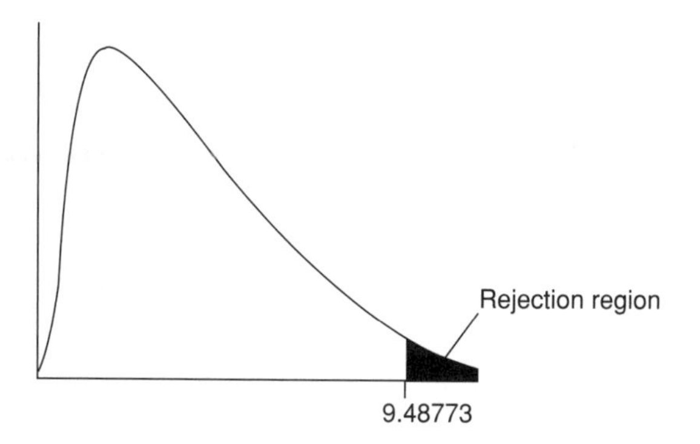

Step 3: Determine the calculated χ^2 The total number of employees selected is $76 + 136 + 89 + 95 + 93 = 489$.

TABLE 14.9

	Selected employees	Expected number of selected employees	Actual proportion
Plant 1	76	$489 \times (1/5) = 97.8$	$76/489 = 0.155$
Plant 2	136	$489 \times (1/5) = 97.8$	$136/489 = 0.278$
Plant 3	89	$489 \times (1/5) = 97.8$	$89/489 = 0.1820$
Plant 4	95	$489 \times (1/5) = 97.8$	$95/489 = 0.194$
Plant 5	93	$489 \times (1/5) = 97.8$	$93/489 = 0.190$
Totals	489	489	1

$$\chi^2 = \frac{(76 - 97.8)^2}{97.8} + \frac{(136 - 97.8)^2}{97.8} + \frac{(89 - 97.8)^2}{97.8}\frac{(95 - 97.8)^2}{97.8} + \frac{(93 - 97.8)^2}{97.8}$$

$$\chi^2 = 20.888$$

Step 3: Decision making Because the calculated χ^2 is greater than the expected $\chi^2_{0.05,4}$ (which is equal to 9.948773), we must reject the null hypothesis and conclude that employees at the different plants did not have an equal probability of being selected.

Using Minitab Open the file *Konakry.mpj* on the included CD. From the State menu, select "Tables and Chi-Square Goodness-of-Fit (One Variable)." Fill out the fields as indicated in Figure 14.3, then select "Graphs. . . "

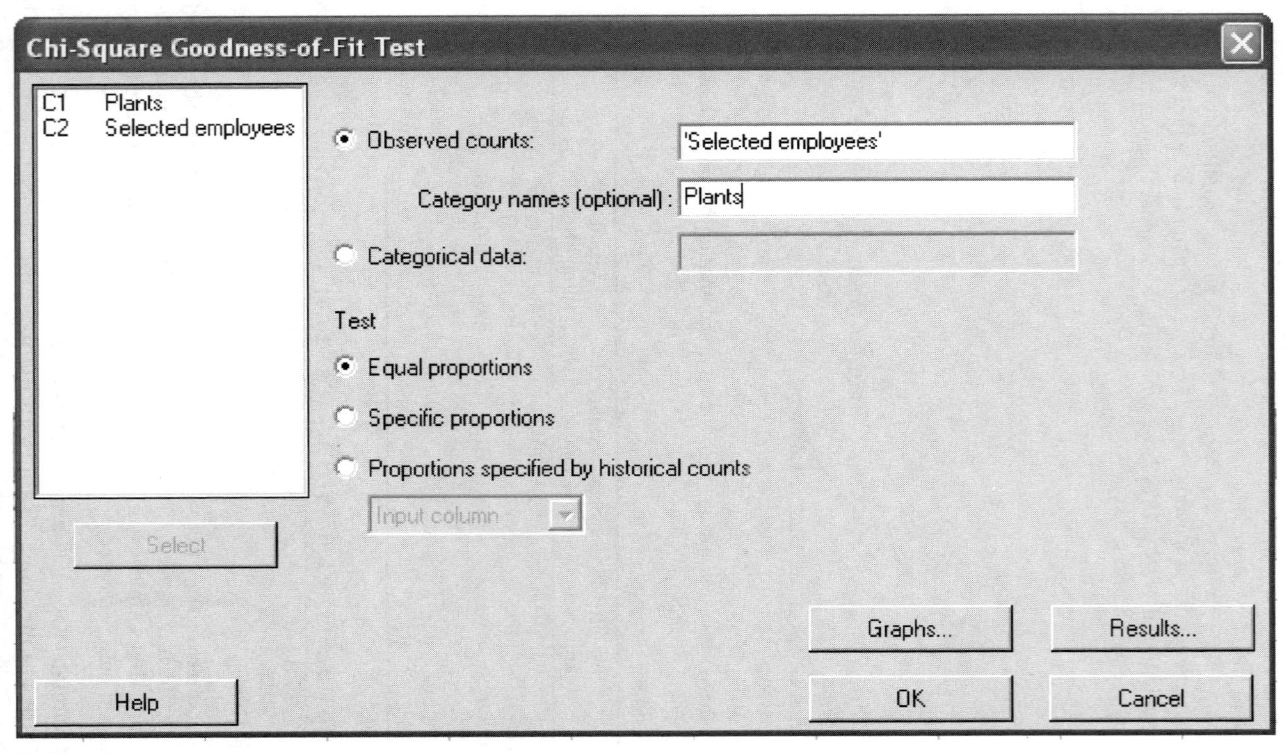
Chi-Square Goodness-of-Fit Test
C1 Plants
C2 Selected employees
Observed counts:
'Selected employees'
Category names (optional) :
Plants
Categorical data:
Test
Equal proportions
Specific proportions
Proportions specified by historical counts
Input column
Select
Graphs...
Results...
Help
OK
Cancel

Figure 14.3

On the "Chi-Square Goodness-of-Fit Graph" dialog box, select each of the options, then select "OK" and select "OK" again. We obtain the results shown in Figure 14.4.

Chi-Square Goodness-of-Fit Test for Observed Counts in Variable: Selected emplo

Using category names in Plants

Category	Observed	Test Proportion	Expected	Contribution to Chi-Sq
Plant 1	76	0.2	97.8	4.8593
Plant 2	136	0.2	97.8	14.9207
Plant 3	89	0.2	97.8	0.7918
Plant 4	95	0.2	97.8	0.0802
Plant 5	93	0.2	97.8	0.2356

N	DF	Chi-Sq	P-Value
489	4	20.8875	0.000

Figure 14.4

The degree of freedom is four, the observed χ^2 is 20.8875, and the P-value equal to zero indicates that there is a statistically significant difference between the samples; therefore, the employees did not have the same probability of being selected.

Interpretation of the graphs The "Chart of Observed and Expected Values" in Figure 14.5 compares the actual number of employees selected

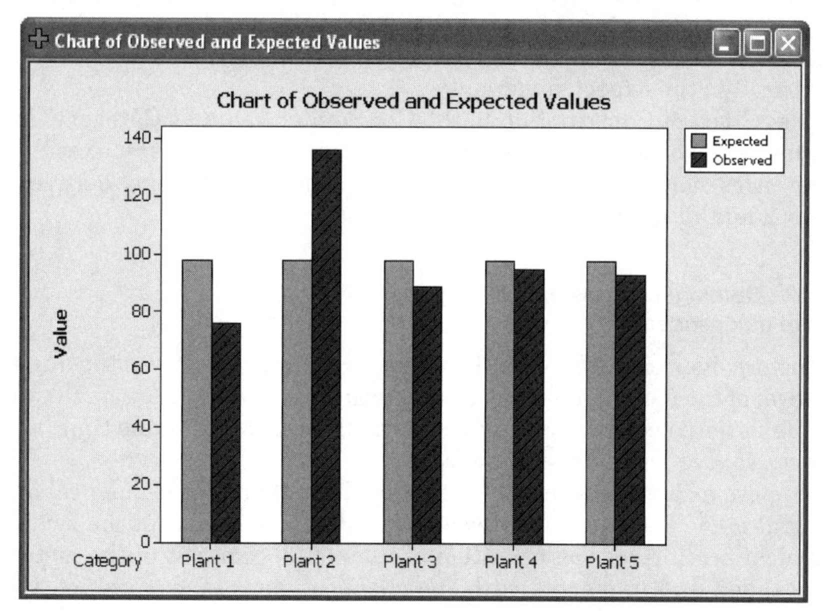

Figure 14.5

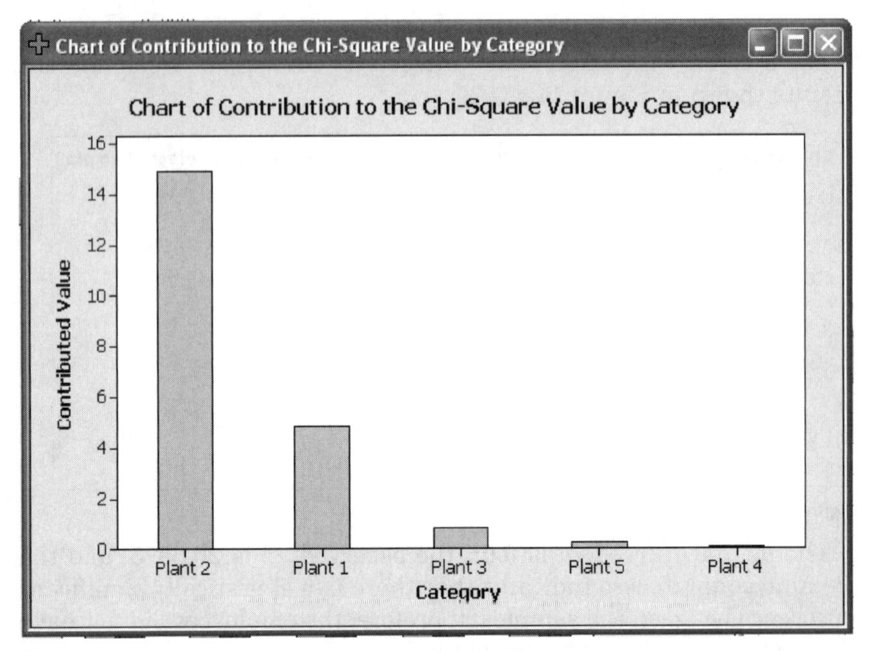

Figure 14.6

to what it should have been for all the employees to have an equal opportunity to be selected. We can clearly see that the number of employees selected from Plant 2 is too high compared to the rest of the plants and compared to the expected value.

The "Chart of Contribution to the Chi-Square Value by Category" in Figure 14.6 shows how the different plants are distributed. It makes the differences more obvious, and it shows how wide the difference between Plant 2 and Plant 4 really is.

14.2.2 Contingency analysis: chi-square test of independence

In the previous example, we only had one variable, which was the quality level of the metal bars measured in terms of strength. If we have two variables with several levels (or categories) to test at the same time, we use *chi-square test of independence.*

Suppose a chemist wants to know the effect of an acidic chemical on a metal alloy. The experimenter wants to know if the use of the acidic chemical accelerates the oxidation of the metal. Samples of the metal were taken and immersed with the chemical, and some were not. Of the samples that were immersed, traces of oxide were found on 79 bars

TABLE 14.10

	Acidic	Non-acidic
Oxide	79	48
No Oxide	1091	1492

and no trace of oxide was found on 1091 bars. For those that were not immersed with the chemical, traces of oxide were found on 48 bars and no oxide was found on 1492 bars. The findings are summarized in Table 14.9.

In this case, if the acidic chemical has no impact on the oxidation level of the metal we should expect that there would be no statistically significant difference between the proportions of the metals with oxidation and the ones without oxidation with respect to their groups.

If we call $\overline{P}_1$ the proportion of the bars with oxide that were immersed in the chemical and $\overline{P}_2$ the proportion of the bars with oxide that were not immersed in the chemical, the null and alternate hypotheses will be as follows:

$$H_0 : \overline{P}_1 = \overline{P}_2$$

$$H_1 : \overline{P}_1 \neq \overline{P}_2$$

Rewrite Table 14.9 by adding the totals.

TABLE 14.11

	Acidic	Non-acidic	Total
Oxide	79	48	127
No Oxide	1091	1492	2583
Total	1170	1540	2710

The grand mean proportion for the bars with traces of oxidation is

$$\overline{P} = \frac{79 + 48}{1170 + 1540} = \frac{127}{2710} = 0.0468635$$

The grand mean proportion of the bars without traces of oxide is

$$\overline{Q} = 1 - \overline{P} = 1 - 0.0468635 = 0.9531365$$

Now we can build the table of the expected frequencies:

TABLE 14.12

	Acidic	Non-acidic	Total
Oxide	0.046864 × 1170 = 54.830295	0.46864 × 1540 = 72.16979	127
No Oxide	0.953137 × 1170 = 1115.169705	0.953137 × 1540 = 1467.83021	2583
Total	1170	1540	2710

Now that we have both the observed data and the expected data, we can use the formula to make the comparison.

The formula that will be used in the case of a contingency table is slightly different from the one of chi-square goodness-of-fit,

$$\chi^2 = \sum\sum \frac{(f_e - f_a)^2}{f_e}$$

with a degree of freedom of

$$df = (r - 1)(c - 1)$$

where c is the number of columns and r is the number of rows. The degree of freedom for this instance will be $(2 - 1)(2 - 1) = 1$. For a significance level of 0.05, the critical $\chi^2_{0.05,1}$ found on the Chi-Square table would be 3.841.

We can now compute the test statistics:

TABLE 14.13

f_e	f_a	$(f_a - f_e)^2$	$(f_a - f_e)^2/f_e$
54.830295	79	584.1746	10.65423
72.16979	48	584.1787	8.094505
1115.169705	1091	584.1746	0.523844
1467.83021	1492	584.1787	0.397988
Totals			19.67057

The calculated χ^2 is 19.67057, which is much higher than the critical $\chi^2_{0.05,1}$, which is 3.841. Therefore, we must reject the null hypothesis. At a confidence level of 0.05, there is enough evidence to suggest that the acidic chemical has an effect on the oxidation of the metal alloy.

Test these findings using Minitab. After pasting the table into a Minitab worksheet, from the Stat menu, select "Tables", then "Chi-Square Test (Table in Worksheet)" as shown in Figure 14.7.

Then the dialog box of Figure 14.8 appears.

Figure 14.7

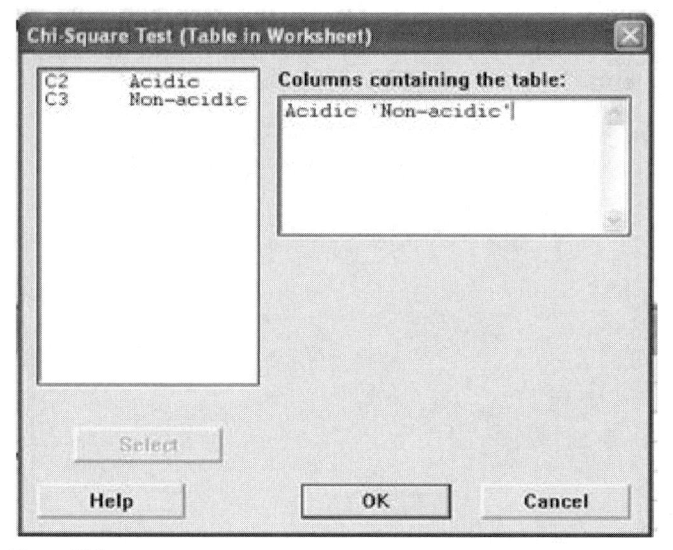

Figure 14.8

Select "Acidic" and "Non-Acidic" and then select "OK."

MINITAB - Untitled - [Session]

File Edit Data Calc Stat Graph Editor Tools Window Help

5/3/2006 3:27:55 PM

Welcome to Minitab, press F1 for help.

Chi-Square Test: Acidic, Non-acidic

```
Expected counts are printed below observed counts
Chi-Square contributions are printed below expected counts

        Acidic  Non-acidic  Total
    1       79          48    127
         54.83       72.17
        10.654       8.094

    2     1091        1492   2583
       1115.17     1467.83
         0.524       0.398

Total     1170        1540   2710

Chi-Sq = 19.671, DF = 1, P-Value = 0.000
```

The degree of freedom is one, and the calculated χ^2 is 19.671. The P-value of zero suggests that there is a statistically significant difference, and therefore we must reject the null hypothesis.

Chapter

15

Pinpointing the Vital Few Root Causes

15.1 Pareto Analysis

Pareto analysis is simple. It is based on the principle that 80 percent of problems find their roots in 20 percent of causes. This principle was established by Vilfredo Pareto, a nineteenth-century Italian economist who discovered that 80 percent of the land in Italy was owned by only 20 percent of the population. Later empirical evidence showed that the 80/20 ratio was determined to have a universal application.

- 80 percent of customer dissatisfaction stems from 20 percent of defects.
- 80 percent of the wealth is in the hands of 20 percent of the people.
- 20 percent of customers account for 80 percent of a business.

When applied to management, the Pareto rule becomes an invaluable tool. For instance, in the case of problem-solving the objective should be to find and eliminate the circumstances that make the 20 percent "vital few" possible so that 80 percent of the problems are eliminated. It is worthy to note that Pareto analysis is a better tool to detect and eliminate sources of problems when those sources are independent variables. If the different causes of a problem are highly correlated, the Pareto principle may not be applicable.

The first step in Pareto analysis will be to clearly define the goals of the analysis. What is it that we are trying to achieve? What is the nature of the problem we are facing?

The next step in Pareto analysis is the data collection. All the data pertaining to the factors that can potentially affect the problem being addressed must be quantified and stratified. In most cases, a sophisticated statistical analysis is not necessary; a simple tally of the numbers suffices to prioritize the different factors. But in some cases, the quantification might require statistical analysis to determine the level of correlation between the cause and the effect. A regression analysis can be used for that purpose, or a correlation coefficient or a coefficient of determination can be derived to estimate the level of association of the different factors to the problem being analyzed. Then a categorization can be made: the factors are arranged according to how much they contribute to the problem. The data generated is used to build a cumulative frequency distribution.

The next step will be to create a *Pareto diagram* or *Pareto chart* to visualize the main factors that contribute to the problem and therefore concentrate focus on the "vital few." The Pareto chart is a simple histogram; the horizontal axis shows the different factors whereas the vertical axis represents the frequencies. Because all the different causes will be listed on the same diagram, it is necessary to standardize the unit of measurement and set the timeframe for the occurrences.

The building of the chart requires a data organization. A four-column data summary must be created to organize the information collected. The first column will list the different factors that cause the problem, the second column will list the frequency of occurrence of the problem during a given timeframe, the third column records the relative frequencies (in other words, the percentage of the total), and the last column will record the cumulative frequencies—keeping in mind that the data are listed from the most important factor to the least.

The data of Table 15.1 was gathered during a period of one month to analyze the reasons behind a high volume of customer returns of cellular phones ordered online.

TABLE 15.1

Factors	Frequency	Relative Frequency	Cumulative Frequency
Misinformed about the contract	165	58%	58%
Wrong products shipped	37	13%	71%
Took too long to receive	30	11%	82%
Defective product	26	9.2%	91.2%
Changed my mind	13	4.6%	95.8%
Never received the phone	12	4.2%	100%
Totals	283	100%	

The diagram itself will consist of three axes. The horizontal axis lists the factors, the left vertical axis lists frequency of occurrence and is graded from zero to at least the highest frequency. The right vertical line is not always present on Pareto charts; it represents the percentage of occurrences and is graded from zero to 100 percent.

The *breaking point* (the point on the cumulative frequency line at which the curve is no longer steep) on the graph of Figure 15.1 occurs at around "Wrong products." Because the breaking point divides the "vital few" from the " trivial many," the two first factors, "Misinformed about contract" and "Wrong products" are the factors that need more attention. By eliminating the circumstances that make them possible, we will eliminate about 71 percent of our problems.

Creating a Pareto chart using Minitab. Open the file *Cell phone.mpj* from the included CD. From the Stat menu, select "Quality Tools" and then select "Pareto Chart."

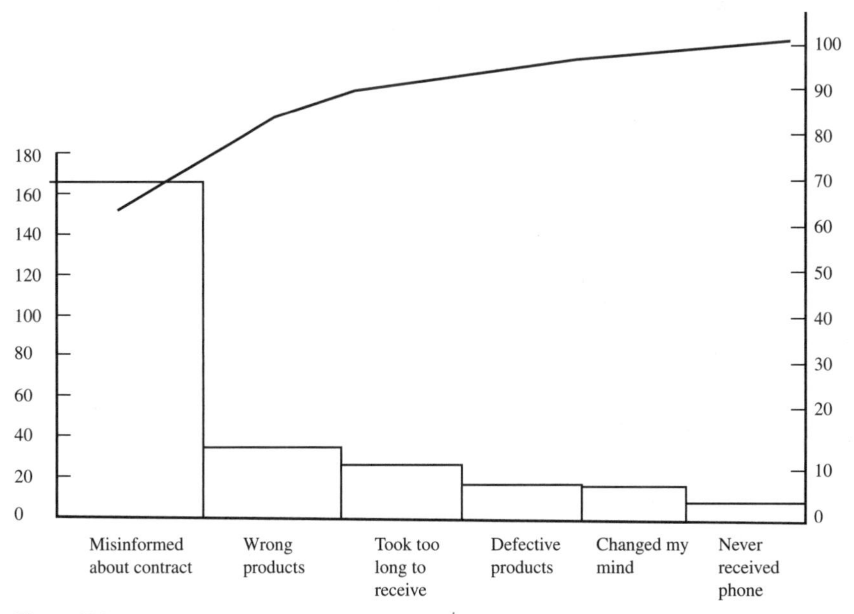

Figure 15.1

Select the option "Chart Defect Table" from the "Pareto Chart" dialog box and change the "Combine remaining defects into one category after this percent" to "99," as indicated in Figure 15.2.

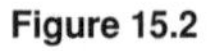

Figure 15.2

Select "OK" to get the output shown in Figure 15.3.

15.2 Cause and Effect Analysis

The *cause-and-effect diagram*—also known as a *fishbone* (because of its shape) or *Ishikawa diagram* (after its creator)—is used to synthesize the different causes of an outcome. It is an analytical tool that provides a

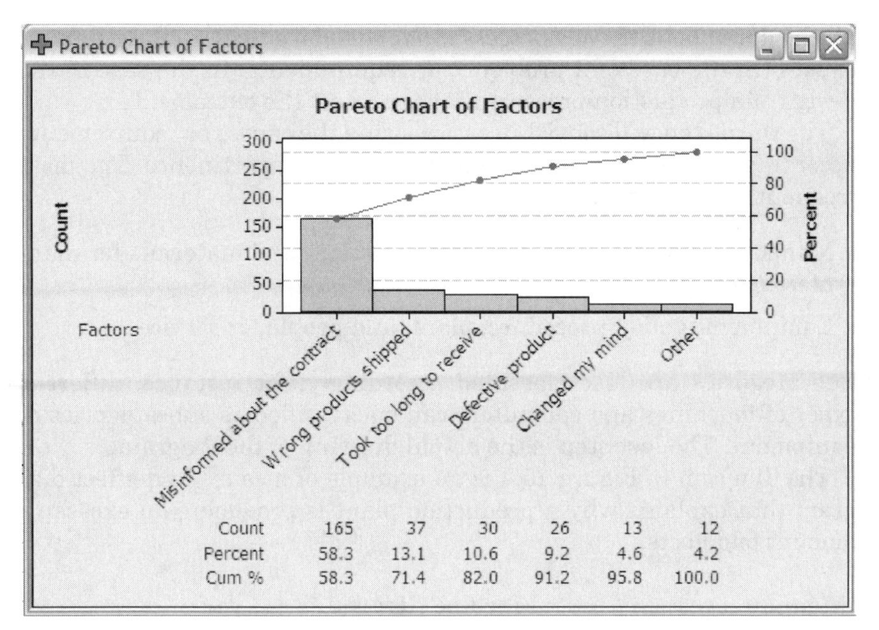

Figure 15.3

visual and systematic way of linking different causes (input) to an effect (output). It can be used in the "Design" phase of a production process as well as in an attempt to identify the root causes of a problem. The effect is considered positive when it is an objective to be reached, as in the case of a manufacturing design. It is negative when it addresses a problem being investigated

The building of the diagram is based on the sequence of events. "Sub-causes" are classified according to how they generate "sub-effects," and those "sub-effects" become the causes of the outcome being addressed.

The fishbone diagram does help visually identify the root causes of an outcome, but it does not quantify the level of correlation between the different causes and the outcome. Further statistical analysis is needed to determine which factors contribute the most to creating the effect. Pareto analysis is a good tool for that purpose but it still requires data gathering. Regression analysis allows the quantification and the determination of the level of association between causes and effects. A combination of Pareto and regression analysis can help not only determine the level of correlation but also stratify the root causes. The causes are stratified hierarchically according to their level of importance and their areas of occurrence

The first step in constructing a fishbone diagram is to clearly define the effect being analyzed. The second step will consist into gathering all the data about the *key process input variables* (KPIV), the potential causes (in the case of a problem), or requirements (in the case of the design of a production process) that can affect the outcome.

The third step will consist in categorizing the causes or requirements according to their level of importance or areas of pertinence. The most frequently used categories are:

- Manpower, machine, method, measurement, and materials for manufacturing
- Equipment, policy, procedure, plant, and people for services

Subcategories are also classified accordingly; for instance, different types of machines and computers can be classified as subcategories of equipment. The last step is the actual drawing of the diagram.

The diagram in Figure 15.4 is an example of a cause-and-effect diagram that explains why a production plant is producing an excessive amount of defects.

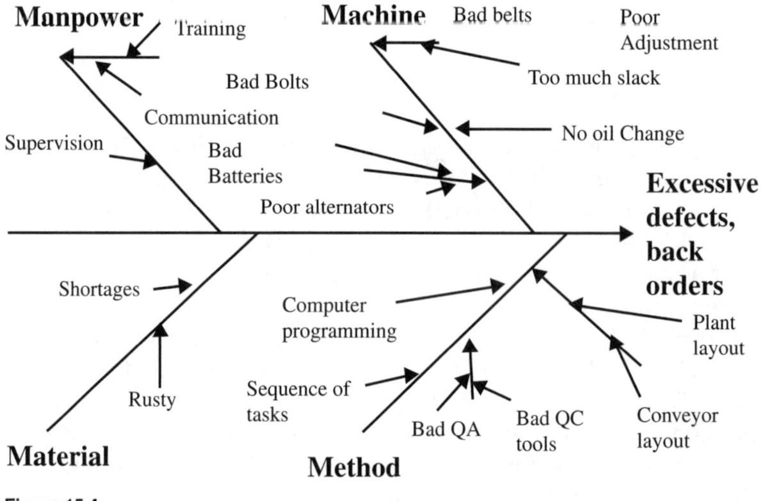

Figure 15.4

Appendices

Appendix 1 Binomial Table $P(x) = {}_nC_x p^x q^{n-x}$

n	s	0	0.02	0.04	0.05	0.06	0.08	0.1	0.12	0.14	0.15	0.16	0.18	0.2	0.22	0.24	0.25
2	0	0.98	0.96	0.92	0.903	0.884	0.846	0.81	0.774	0.74	0.72	0.706	0.672	0.64	0.608	0.578	0.563
2	1	0.02	0.039	0.08	0.095	0.113	0.147	0.18	0.211	0.241	0.26	0.269	0.295	0.32	0.343	0.365	0.375
2	2			0	0.003	0.004	0.006	0.01	0.014	0.02	0.02	0.026	0.032	0.04	0.048	0.058	0.063
3	0	0.97	0.941	0.89	0.857	0.831	0.779	0.729	0.681	0.636	0.61	0.593	0.551	0.512	0.475	0.439	0.422
3	1	0.03	0.058	0.11	0.135	0.159	0.203	0.243	0.279	0.311	0.33	0.339	0.363	0.384	0.402	0.416	0.422
3	2		0.001	0.01	0.007	0.01	0.018	0.027	0.038	0.051	0.06	0.065	0.08	0.096	0.113	0.131	0.141
3	3						0.001	0.001	0.002	0.003	0	0.004	0.006	0.008	0.011	0.014	0.016
4	0	0.96	0.922	0.85	0.815	0.781	0.716	0.656	0.6	0.547	0.52	0.498	0.452	0.41	0.37	0.334	0.316
4	1	0.04	0.075	0.14	0.171	0.199	0.249	0.292	0.327	0.356	0.37	0.379	0.397	0.41	0.418	0.421	0.422
4	2	0	0.002	0.01	0.014	0.019	0.033	0.049	0.067	0.087	0.1	0.108	0.131	0.154	0.177	0.2	0.211
4	3					0.001	0.002	0.004	0.006	0.009	0.01	0.014	0.019	0.026	0.033	0.042	0.047
4	4										0	0.001	0.001	0.002	0.002	0.003	0.004
		0	**0.02**	**0.04**	**0.05**	**0.06**	**0.08**	**0.1**	**0.12**	**0.14**	**0.15**	**0.16**	**0.18**	**0.2**	**0.22**	**0.24**	**0.25**
5	0	0.95	0.904	0.82	0.774	0.734	0.659	0.59	0.528	0.47	0.44	0.418	0.371	0.328	0.289	0.254	0.237
5	1	0.05	0.092	0.17	0.204	0.234	0.287	0.328	0.36	0.383	0.39	0.398	0.407	0.41	0.407	0.4	0.396
5	2	0	0.004	0.01	0.021	0.03	0.05	0.073	0.098	0.125	0.14	0.152	0.179	0.205	0.23	0.253	0.264
5	3			0	0.001	0.002	0.004	0.008	0.013	0.02	0.02	0.029	0.039	0.051	0.065	0.08	0.088
5	4								0.001	0.002	0	0.003	0.004	0.006	0.009	0.013	0.015
5	5														0.001	0.001	0.001

		0	**0.02**	**0.04**	**0.05**	**0.06**	**0.08**	**0.1**	**0.12**	**0.14**	**0.15**	**0.16**	**0.18**	**0.2**	**0.22**	**0.24**	**0.25**
6	0	0.94	0.886	0.78	0.735	0.69	0.606	0.531	0.464	0.405	0.38	0.351	0.304	0.262	0.225	0.193	0.178
6	1	0.06	0.108	0.2	0.232	0.264	0.316	0.354	0.38	0.395	0.4	0.401	0.4	0.393	0.381	0.365	0.356
6	2	0	0.006	0.02	0.031	0.042	0.069	0.098	0.13	0.161	0.18	0.191	0.22	0.246	0.269	0.288	0.297
6	3			0	0.002	0.004	0.008	0.015	0.024	0.035	0.04	0.049	0.064	0.082	0.101	0.121	0.132
6	4						0.001	0.001	0.002	0.004	0.01	0.007	0.011	0.015	0.021	0.029	0.033
6	5											0.001	0.001	0.002	0.002	0.004	0.004
6	6																
		0	**0.02**	**0.04**	**0.05**	**0.06**	**0.08**	**0.1**	**0.12**	**0.14**	**0.15**	**0.16**	**0.18**	**0.2**	**0.22**	**0.24**	**0.25**
7	0	0.93	0.868	0.75	0.698	0.648	0.558	0.478	0.409	0.348	0.32	0.295	0.249	0.21	0.176	0.146	0.133
7	1	0.07	0.124	0.22	0.257	0.29	0.34	0.372	0.39	0.396	0.4	0.393	0.383	0.367	0.347	0.324	0.311
7	2	0	0.008	0.03	0.041	0.055	0.089	0.124	0.16	0.194	0.21	0.225	0.252	0.275	0.293	0.307	0.311
7	3			0	0.004	0.006	0.013	0.023	0.036	0.053	0.06	0.071	0.092	0.115	0.138	0.161	0.173
7	4						0.001	0.003	0.005	0.009	0.01	0.014	0.02	0.029	0.039	0.051	0.058
7	5									0.001	0	0.002	0.003	0.004	0.007	0.01	0.012
7	6														0.001	0.001	0.001
7	7																
		0	**0.02**	**0**	**0.05**	**0.06**	**0.08**	**0.1**	**0.12**	**0.14**	**0.2**	**0.16**	**0.2**	**0.2**	**0.22**	**0.24**	**0.3**
8	0	0.9	0.85	0.72	0.663	0.61	0.513	0.43	0.36	0.299	0.27	0.25	0.2	0.17	0.14	0.111	0.1
8	1	0.1	0.14	0.24	0.279	0.311	0.357	0.38	0.392	0.39	0.39	0.38	0.36	0.34	0.31	0.281	0.27
8	2	0	0.01	0.04	0.051	0.07	0.109	0.15	0.187	0.222	0.24	0.25	0.28	0.29	0.31	0.311	0.31
8	3			0	0.005	0.009	0.019	0.03	0.051	0.072	0.08	0.1	0.12	0.15	0.17	0.196	0.21
8	4					0.001	0.002	0.01	0.009	0.015	0.02	0.02	0.03	0.05	0.06	0.077	0.09
8	5								0.001	0.002	0	0	0.01	0.01	0.01	0.02	0.02
8	6												0	0	0	0.003	0
8	7																
8	8																

(continues)

Appendix 1 Binomial Table $P(x) = {}_nC_x p^x q^{n-x}$ (*Continued*)

9	0	0.9	0.83	0.69	0.63	0.573	0.472	0.39	0.316	0.257	0.23	0.21	0.17	0.13	0.11	0.085	0.08
9	1	0.1	0.15	0.26	0.299	0.329	0.37	0.39	0.388	0.377	0.37	0.36	0.33	0.3	0.27	0.24	0.23
9	2	0	0.01	0.04	0.063	0.084	0.129	0.17	0.212	0.245	0.26	0.27	0.29	0.3	0.31	0.304	0.3
9	3		0	0	0.008	0.013	0.026	0.05	0.067	0.093	0.11	0.12	0.15	0.18	0.2	0.224	0.23
9	4				0.001	0.001	0.003	0.01	0.014	0.023	0.03	0.04	0.05	0.07	0.09	0.106	0.12
9	5							0	0.002	0.004	0.01	0.01	0.01	0.02	0.02	0.033	0.04
9	6										0	0	0	0	0.01	0.007	0.01
9	7														0	0.001	0
9	8																
9	9																
		0	**0.02**	**0**	**0.05**	**0.06**	**0.08**	**0.1**	**0.12**	**0.14**	**0.2**	**0.16**	**0.2**	**0.2**	**0.22**	**0.24**	**0.3**
10	0	0.9	0.82	0.67	0.599	0.539	0.434	0.35	0.279	0.221	0.2	0.18	0.14	0.11	0.08	0.064	0.06
10	1	0.1	0.17	0.28	0.315	0.344	0.378	0.39	0.38	0.36	0.35	0.33	0.3	0.27	0.24	0.203	0.19
10	2	0	0.02	0.05	0.075	0.099	0.148	0.19	0.233	0.264	0.28	0.29	0.3	0.3	0.3	0.288	0.28
10	3		0	0.01	0.01	0.017	0.034	0.06	0.085	0.115	0.13	0.15	0.17	0.2	0.22	0.243	0.25
10	4				0.001	0.002	0.005	0.01	0.02	0.033	0.04	0.05	0.07	0.09	0.11	0.134	0.15
10	5						0.001	0	0.003	0.006	0.01	0.01	0.02	0.03	0.04	0.051	0.06
10	6									0.001	0	0	0	0.01	0.01	0.013	0.02
10	7													0	0	0.002	0

Appendix 2 Poisson Table $P(x) = \lambda^x e^{-\lambda}/x!$

Events	Mean 0.1	0.2	0.3	0.4	0.5	0.6	0.7	0.8	0.9	1
0	0.90484	0.81873	0.74082	0.67032	0.60653	0.54881	0.49659	0.4493	0.40657	0.36788
1	0.09048	0.16375	0.22225	0.26813	0.30327	0.32929	0.34761	0.3595	0.36591	0.36788
2	0.00452	0.01637	0.03334	0.05363	0.07582	0.09879	0.12166	0.1438	0.16466	0.18394
3	0.00015	0.00109	0.00333	0.00715	0.01264	0.01976	0.02839	0.0383	0.0494	0.06131
4	0	0.00005	0.00025	0.00072	0.00158	0.00296	0.00497	0.0077	0.01111	0.01533
5	0	0	0.00002	0.00006	0.00016	0.00036	0.0007	0.0012	0.002	0.00307
	1.1	**1.2**	**1.3**	**1.4**	**1.5**	**1.6**	**1.7**	**1.8**	**1.9**	**2**
0	0.33287	0.30119	0.27253	0.2466	0.22313	0.2019	0.18268	0.1653	0.14957	0.13534
1	0.36616	0.36143	0.35429	0.34524	0.3347	0.32303	0.31056	0.2975	0.28418	0.27067
2	0.20139	0.21686	0.23029	0.24167	0.25102	0.25843	0.26398	0.2678	0.26997	0.27067
3	0.07384	0.08674	0.09979	0.11278	0.12551	0.13783	0.14959	0.1607	0.17098	0.18045
4	0.02031	0.02602	0.03243	0.03947	0.04707	0.05513	0.06357	0.0723	0.08122	0.09022
5	0.00447	0.00625	0.00843	0.01105	0.01412	0.01764	0.02162	0.026	0.03086	0.03609
6	0.00082	0.00125	0.00183	0.00258	0.00353	0.0047	0.00612	0.0078	0.00977	0.01203
7	0.00013	0.00021	0.00034	0.00052	0.00076	0.00108	0.00149	0.002	0.00265	0.00344
8	0.00002	0.00003	0.00006	0.00009	0.00014	0.00022	0.00032	0.0005	0.00063	0.00086
9	0	0	0.00001	0.00001	0.00002	0.00004	0.00006	9E-05	0.00013	0.00019
10	0	0	0	0	0	0.00001	0.00001	2E-05	0.00003	0.00004

(continues)

Appendix 2 Poisson Table $P(x) = \lambda^x e^{-\lambda}/x!$ (*Continued*)

	2.1	**2.2**	**2.3**	**2.4**	**2.5**	**2.6**	**2.7**	**2.8**	**2.9**	**3**
0	0.12246	0.1108	0.10026	0.09072	0.08208	0.07427	0.06721	0.0608	0.05502	0.04979
1	0.25716	0.24377	0.2306	0.21772	0.20521	0.19311	0.18145	0.1703	0.15957	0.14936
2	0.27002	0.26814	0.26518	0.26127	0.25652	0.25104	0.24496	0.2384	0.23137	0.22404
3	0.18901	0.19664	0.20331	0.20901	0.21376	0.21757	0.22047	0.2225	0.22366	0.22404
4	0.09923	0.10815	0.1169	0.12541	0.1336	0.14142	0.14882	0.1557	0.16215	0.16803
5	0.04168	0.04759	0.05378	0.0602	0.0668	0.07354	0.08036	0.0872	0.09405	0.10082
6	0.01459	0.01745	0.02061	0.02408	0.02783	0.03187	0.03616	0.0407	0.04546	0.05041
7	0.00438	0.00548	0.00677	0.00826	0.00994	0.01184	0.01395	0.0163	0.01883	0.0216
8	0.00115	0.00151	0.00195	0.00248	0.00311	0.00385	0.00471	0.0057	0.00683	0.0081
9	0.00027	0.00037	0.0005	0.00066	0.00086	0.00111	0.00141	0.0018	0.0022	0.0027
10	0.00006	0.00008	0.00011	0.00016	0.00022	0.00029	0.00038	0.0005	0.00064	0.00081
11	0.00001	0.00002	0.00002	0.00003	0.00005	0.00007	0.00009	0.0001	0.00017	0.00022
12	0	0	0	0.00001	0.00001	0.00001	0.00002	3E-05	0.00004	0.00006
13	0	0	0	0	0	0	0	1E-05	0.00001	0.00001
14	0	0	0	0	0	0	0	0	0	0
15	0	0	0	0	0	0	0	0	0	0
	3.1	**3.2**	**3.3**	**3.4**	**3.5**	**3.6**	**3.7**	**3.8**	**3.9**	**4**
0	0.04505	0.04076	0.03688	0.03337	0.0302	0.02732	0.02472	0.0224	0.02024	0.01832
1	0.13965	0.13044	0.12171	0.11347	0.10569	0.09837	0.09148	0.085	0.07894	0.07326
2	0.21646	0.2087	0.20083	0.1929	0.18496	0.17706	0.16923	0.1615	0.15394	0.14653
3	0.22368	0.22262	0.22091	0.21862	0.21579	0.21247	0.20872	0.2046	0.20012	0.19537
4	0.17335	0.17809	0.18225	0.18582	0.18881	0.19122	0.19307	0.1944	0.19512	0.19537
5	0.10748	0.11398	0.12029	0.12636	0.13217	0.13768	0.14287	0.1477	0.15219	0.15629
6	0.05553	0.06079	0.06616	0.0716	0.0771	0.08261	0.0881	0.0936	0.09893	0.1042
7	0.02459	0.02779	0.03119	0.03478	0.03855	0.04248	0.04657	0.0508	0.05512	0.05954
8	0.00953	0.01112	0.01287	0.01478	0.01687	0.01912	0.02154	0.0241	0.02687	0.02977
9	0.00328	0.00395	0.00472	0.00558	0.00656	0.00765	0.00885	0.0102	0.01164	0.01323
10	0.00102	0.00126	0.00156	0.0019	0.0023	0.00275	0.00328	0.0039	0.00454	0.00529
11	0.00029	0.00037	0.00047	0.00059	0.00073	0.0009	0.0011	0.0013	0.00161	0.00192
12	0.00007	0.0001	0.00013	0.00017	0.00021	0.00027	0.00034	0.0004	0.00052	0.00064
13	0.00002	0.00002	0.00003	0.00004	0.00006	0.00007	0.0001	0.0001	0.00016	0.0002
14	0	0.00001	0.00001	0.00001	0.00001	0.00002	0.00003	3E-05	0.00004	0.00006
15	0	0	0	0	0	0	0.00001	1E-05	0.00001	0.00002

	4.1	4.2	4.3	4.4	4.5	4.6	4.7	4.8	4.9	5
0	0.01657	0.015	0.01357	0.01228	0.01111	0.01005	0.0091	0.0082	0.00745	0.00674
1	0.06795	0.06298	0.05834	0.05402	0.04999	0.04624	0.04275	0.0395	0.03649	0.03369
2	0.13929	0.13226	0.12544	0.11884	0.11248	0.10635	0.10046	0.0948	0.0894	0.08422
3	0.19037	0.18517	0.1798	0.17431	0.16872	0.16307	0.15738	0.1517	0.14601	0.14037
4	0.19513	0.19442	0.19328	0.19174	0.18981	0.18753	0.18493	0.182	0.17887	0.17547
5	0.16	0.16332	0.16622	0.16873	0.17083	0.17253	0.17383	0.1748	0.17529	0.17547
6	0.10934	0.11432	0.11913	0.12373	0.12812	0.13227	0.13617	0.1398	0.14315	0.14622
7	0.06404	0.06859	0.07318	0.07778	0.08236	0.08692	0.09143	0.0959	0.10021	0.10444
8	0.03282	0.03601	0.03933	0.04278	0.04633	0.04998	0.05371	0.0575	0.06138	0.06528
9	0.01495	0.01681	0.01879	0.02091	0.02316	0.02554	0.02805	0.0307	0.03342	0.03627
10	0.00613	0.00706	0.00808	0.0092	0.01042	0.01175	0.01318	0.0147	0.01637	0.01813
11	0.00228	0.00269	0.00316	0.00368	0.00426	0.00491	0.00563	0.0064	0.00729	0.00824
12	0.00078	0.00094	0.00113	0.00135	0.0016	0.00188	0.00221	0.0026	0.00298	0.00343
13	0.00025	0.0003	0.00037	0.00046	0.00055	0.00067	0.0008	0.001	0.00112	0.00132
14	0.00007	0.00009	0.00011	0.00014	0.00018	0.00022	0.00027	0.0003	0.00039	0.00047
15	0.00002	0.00003	0.00003	0.00004	0.00005	0.00007	0.00008	0.0001	0.00013	0.00016

	5.1	5.2	5.3	5.4	5.5	5.6	5.7	5.8	5.9	6
0	0.0061	0.00552	0.00499	0.00452	0.00409	0.0037	0.00335	0.003	0.00274	0.00248
1	0.03109	0.02869	0.02646	0.02439	0.02248	0.02071	0.01907	0.0176	0.01616	0.01487
2	0.07929	0.07458	0.07011	0.06585	0.06181	0.05798	0.05436	0.0509	0.04768	0.04462
3	0.13479	0.12928	0.12386	0.11853	0.11332	0.10823	0.10327	0.0985	0.09377	0.08924
4	0.17186	0.16806	0.16411	0.16002	0.15582	0.15153	0.14717	0.1428	0.13831	0.13385
5	0.17529	0.17479	0.17396	0.17282	0.1714	0.16971	0.16777	0.1656	0.16321	0.16062
6	0.149	0.15148	0.15366	0.15554	0.15712	0.1584	0.15938	0.1601	0.16049	0.16062
7	0.10856	0.11253	0.11634	0.11999	0.12345	0.12672	0.12978	0.1326	0.13527	0.13768
8	0.06921	0.07314	0.07708	0.08099	0.08487	0.0887	0.09247	0.0962	0.09976	0.10326
9	0.03922	0.04226	0.04539	0.04859	0.05187	0.05519	0.05856	0.062	0.0654	0.06884
10	0.02	0.02198	0.02406	0.02624	0.02853	0.03091	0.03338	0.0359	0.03859	0.0413
11	0.00927	0.01039	0.01159	0.01288	0.01426	0.01573	0.0173	0.019	0.0207	0.02253
12	0.00394	0.0045	0.00512	0.0058	0.00654	0.00734	0.00822	0.0092	0.01018	0.01126
13	0.00155	0.0018	0.00209	0.00241	0.00277	0.00316	0.0036	0.0041	0.00462	0.0052
14	0.00056	0.00067	0.00079	0.00093	0.00109	0.00127	0.00147	0.0017	0.00195	0.00223
15	0.00019	0.00023	0.00028	0.00033	0.0004	0.00047	0.00056	0.0007	0.00077	0.00089

(continues)

Appendix 2 Poisson Table $P(x) = \lambda^x e^{-\lambda}/x!$ (*Continued*)

	6.1	6.2	6.3	6.4	6.5	6.6	6.7	6.8	6.9	7
0	0.00224	0.00203	0.00184	0.00166	0.0015	0.00136	0.00123	0.0011	0.00101	0.00091
1	0.01368	0.01258	0.01157	0.01063	0.00977	0.00898	0.00825	0.0076	0.00695	0.00638
2	0.04173	0.03901	0.03644	0.03403	0.03176	0.02963	0.02763	0.0258	0.02399	0.02234
3	0.08485	0.08061	0.07653	0.07259	0.06881	0.06518	0.0617	0.0584	0.05518	0.05213
4	0.12939	0.12495	0.12053	0.11615	0.11182	0.10755	0.10335	0.0992	0.09518	0.09123
5	0.15786	0.15494	0.15187	0.14867	0.14537	0.14197	0.13849	0.135	0.13135	0.12772
6	0.16049	0.1601	0.15946	0.15859	0.15748	0.15617	0.15465	0.1529	0.15105	0.149
7	0.13986	0.1418	0.14352	0.14499	0.14623	0.14724	0.14802	0.1486	0.1489	0.149
8	0.10664	0.1099	0.11302	0.11599	0.11882	0.12148	0.12397	0.1263	0.12842	0.13038
9	0.07228	0.07571	0.07911	0.08248	0.08581	0.08908	0.09229	0.0954	0.09846	0.1014
10	0.04409	0.04694	0.04984	0.05279	0.05578	0.05879	0.06183	0.0649	0.06794	0.07098
11	0.02445	0.02646	0.02855	0.03071	0.03296	0.03528	0.03766	0.0401	0.04261	0.04517
12	0.01243	0.01367	0.01499	0.01638	0.01785	0.0194	0.02103	0.0227	0.0245	0.02635
13	0.00583	0.00652	0.00726	0.00806	0.00893	0.00985	0.01084	0.0119	0.01301	0.01419
14	0.00254	0.00289	0.00327	0.00369	0.00414	0.00464	0.00519	0.0058	0.00641	0.00709
15	0.00103	0.00119	0.00137	0.00157	0.0018	0.00204	0.00232	0.0026	0.00295	0.00331
16	0.00039	0.00046	0.00054	0.00063	0.00073	0.00084	0.00097	0.0011	0.00127	0.00145
17	0.00014	0.00017	0.0002	0.00024	0.00028	0.00033	0.00038	0.0005	0.00052	0.0006
18	0.00005	0.00006	0.00007	0.00008	0.0001	0.00012	0.00014	0.0002	0.0002	0.00023
19	0.00002	0.00002	0.00002	0.00003	0.00003	0.00004	0.00005	6E-05	0.00007	0.00009
20	0	0.00001	0.00001	0.00001	0.00001	0.00001	0.00002	2E-05	0.00002	0.00003

	7.1	**7.2**	**7.3**	**7.4**	**7.5**	**7.6**	**7.7**	**7.8**	**7.9**	**8**
0	0.00083	0.00075	0.00068	0.00061	0.00055	0.0005	0.00045	0.0004	0.00037	0.00034
1	0.00586	0.00538	0.00493	0.00452	0.00415	0.0038	0.00349	0.0032	0.00293	0.00268
2	0.0208	0.01935	0.018	0.01674	0.01556	0.01445	0.01342	0.0125	0.01157	0.01073
3	0.04922	0.04644	0.0438	0.04128	0.03889	0.03661	0.03446	0.0324	0.03047	0.02863
4	0.08736	0.0836	0.07993	0.07637	0.07292	0.06957	0.06633	0.0632	0.06017	0.05725
5	0.12406	0.12038	0.1167	0.11303	0.10937	0.10574	0.10214	0.0986	0.09507	0.0916
6	0.1468	0.14446	0.14199	0.13941	0.13672	0.13394	0.13108	0.1282	0.12517	0.12214
7	0.1489	0.14859	0.14807	0.14737	0.14648	0.14542	0.14419	0.1428	0.14126	0.13959
8	0.13215	0.13373	0.13512	0.13632	0.13733	0.13815	0.13878	0.1392	0.1395	0.13959
9	0.10425	0.10698	0.1096	0.11208	0.11444	0.11666	0.11874	0.1207	0.12245	0.12408
10	0.07402	0.07703	0.08	0.08294	0.08583	0.08866	0.09143	0.0941	0.09673	0.09926
11	0.04777	0.05042	0.05309	0.0558	0.05852	0.06126	0.064	0.0667	0.06947	0.07219
12	0.02827	0.03025	0.0323	0.03441	0.03658	0.0388	0.04107	0.0434	0.04574	0.04813
13	0.01544	0.01675	0.01814	0.01959	0.0211	0.02268	0.02432	0.026	0.02779	0.02962
14	0.00783	0.00862	0.00946	0.01035	0.0113	0.01231	0.01338	0.0145	0.01568	0.01692
15	0.00371	0.00414	0.0046	0.00511	0.00565	0.00624	0.00687	0.0075	0.00826	0.00903
16	0.00164	0.00186	0.0021	0.00236	0.00265	0.00296	0.0033	0.0037	0.00408	0.00451
17	0.00069	0.00079	0.0009	0.00103	0.00117	0.00132	0.0015	0.0017	0.0019	0.00212
18	0.00027	0.00032	0.00037	0.00042	0.00049	0.00056	0.00064	0.0007	0.00083	0.00094
19	0.0001	0.00012	0.00014	0.00016	0.00019	0.00022	0.00026	0.0003	0.00035	0.0004
20	0.00004	0.00004	0.00005	0.00006	0.00007	0.00009	0.0001	0.0001	0.00014	0.00016

(continues)

Appendix 2 Poisson Table $P(x) = \lambda^x e^{-\lambda}/x!$ (*Continued*)

	8.1	**8.2**	**8.3**	**8.4**	**8.5**	**8.6**	**8.7**	**8.8**	**8.9**	**9**
0	0.0003	0.00027	0.00025	0.00022	0.0002	0.00018	0.00017	0.0002	0.00014	0.00012
1	0.00246	0.00225	0.00206	0.00189	0.00173	0.00158	0.00145	0.0013	0.00121	0.00111
2	0.00996	0.00923	0.00856	0.00793	0.00735	0.00681	0.0063	0.0058	0.0054	0.005
3	0.02689	0.02524	0.02368	0.02221	0.02083	0.01952	0.01828	0.0171	0.01602	0.01499
4	0.05444	0.05174	0.04914	0.04665	0.04425	0.04196	0.03977	0.0377	0.03566	0.03374
5	0.0882	0.08485	0.08158	0.07837	0.07523	0.07217	0.06919	0.0663	0.06347	0.06073
6	0.11907	0.11597	0.11285	0.10972	0.10658	0.10345	0.10033	0.0972	0.09414	0.09109
7	0.13778	0.13585	0.1338	0.13166	0.12942	0.12709	0.12469	0.1222	0.1197	0.11712
8	0.1395	0.13924	0.13882	0.13824	0.13751	0.13663	0.1356	0.1345	0.13316	0.13176
9	0.12555	0.12687	0.12803	0.12903	0.12987	0.13055	0.13108	0.1315	0.13168	0.13176
10	0.1017	0.10403	0.10626	0.10838	0.11039	0.11228	0.11404	0.1157	0.1172	0.11858
11	0.07488	0.07755	0.08018	0.08276	0.0853	0.08778	0.0902	0.0926	0.09482	0.09702
12	0.05055	0.05299	0.05546	0.05793	0.06042	0.06291	0.06539	0.0679	0.07033	0.07277
13	0.03149	0.03343	0.03541	0.03743	0.03951	0.04162	0.04376	0.0459	0.04815	0.05038
14	0.01822	0.01958	0.02099	0.02246	0.02399	0.02556	0.0272	0.0289	0.03061	0.03238
15	0.00984	0.0107	0.01162	0.01258	0.01359	0.01466	0.01577	0.0169	0.01816	0.01943
16	0.00498	0.00549	0.00603	0.0066	0.00722	0.00788	0.00858	0.0093	0.0101	0.01093
17	0.00237	0.00265	0.00294	0.00326	0.00361	0.00399	0.00439	0.0048	0.00529	0.00579
18	0.00107	0.00121	0.00136	0.00152	0.0017	0.0019	0.00212	0.0024	0.00261	0.00289
19	0.00046	0.00052	0.00059	0.00067	0.00076	0.00086	0.00097	0.0011	0.00122	0.00137
20	0.00018	0.00021	0.00025	0.00028	0.00032	0.00037	0.00042	0.0005	0.00055	0.00062

	9.1	9.2	9.3	9.4	9.5	9.6	9.7	9.8	9.9	10
0	0.00011	0.0001	0.00009	0.00008	0.00007	0.00007	0.00006	6E-05	0.00005	0.00005
1	0.00102	0.00093	0.00085	0.00078	0.00071	0.00065	0.00059	0.0005	0.0005	0.00045
2	0.00462	0.00428	0.00395	0.00365	0.00338	0.00312	0.00288	0.0027	0.00246	0.00227
3	0.01402	0.01311	0.01226	0.01145	0.0107	0.00999	0.00932	0.0087	0.00811	0.00757
4	0.03191	0.03016	0.0285	0.02691	0.0254	0.02397	0.02261	0.0213	0.02008	0.01892
5	0.05807	0.05549	0.053	0.05059	0.04827	0.04602	0.04386	0.0418	0.03976	0.03783
6	0.08807	0.08509	0.08215	0.07926	0.07642	0.07363	0.0709	0.0682	0.06561	0.06306
7	0.11449	0.11183	0.10915	0.10644	0.10371	0.10098	0.09825	0.0955	0.09279	0.09008
8	0.13024	0.12861	0.12688	0.12506	0.12316	0.12118	0.11912	0.117	0.11483	0.1126
9	0.13168	0.13147	0.13111	0.13062	0.13	0.12926	0.12839	0.1274	0.12631	0.12511
10	0.11983	0.12095	0.12193	0.12279	0.1235	0.12409	0.12454	0.1249	0.12505	0.12511
11	0.09913	0.10116	0.10309	0.10493	0.10666	0.10829	0.10982	0.1112	0.11254	0.11374
12	0.07518	0.07755	0.0799	0.08219	0.08444	0.08663	0.08877	0.0908	0.09285	0.09478
13	0.05262	0.05488	0.05716	0.05943	0.06171	0.06398	0.06624	0.0685	0.07071	0.07291
14	0.03421	0.03607	0.03797	0.0399	0.04187	0.04387	0.04589	0.0479	0.05	0.05208
15	0.02075	0.02212	0.02354	0.02501	0.02652	0.02808	0.02968	0.0313	0.033	0.03472
16	0.0118	0.01272	0.01368	0.01469	0.01575	0.01685	0.01799	0.0192	0.02042	0.0217
17	0.00632	0.00688	0.00749	0.00812	0.0088	0.00951	0.01027	0.0111	0.01189	0.01276
18	0.00319	0.00352	0.00387	0.00424	0.00464	0.00507	0.00553	0.006	0.00654	0.00709
19	0.00153	0.0017	0.00189	0.0021	0.00232	0.00256	0.00282	0.0031	0.00341	0.00373
20	0.0007	0.00078	0.00088	0.00099	0.0011	0.00123	0.00137	0.0015	0.00169	0.00187

Appendix 3 Normal Z table

Z	**0**	**0.01**	**0.02**	**0.03**	**0.04**	**0.05**	**0.06**	**0.07**	**0.08**	**0.09**
0	0	0.004	0.008	0.012	0.016	0.0199	0.0239	0.0279	0.0319	0.0359
0.1	0.0398	0.0438	0.0478	0.0517	0.0557	0.0596	0.0636	0.0675	0.0714	0.0753
0.2	0.0793	0.0832	0.0871	0.091	0.0948	0.0987	0.1026	0.1064	0.1103	0.1141
0.3	0.1179	0.1217	0.1255	0.1293	0.1331	0.1368	0.1406	0.1443	0.148	0.1517
0.4	0.1554	0.1591	0.1628	0.1664	0.17	0.1736	0.1772	0.1808	0.1844	0.1879
0.5	0.1915	0.195	0.1985	0.2019	0.2054	0.2088	0.2123	0.2157	0.219	0.2224
0.6	0.2257	0.2291	0.2324	0.2357	0.2389	0.2422	0.2454	0.2486	0.2517	0.2549
0.7	0.258	0.2611	0.2642	0.2673	0.2704	0.2734	0.2764	0.2794	0.2823	0.2852
0.8	0.2881	0.291	0.2939	0.2967	0.2995	0.3023	0.3051	0.3078	0.3106	0.3133
0.9	0.3159	0.3186	0.3212	0.3238	0.3264	0.3289	0.3315	0.334	0.3365	0.3389
1	0.3413	0.3438	0.3461	0.3485	0.3508	0.3531	0.3554	0.3577	0.3599	0.3621
1.1	0.3643	0.3665	0.3686	0.3708	0.3729	0.3749	0.377	0.379	0.381	0.383
1.2	0.3849	0.3869	0.3888	0.3907	0.3925	0.3944	0.3962	0.398	0.3997	0.4015
1.3	0.4032	0.4049	0.4066	0.4082	0.4099	0.4115	0.4131	0.4147	0.4162	0.4177
1.4	0.4192	0.4207	0.4222	0.4236	0.4251	0.4265	0.4279	0.4292	0.4306	0.4319
1.5	0.4332	0.4345	0.4357	0.437	0.4382	0.4394	0.4406	0.4418	0.4429	0.4441
1.6	0.4452	0.4463	0.4474	0.4484	0.4495	0.4505	0.4515	0.4525	0.4535	0.4545
1.7	0.4554	0.4564	0.4573	0.4582	0.4591	0.4599	0.4608	0.4616	0.4625	0.4633
1.8	0.4641	0.4649	0.4656	0.4664	0.4671	0.4678	0.4686	0.4693	0.4699	0.4706
1.9	0.4713	0.4719	0.4726	0.4732	0.4738	0.4744	0.475	0.4756	0.4761	0.4767
2	0.4772	0.4778	0.4783	0.4788	0.4793	0.4798	0.4803	0.4808	0.4812	0.4817
2.1	0.4821	0.4826	0.483	0.4834	0.4838	0.4842	0.4846	0.485	0.4854	0.4857
2.2	0.4861	0.4864	0.4868	0.4871	0.4875	0.4878	0.4881	0.4884	0.4887	0.489
2.3	0.4893	0.4896	0.4898	0.4901	0.4904	0.4906	0.4909	0.4911	0.4913	0.4916
2.4	0.4918	0.492	0.4922	0.4925	0.4927	0.4929	0.4931	0.4932	0.4934	0.4936
2.5	0.4938	0.494	0.4941	0.4943	0.4945	0.4946	0.4948	0.4949	0.4951	0.4952
2.6	0.4953	0.4955	0.4956	0.4957	0.4959	0.496	0.4961	0.4962	0.4963	0.4964
2.7	0.4965	0.4966	0.4967	0.4968	0.4969	0.497	0.4971	0.4972	0.4973	0.4974
2.8	0.4974	0.4975	0.4976	0.4977	0.4977	0.4978	0.4979	0.4979	0.498	0.4981
2.9	0.4981	0.4982	0.4982	0.4983	0.4984	0.4984	0.4985	0.4985	0.4986	0.4986
3	0.4987	0.4987	0.4987	0.4988	0.4988	0.4989	0.4989	0.4989	0.499	0.499

Appendix 4 Student's *t* table

	0.4	0.25	0.1	0.05	0.025	0.01	0.005	0.0005
1	0.32492	1	3.077684	6.313752	12.7062	31.82052	63.65674	636.6192
2	0.288675	0.816497	1.885618	2.919986	4.30265	6.96456	9.92484	31.5991
3	0.276671	0.764892	1.637744	2.353363	3.18245	4.5407	5.84091	12.924
4	0.270722	0.740697	1.533206	2.131847	2.77645	3.74695	4.60409	8.6103
5	0.267181	0.726687	1.475884	2.015048	2.57058	3.36493	4.03214	6.8688
6	0.264835	0.717558	1.439756	1.94318	2.44691	3.14267	3.70743	5.9588
7	0.263167	0.711142	1.414924	1.894579	2.36462	2.99795	3.49948	5.4079
8	0.261921	0.706387	1.396815	1.859548	2.306	2.89646	3.35539	5.0413
9	0.260955	0.702722	1.383029	1.833113	2.26216	2.82144	3.24984	4.7809
10	0.260185	0.699812	1.372184	1.812461	2.22814	2.76377	3.16927	4.5869
11	0.259556	0.697445	1.36343	1.795885	2.20099	2.71808	3.10581	4.437
12	0.259033	0.695483	1.356217	1.782288	2.17881	2.681	3.05454	4.3178
13	0.258591	0.693829	1.350171	1.770933	2.16037	2.65031	3.01228	4.2208
14	0.258213	0.692417	1.34503	1.76131	2.14479	2.62449	2.97684	4.1405
15	0.257885	0.691197	1.340606	1.75305	2.13145	2.60248	2.94671	4.0728
16	0.257599	0.690132	1.336757	1.745884	2.11991	2.58349	2.92078	4.015
17	0.257347	0.689195	1.333379	1.739607	2.10982	2.56693	2.89823	3.9651
18	0.257123	0.688364	1.330391	1.734064	2.10092	2.55238	2.87844	3.9216
19	0.256923	0.687621	1.327728	1.729133	2.09302	2.53948	2.86093	3.8834
20	0.256743	0.686954	1.325341	1.724718	2.08596	2.52798	2.84534	3.8495
21	0.25658	0.686352	1.323188	1.720743	2.07961	2.51765	2.83136	3.8193
22	0.256432	0.685805	1.321237	1.717144	2.07387	2.50832	2.81876	3.7921
23	0.256297	0.685306	1.31946	1.713872	2.06866	2.49987	2.80734	3.7676
24	0.256173	0.68485	1.317836	1.710882	2.0639	2.49216	2.79694	3.7454
25	0.25606	0.68443	1.316345	1.708141	2.05954	2.48511	2.78744	3.7251
26	0.255955	0.684043	1.314972	1.705618	2.05553	2.47863	2.77871	3.7066
27	0.255858	0.683685	1.313703	1.703288	2.05183	2.47266	2.77068	3.6896
28	0.255768	0.683353	1.312527	1.701131	2.04841	2.46714	2.76326	3.6739
29	0.255684	0.683044	1.311434	1.699127	2.04523	2.46202	2.75639	3.6594
30	0.255605	0.682756	1.310415	1.697261	2.04227	2.45726	2.75	3.646
inf	0.253347	0.67449	1.281552	1.644854	1.95996	2.32635	2.57583	3.2905

Appendix 5 Chi-Square Table

area df	0.995	0.990	0.975	0.950	0.900	0.750	0.500	0.250	0.100	0.050
1	0.0000	0.0002	0.0010	0.0039	0.0158	0.1015	0.4549	1.3233	2.7055	3.8415
2	0.0100	0.0201	0.0506	0.1026	0.2107	0.5754	1.3863	2.7726	4.6052	5.9915
3	0.0717	0.1148	0.2158	0.3519	0.5844	1.2125	2.3660	4.1083	6.2514	7.8147
4	0.2070	0.2971	0.4844	0.7107	1.0636	1.9226	3.3567	5.3853	7.7794	9.4877
5	0.4117	0.5543	0.8312	1.1455	1.6103	2.6746	4.3515	6.6257	9.2364	11.0705
6	0.6757	0.8721	1.2373	1.6354	2.2041	3.4546	5.3481	7.8408	10.6446	12.5916
7	0.9893	1.2390	1.6899	2.1674	2.8331	4.2549	6.3458	9.0372	12.0170	14.0671
8	1.3444	1.6465	2.1797	2.7326	3.4895	5.0706	7.3441	10.2189	13.3616	15.5073
9	1.7349	2.0879	2.7004	3.3251	4.1682	5.8988	8.3428	11.3888	14.6837	16.9190
10	2.1559	2.5582	3.2470	3.9403	4.8652	6.7372	9.3418	12.5489	15.9872	18.3070
11	2.6032	3.0535	3.8158	4.5748	5.5778	7.5841	10.3410	13.7007	17.2750	19.6751
12	3.0738	3.5706	4.4038	5.2260	6.3038	8.4384	11.3403	14.8454	18.5494	21.0261
13	3.5650	4.1069	5.0088	5.8919	7.0415	9.2991	12.3398	15.9839	19.8119	22.3620
14	4.0747	4.6604	5.6287	6.5706	7.7895	10.1653	13.3393	17.1169	21.0641	23.6848
15	4.6009	5.2294	6.2621	7.2609	8.5468	11.0365	14.3389	18.2451	22.3071	24.9958
16	5.1422	5.8122	6.9077	7.9617	9.3122	11.9122	15.3385	19.3689	23.5418	26.2962
17	5.6972	6.4078	7.5642	8.6718	10.0852	12.7919	16.3382	20.4887	24.7690	27.5871
18	6.2648	7.0149	8.2308	9.3905	10.8649	13.6753	17.3379	21.6049	25.9894	28.8693
19	6.8440	7.6327	8.9065	10.1170	11.6509	14.5620	18.3377	22.7178	27.2036	30.1435
20	7.4338	8.2604	9.5908	10.8508	12.4426	15.4518	19.3374	23.8277	28.4120	31.4104
21	8.0337	8.8972	10.2829	11.5913	13.2396	16.3444	20.3372	24.9348	29.6151	32.6706
22	8.6427	9.5425	10.9823	12.3380	14.0415	17.2396	21.3370	26.0393	30.8133	33.9244
23	9.2604	10.1957	11.6886	13.0905	14.8480	18.1373	22.3369	27.1413	32.0069	35.1725
24	9.8862	10.8564	12.4012	13.8484	15.6587	19.0373	23.3367	28.2412	33.1962	36.4150
25	10.5197	11.5240	13.1197	14.6114	16.4734	19.9393	24.3366	29.3389	34.3816	37.6525
26	11.1602	12.1982	13.8439	15.3792	17.2919	20.8434	25.3365	30.4346	35.5632	38.8851
27	11.8076	12.8785	14.5734	16.1514	18.1139	21.7494	26.3363	31.5284	36.7412	40.1133
28	12.4613	13.5647	15.3079	16.9279	18.9392	22.6572	27.3362	32.6205	37.9159	41.3371
29	13.1212	14.2565	16.0471	17.7084	19.7677	23.5666	28.3361	33.7109	39.0875	42.5570
30	13.7867	14.9535	16.7908	18.4927	20.5992	24.4776	29.3360	34.7997	40.2560	43.7730

Appendix 6 *F* table ($\alpha = 0.05$)

df2/df1	1	2	3	4	5	6	7	8	9	10
1	39.863	49.5	53.593	55.833	57.24	58.2044	58.906	59.439	59.858	60.195
2	8.5263	9	9.1618	9.2434	9.2926	9.32553	9.3491	9.3668	9.3805	9.3916
3	5.5383	5.4624	5.3908	5.3426	5.3092	5.28473	5.2662	5.2517	5.24	5.2304
4	4.5448	4.3246	4.1909	4.1073	4.0506	4.00975	3.979	3.9549	3.9357	3.9199
5	4.0604	3.7797	3.6195	3.5202	3.453	3.40451	3.3679	3.3393	3.3163	3.2974
6	3.776	3.4633	3.2888	3.1808	3.1075	3.05455	3.0145	2.983	2.9577	2.9369
7	3.5894	3.2574	3.0741	2.9605	2.8833	2.82739	2.7849	2.7516	2.7247	2.7025
8	3.4579	3.1131	2.9238	2.8064	2.7265	2.66833	2.6241	2.5894	2.5612	2.538
9	3.3603	3.0065	2.8129	2.6927	2.6106	2.55086	2.5053	2.4694	2.4403	2.4163
10	3.285	2.9245	2.7277	2.6053	2.5216	2.46058	2.414	2.3772	2.3473	2.3226
11	3.2252	2.8595	2.6602	2.5362	2.4512	2.38907	2.3416	2.304	2.2735	2.2482
12	3.1766	2.8068	2.6055	2.4801	2.394	2.33102	2.2828	2.2446	2.2135	2.1878
13	3.1362	2.7632	2.5603	2.4337	2.3467	2.28298	2.2341	2.1954	2.1638	2.1376
14	3.1022	2.7265	2.5222	2.3947	2.3069	2.24256	2.1931	2.1539	2.122	2.0954
15	3.0732	2.6952	2.4898	2.3614	2.273	2.20808	2.1582	2.1185	2.0862	2.0593
16	3.0481	2.6682	2.4618	2.3327	2.2438	2.17833	2.128	2.088	2.0553	2.0282
17	3.0262	2.6446	2.4374	2.3078	2.2183	2.15239	2.1017	2.0613	2.0284	2.0009
18	3.007	2.624	2.416	2.2858	2.1958	2.12958	2.0785	2.0379	2.0047	1.977
19	2.9899	2.6056	2.397	2.2663	2.176	2.10936	2.058	2.0171	1.9836	1.9557
20	2.9747	2.5893	2.3801	2.2489	2.1582	2.09132	2.0397	1.9985	1.9649	1.9367
21	2.961	2.5746	2.3649	2.2333	2.1423	2.07512	2.0233	1.9819	1.948	1.9197
22	2.9486	2.5613	2.3512	2.2193	2.1279	2.0605	2.0084	1.9668	1.9327	1.9043
23	2.9374	2.5493	2.3387	2.2065	2.1149	2.04723	1.9949	1.9531	1.9189	1.8903
24	2.9271	2.5383	2.3274	2.1949	2.103	2.03513	1.9826	1.9407	1.9063	1.8775
25	2.9177	2.5283	2.317	2.1842	2.0922	2.02406	1.9714	1.9293	1.8947	1.8658
26	2.9091	2.5191	2.3075	2.1745	2.0822	2.01389	1.961	1.9188	1.8841	1.855
27	2.9012	2.5106	2.2987	2.1655	2.073	2.00452	1.9515	1.9091	1.8743	1.8451
28	2.8939	2.5028	2.2906	2.1571	2.0645	1.99585	1.9427	1.9001	1.8652	1.8359
29	2.887	2.4955	2.2831	2.1494	2.0566	1.98781	1.9345	1.8918	1.8568	1.8274
30	2.8807	2.4887	2.2761	2.1422	2.0493	1.98033	1.9269	1.8841	1.849	1.8195

Index